AF664499

Key Notes on

Genetics and Plant Breeding

For Ready Reference to the Students, Teachers, Researchers & Aspirants of Competitive Examinations

3rd Revised and Enlarged Edition

THE AUTHOR

Venkata R. Prakash Reddy received his B.Sc. (Ag) degree from Acharya NG Ranga Agricultural University, Hyderabad, Andhra Pradesh and M.Sc. (Ag) in Genetics and Plant Breeding from University of Agricultural Sciences, Dharwad, Karnataka. Where he received University Gold Medal. He is a recipient of Inspire Fellowship of Department of Science and Technology for Ph.D programme and doing Ph.D in Acharya NG Ranga Agricultural University. Currently working as an Assistant Professor at Acharya NG Ranga Agricultural University, Agricultural College, Mahanandi. Dr. Reddy has pubished several research, review papers in various national and international journals of repute. He has already authored on book titled Heterosis and Combining ability studies in Sunflower. His research interests include Plant Breeding and Molecular Biology.

Key Notes on

Genetics and Plant Breeding

For Ready Reference to the Students, Teachers, Researchers & Aspirants of Competitive Examinations

3rd Revised and Enlarged Edition

Author

Venkata R. Prakash Reddy

2026

Daya Publishing House®

A Division of

Astral International (P) Ltd

New Delhi 110 002

First Edition 2016
Third Revised Edition 2026

ISBN: 978-93-5919-983-2 (PB)

Published by : **Daya Publishing House®**
A Division of
Astral International Pvt. Ltd.
– ISO 9001:2015 Certified Company –
4736/23, Ansari Road, Darya Ganj
New Delhi-110 002
Ph. 011-43549197
E-mail: info@astralint.com
Website: www.astralint.com

Printed at GK Fine Art, Delhi

DEDICATED TO

My Grand Father Late B. Pedda Venkata Reddy

Preface to the 3rd Revised Edition

The pursuit of knowledge is a journey of continuous evolution—shaped by exploration, discovery, and refinement. As we present the third revised edition of Genetics and Plant Breeding, we remain steadfast in our original goal: to provide relevant, accessible, and student-friendly content tailored to the needs of aspirants preparing for various competitive examinations. The overwhelming success and sustained demand for this book reflect its clarity, simplicity, and effectiveness in communicating complex concepts. In this new edition, we have worked diligently to preserve the core objectives of the previous editions while integrating valuable feedback from our readers and adapting the material to reflect the dynamic educational environment.

Each chapter has been meticulously reviewed and revised to correct errors and enhance accuracy, guided by constructive suggestions from readers across India. The disciplines of Genetics and Plant Breeding continue to advance rapidly, driven by breakthroughs in genetic understanding, technological innovations, and modern methodologies. In response, this edition incorporates updated scientific knowledge, contributions from renowned scientists, and the achievements of distinguished Indian plant breeders. The scope of the book covers a wide range of topics, including recent advancements in plant breeding and genetics, general genetics, and genome organization in crop plants. It also addresses cytogenetics, reproductive and pollination biology, modern breeding techniques, and the foundational principles of population and quantitative genetics. Additional subjects include biometrical genetics, breeding for resistance to biotic and abiotic stresses, improvement of nutritional quality, genetic engineering, and the application of biotechnological tools in plant breeding. The book also explores the conservation of plant genetic resources, regulatory systems, seed classification and certification procedures, as well as economic botany and medicinal plants. To support applied learning and research, the book includes chapters on statistical methods and field plot techniques, ensuring students gain both theoretical knowledge and practical insight. This edition is particularly designed to benefit graduate and postgraduate students in Agriculture and Biological Sciences across Indian universities. Furthermore, it is an invaluable resource for aspirants preparing for competitive examinations, including the Agricultural Research Services (ARS) of the ICAR, the National Eligibility Test (NET), the Civil Services Examination, and other related assessments.

These enhancements aim to ensure that the book remains an indispensable resource for students, educators, researchers, and enthusiasts in the field of agriculture. While every effort has been made to improve the content, I humbly acknowledge that no work is beyond refinement. I warmly welcome further feedback and suggestions that can help us continue improving future editions. To all our readers—thank you for your continued support. It is my sincere hope that this revised edition of Genetics and Plant Breeding will remain a trusted and valuable companion on your academic and professional journey.

Venkata R. Prakash Reddy

PREFACE

This book has been written primarily for undergraduate and post graduate students of Agriculture. Genetics and Plant Breeding have practical applications in improving the genetic makeup plants in relation to their economic use for mankind. Students of these disciplines have to answer both descriptive and objective questions in various competitive examinations. This book will help the students in the development of this faculty through the presentation of typical questions and answers and help them gain an understanding of basic genetic principles. I am conscious that all attempts leave scope for further improvements and shall be most grateful to the readers to give me the benefit of their advice and suggestions so that these can be incorporated in a later edition.

Venkata R. Prakash Reddy

CONTENTS

Scientific Contributions of Eminent Scientists

Name of the Scientists		Contributions
Allard and Bradshaw	-	G x E interaction
Recurrent Selection for SCA	-	Hull
Recurrent Selection for GCA	-	Jenkins
Dominance hypothesis	–	Davenport
Gene for gene hypothesis	-	Flor
Pure line concept	-	Johannsen
Backcross method	-	Harlan and Pope
Double cross scheme	-	Jones
Cytoplasmic Genetic Male sterility	-	Jones and Davis
Ear to row method	-	Hopkins
Colchicine	-	Blackslee and Nebel
Single Seed Descent Method	-	Goulden
Self-incompatibility	-	Lewis
Vertifolia effect	-	Van Der Plank
Centres of diversity, Law of homologus series	-	Vavilov
Grater initial capital hypothesis	-	Ashby
Progeny test	-	Vilmorin
First artificial hybrid	-	Thomas Fairchild
Triticale	-	Rimpau
Mutation	-	Hugo de Vries
Sprophytic System of self-incompatibility	-	Hughes and Babcock
Bulk method	-	Nilsson & Ehle
Raphano brassica	-	Karpenchenko

Name of the Scientists		Contributions
Heterosis	-	Shull
Male sterility	-	Jones and Davis
Father of hybrid rice	-	Yuan Long Ping
Self-incompatibility classification	-	Lewis
Mechanism of insect resistance	-	Painter
Modified bulk method	-	Atkins
Components of genetic variance classification	-	Fischer
Male sterility in maize	-	Rhoades
Microcentre	-	Harlan
Chemical mutagen	-	Aurbach
Multiline concept	-	Jenson
Green revolution in India	-	M.S. Swaminathan
Semidwarf rice varieties at IRRI	-	T.T. Chang
Forage breeder	-	G.W. Burton
Forage breeder	-	T.J. Jenkin
Soyabean breeder	-	E.E. Hartwig

Model Test Paper (Solved)

1. **Match the following columns:**

Column I (Parental species)	Column II (Resulting allopolyploid species)
a. Raphanus sativus x Brassica oleracea	(a) Nicotiana digluta
b. Nicotianan tabacum x Nicotiana glutinosa	(b) Raphanobrassica
c. Triticum aestivum x Secale cerele	(c) Gossypium hirsutum
d. Gossypium herbaceum x Gossypium raimondii	(d) Triticale

(a) a-2, b-1, c-4, d-3 (b) a-1, b-2, c-3, d-4
(c) a-4, b-3, c-2, d-1 (d) a-3, b-4, c-2, d-1

2. **Diffused centromere means:**
(a) Centromere is at end of the chromosome
(b) Centromer is at the middle of the chromosome
(c) Centromere is nonlocalised and spread over chromosome
(d) Centromere is localized

3. **A pure tall plant was grown in nutrient limited soil and became dwarf. If this plant was selfed then the genotype in the F_1 generation are**
(a) All dwarf plant (b) 75% tall and 25% dwarf plants
(c) All tall plants (d) 50% tall and 50% dwarf plants

4. **Consider the following statements Hybrid seed production process induces**

A. Production of inbreed lines **B. Selection of inbreed lines**
C. Production of F1 seed. **D. Testing of combining ability**

Select the correct answer
(a) 1, 2, 3 and 4 (b) 2, 1, 4 and 3
(c) 1, 2, 4 and 3 (d) 1, 4, 2 and 3

5. **Match the following columns:**

Column I (Crop)	Column II (Method of hybrid seed production)
a. Brinjal	(a) Genetic male sterility
b. Chilli	(b) Self incompatibility
c. Cucumber	(c) Functional male sterility
d. Cabbage	(d) Gynocracy

(a) a-2, b-4, c-1, d-3 (b) a-3, b-1, c-4, d-2
(c) O a-2, b-1, c-4, d-3 (d) a-3, b-4, c-1, d-2

6. **All possible phenotypic correlation among 'n' variables are required for estimation of**
(a) Genotypic path (b) Phenotypic path
(c) Environmental path (d) All of these

7. **Better performance of segregating generations over the parents is called:**
(a) Heterosis (b) Inbreeding depression
(c) Transgressive segregation (d) Hybrid vigour

8. **The term "linkage drag" is used to explain**
(a) Transfer of unfavourabel genes along with favourable genes
(b) Combining recessive alleles of different genes together
(c) Epistatic gene masks the effect of lined hyhpostatic gene
(d) Process of breaking linkage groups

9. **If the sequence of nucleotides on one strand of DNA is A-T-A-G-C, the sequence in the other strand will be:**
(a) T-T-C-G-G (b) C-A-T-A-T
(c) G-T-A-C-C (d) T-A-T-C-G

10. **Which of the following acids is not a polyprotic acid?**
(a) Chromic acid (b) Malonic acid
(c) Carbonic acid (d) Formic acid

11. **The pH of a glycine solution, in which the α-NH group (having pKa 9.6) is one third dissociated would be equal to :**
(a) 9.3 (b) 9.9
(c) 10.1 (d) 9.1

12. **Most soluble Endoplasmic Reticulum-resident proteins have a C-terminal motif/tag viz. KDEL. The sequence of amino acids in this motif/tag is ?**
(a) Lysine-Aspartic acid-Glutamic acid-Leucine
(b) Leucine-Aspartic acid-Glutamic acid-Lysine
(c) Lysine-Glutamic acid-Aspartic acid-Leucine
(d) Leucine-Glutamic acid-Aspartic acid-Lysine

13 In ranz Knoop's classic experiment, which indicated that fatty acids are metabolically oxidized at the β-carbon atom, ω- phenyl labeled fatty acids containing odd number of carbon atoms are broken down to which of the following metabolite?

(a) Phenylacetic acid (b) Benzoic acid

(c) Acetoacetatic acid (d) Propionic acid

14. Which of the following statements about the enzyme acetyl CoA carboxylase (ACCase), involved in the biosynthesis of fatty acids, is incorrect?

(a) It catalyses ATP-dependent conversion of acetyl CoA into malonyl CoA

(b) It is a member of a family of biotin-dependent carboxylases

(c) Reaction catalysed by ACCase is a single-step reaction

(d) The ACCase in plants could be multifunctional or multisubunit type

15. Match List I with List II

List I	List II
Name of the fatty acid	**Type of the fatty acid**
A. Ricinoleic acid	I. ω-3
B. Linolenic acid	II. ω-6
C. Palmitoleic acid	III. ω-7
D. Erucic acid	IV. ω-9
	V. Hydroxylated

Choose the correct answer from the options given below:

(a) A-V, B-I, C-III, D-IV (b) A-V, B-II, C-I, D-IV

(c) A-III, B-I, C-V, D-II (d) A-IV, B-II, C-III, D-I

16. Which of the following molecules functions as provitamin D ?

(a) 7-Dehydrocholesterol (b) Ergosterol

(c) Ergocalciferol (d) 1, 25-Dihydroxy cholecalciferol

17. The vitamin B deficiency will lead to the accumulation of which of the following metabolites?

(a) Pyruvic acid (b) Methyl malonic acid

(c) α-Keto glutarate (d) Biotin

18. The enzyme Papain, derived from papaya, is commonly used in food industry as a meat tenderizer. It is a/an:

(a) Serine protease (b) Metalloprotease

(c) Cysteine protease (d) Aspartic protease

19. How does the effectiveness of a competitive inhibitor gets affected under high physiological concentration of the substrate?

(a) The inhibitor becomes more effective.

(b) The effectiveness of the inhibitor gets reduced.

(c) The inhibitor remains equally effective

(d) The inhibitor gets irreversibly inactivated

20. Which of the followings is not true about the DNA denaturation process?

(a) DNA denaturation causes decrease in UV absorption at all the wavelengths

(b) DNA denaturation is a cooperative process

(c) The melting temperature of the DNA decreases under alkaline conditions.

(d) The melting temperature of the DNA increases linearly with its GC content

21. In globular proteins, α-helices have an average span of 12 residues, which would correspond to how much length of the helix?

(a) 5.4 A (b) 8 A

(c) 3.6 A (d) 36 A

22. Given below are two statements, one is labelled as Assertion A and the other is labelled as Reason R

Assertion A: Although most of the animals cannot digest α-keratin but clothes moth larvae can very easily do so.

Reason R: The clothes moth larvae has a very high concentration of mercaptans in their digestive tract that helps them to cleave disulfide bonds abundantly present in α-keratin.

In light of the above statements, choose the *correct* answer from the options given below

(a) Both A and R are true and R is the correct explanation of A

(b) Both A and R are true but R is NOT the correct explanation of A

(c) A is true but R is false

(d) A is false but R is true

23. Which of the following molybdenum-containing enzymes do not contain molybdopterin as its cofactor?

(a) Xanthine oxidase (b) Nitrate reductase

(c) Sulfite oxidase (d) Nitrogenase

24. Match List I with List II

List I Enzyme	List II (Reaction catalyzed
A. Glutamate dehydrogenase	I. Catalyzes ATP-dependent conversion of glutamate to glutamine using ammonium as substrate.
B. GS	II. Catalyzes reductive amination of 2-oxoglutarate to form glutamate using ammonium as substrate.
C. GOGAT	III. Catalyzes the synthesis of two molecules of glutamate from glutamine and α-ketoglutarate.
D. Glutamate decarboxylase	IV. Catayzes the synthesis of GABA from glutamate.
	V. Catalyses synthesis of aminolevulinic acid in the chlorophyll biosynthetic pathway.

Choose the correct answer from the options given below:

(a) A -I , B -II , C -III , D -V (b) A -II , B -I , C -III , D -IV

(c) A -I , B -III , C -II , D -V (d) A -III , B -I , C -II , D -IV

25. If a 100 μM solution of NADH exhibits an absorbance of 0.62 at a wavelength of 340 nm in a 1 cm cell, then its molar extinction coefficient would be equal to:

(a) 6.2 $cm^2 mol^{-1}$ (b) 6.2×10^2 $cm^2 mol^{-1}$

(c) 6.2×10^4 $cm^2 mol^{-1}$ (d) 6.2×10^6 $cm^2 mol^{-1}$

26. The mitochondria play a direct role in which of the following subtype(s) of C photosynthesis?

(a) NADP-dependent malic enzyme (NADP-ME) subtype

(b) NADP-ME and NAD-dependent malic enzyme (NAD-ME) subtypes

(c) PEP-carboxykinase (PCK) subtype

(d) NAD-ME and PCK subtypes

27. Which one of the following statements is *not true* about Z-DNA structure?

(a) Phosphate backbone in Z-DNA zig-zag left rather than to the right as in B-DNA.

(b) Z-DNA has only minor groove and major groove is absent.

(c) Z-DNA is much broader than B-DNA.

(d) There is alternate stacking of bases in *anti-* and *syn-* conformation.

28. Match List I with List II

List I Inhibitor	List II Changes observed in the Kinetic parameters
A. Competitive	**I. No effect on the apparent value of the V /K ratio but apparent value of V gets affected.**
B. Uncompetitive	**II. The apparent values of both the V /K ratio and V are altered but by the same magnitude.**
C. Noncompetitive	**III. The apparent values of both the V /K ratio and V are altered but by the different magnitude.**
D. Mixed	**IV. The Vmax/ Km ratio decreases but no effect on the apparent value of V max**
	V. No changes observed in any of the kinetic parameter

Choose the correct answer from the options given below:

(a) A -IV , B -I , C -II , D -III (b) A -I , B -II , C -IV , D -V

(c) A -III , B -I , C -II , D -V (d) A -IV , B -III , C -I , D -II

29. Given below are two statements, one is labelled as Assertion A and the other is labelled as Reason R

Assertion A : The uncompetitive inhibitors bind only to the enzyme-substrate (ES) complex and not to the free enzyme.

Reason R: Binding of the substrate to the enzyme causes a change in its conformation, which reveals the inhibitor-binding site.

In light of the above statements, choose the *correct* answer from the options given below:

(a) Both A and R are true and R is the correct explanation of A

(b) Both A and R are true but R is NOT the correct explanation of A

(c) A is true but R is false

(d) A is false but R is true

30. Given below are two statements, one is labelled as Assertion A and the other is labelled as Reason R

Assertion A: The presence of thick and fleshy water-storing leaves or stems in CAM plants enhances their photosynthetic efficiency.

Reason R: The photosynthetic rates in CAM plants are limited by their vacuolar storage capacity, as malate synthesized at night through carboxylation of PEP has to be stored in the large central vacuole until the daytime.

In light of the above statements, choose the *correct* answer from the options given below

(a) Both A and R are true and R is the correct explanation of A

(b) Both A and R are true but R is NOT the correct explanation of A

(c) A is true but R is false

(d) A is false but R is true

31. Given below are two statements, one is labelled as Assertion A and the other is labelled as Reason R

Assertion A: In Gas-Liquid Chromatography (GLC) the Fatty Acid Methyl Esters (FAMEs) with long fatty acyl chains come out first from the column.

Reason R: Increased chain length of fatty acids leads to lower volatility and their increased retention on the GLC column.

In light of the above statements, choose the *correct* answer from the options given below

(a) Both A and R are true and R is the correct explanation of A

(b) Both A and R are true but R is NOT the correct explanation of A

(c) A is true but R is false

(d) A is false but R is true

32. The antibiotic oligomycin inhibits the oxidative phosphorylation by which of the following mechanisms?

(a) By binding and blocking F component of the mitochondrial ATP synthase complex

(b) By binding and blocking F component of the mitochondrial ATP synthase complex.

(c) By inactivating both the F and F components of the ATP Synthase.

(d) By collapsing the proton motive force

33. Six enzyme classes have been recognized and approved by International Union of Biochemistry since 196(a) In 2018, which one of the following group of enzymes has been added as the seventh class, with Enzyme Commission number 7 (EC 7), by the Nomenclature Committee of the International Union of Biochemistry and Molecular Biology (NC-IUBMB)?

(a) Racemases (b) Translocases

(c) GTPases (d) Ubiquitinases

34. Components of thc mitochondrial Electron Transport Chain (ETC):

A. NADH-Coenzyme Q oxidoreductase

B. Cytochrome bc complex

C. Ubiquinone

D. Cytochrome C

E. Cytochrome C Oxidase

Choose the *correct* sequence of electron tranfer through the aforementioned components of the mitochondrial ETC

(a) A, B, C, D, E (b) A, C, B, D, E

(c) A, D, B, C, E (d) A, B, D, C, E

35. Enzymes involved in the complete oxidation and utilization of odd-chain fatty acids:

A. Propionyl CoA carboxylase

B. Thiolase

C. Fatty acyl CoA synthetase

D. Methylmalonyl CoA mutase

E. Methylmalonyl CoA racemase

Choose the *correct* sequence in which aforementioned enzymes act for the purpose of oxidation and utilization of odd-chain fatty acids as a source of energy.

(a) C, B, A, E, D (b) B, C, A, D, E

(c) A, D, E, C, B (d) A, E, D, B, C

36. Enzymes of the "Kennedy pathway" for the triacylglycerol biosynthesis in plants:

A. Glycerol-3-P dehydrogenase

B. Glycerol-3-phosphate acyltransferase

C. Lysophosphatidic acid acyltransferase

D. Phosphatidic acid phosphatase

E. Diacylglycerol acyltransferase

Choose the *correct* sequence of the action of aforementioned enzymes of the "Kennedy pathway" in the triacylglycerol biosynthesis.

(a) A, B, C, D, E (b) B, A, D, C, E

(c) B, A, C, D, E (d) B, A, C, E, D

37. **Names of the coenzymes and prosthetic groups:**

A. Thiamine pyrophosphate

B. Lipoic acid

C. Pyridoxal phosphate

D. Coenzyme A

E. Nicotinamide adenine dinucleotide (NAD^+)

Choose the *correct* coenzymes and prosthetic groups, from the above list, which are required by the Krebs cycle enzyme Pyruvate dehydrogenase:

(a) A, D and E only (b) A, C and E only

(c) A, B, D and E only (d) A, B, C, D and E only

38. **The biochemical nature of Nuclear Localization Signal is?**

(a) Sequence of amino acids

(b) Sequence of nucleotides similar to DNA

(c) Sequence of nucleotides similar to RNA

(d) Sequence of nucleosides

39. **Which of the following statement is correct for DNA polymerase I of *E. coli*?**

A. Also known as Kornberg polymerase

B. Possesses 5'exonuclease activity

C. Possesses 3'exonuclease activity

D. Plays central role in replication of its genome

Choose the *correct* answer from the options given below

(a) A, B and C only (b) A, C and D only

(c) B, C and D only (d) A, B, C and D

40. **Which of the following combination is appropriate for hybridoma production?**

(a) Lymphocytes from immune mice, HAT, Myeloma cells

(b) Lymphocytes from immune mice, HAT, PEG

(c) Sendai virus, HAT, Myeloma cells

(d) Lymphocytes from immune mice, Aminopterin, Myeloma cells

41. **In order to protect DNA from DNase, which of the following chemical is used during DNA extraction?**

(a) EDTA (b) DEPC

(c) EtOH (d) CTAB

42. Which of the following is most critical temperature, to be decided by primer sequence, in any PCR experiment?

(a) Annealing temperature (b) Denaturation temperature

(c) Primer extension temperature (d) Final holding temperature

43. Which of the following is correct extended form of URL?

(a) Uniform Resource Location (b) Uniform Resource Locator

(c) Uniform Research Linkage (d) Uniform Resource Learning

44. Which of the following technique(s) is used for protein-protein interaction studies?

(a) Co-immunoprecipitation (b) Yeast one hybrid system

(c) Western blotting (d) 2D gel electrophoresis

45. Which of the following nuclease is/are used in genetic engineering?

(a) ZFN (b) *Eco*RI

(c) Cas9 (d) More than one answer are correct

46. Which type of following primers may be used in Reverse Transcriptase-PCR?

(a) oligo-dT (b) Random

(c) Gene specific (d) More than one answer are correct

47. Which of the following technique(s) can be used to study differential gene expression in two tissues under two different external conditions?

A. Northern hybridization

B. Whole genome RNA sequencing

C. qPCR

D. Semi-quantitative PCR

E. Whole Genome Sequencing

Choose the *correct* answer from the options given below

(a) A, B, C and D only (b) A, C, D and E only

(c) B, C, D and E only (d) A, B, D and E only

48. Which of the following is correct statement for RNA synthesis?

A. A single RNA polymerase can synthesize all types of RNA in prokaryotes

B. RNA polymerase II can synthesize mRNA in eukaryotes

C. A single RNA polymerase can synthesize all types of RNA in eukaryotes

D. RNA polymerase II can synthesize all types of RNA in prokaryotes

E. ***E. coli*** **cell has only one type of RNA polymerase**

Choose the *correct* answer from the options given below

(a) A, B and E only (b) A, B and C only

(c) B, C and E only (d) B, C and D only

49. Bacterial conjugation can lead to gain of new alleles in recipient cell

Statement I: Through involving crossing over between DNA fragment of donor and recipient cell

Statement II: Through gain of plasmid from donor bacterium to recipient bacterium In light of the above statements, choose the *correct* answer from the options given below

(a) Both Statement I and Statement II are true

(b) Both Statement I and Statement II are false

(c) Statement I is true but Statement II is false

(d) Statement I is false but Statement II is true

50. Match List I with List II

List I	List II
A. Gene	I. Barbara McClintock
B. Jumping gene	II. Johannsen
C. Cistron	III. Benzer
D. Rolling circle θX DNA replication	IV. Dressler
	V. Cairns

Choose the correct answer from the options given below:

(a) A-V, B-IV, C-I, D-II (b) A-IV, B-V, C-I, D-II

(c) A-II, B-I, C-III, D-IV (d) A-I, B-V, C-III, D-II

51. Match List I with List II

List I	List II
A. Functional genomics	I. Homologous recombination
B. Gene inactivation	II. Protein-Protein interaction
C. In-vivo protein-protein interaction	III. Mass spectrometry
D. Identification of unknown protein	IV. Bimolecular Fluorescence Complementation
V. Yeast 2-hybrid assay	

Choose the correct answer from the options given below:

(a) A-I, B-V, C-II, D-IV (b) A-V, B-IV, C-I, D-II

(c) A-II, B-I, C-IV, D-III (d) A-II, B-III, C-IV, D-V

52. The major groups of multicellular eukaryotes - plants, fungi and animals -each evolved from a different group of the eukaryotes generally referred to as *protists*.

A. Most protists are unicellular.

B. Chloroplast containing protists gave rise to plants.

C. Multicellularity evolved in several protists lineages.

D. Archaea are also protists

Choose the *correct* answer describing *Protists* from the options given below

(a) A and B only (b) B and C only

(c) A and C only (d) A, B and C only

53. Blue-white colonies screening of transformed *E. coli* cells involved disruption of which of the following gene of cloning vector?

(a) X gal (b) Kanamycin

(c) Ampicillin (d) *Lac Z'*

54. Which of the following is NOT correct for vacuole?

(a) Storage of reserves

(b) Contains different types of enzymes involved in breakdown and recycling

(c) Accumulate toxic compounds used in defense against microbial pathogens

(d) Present in a fixed number and size

55. Which of the following usually have less dependency on nitrogenous fertilizer for its commercial cultivation?

(a) Wheat (b) Chickpea

(c) Rice (d) Maize

56. Which of the following statement(s) are correct in describing the gene in DNA?

StatementI: Thestart/initiatorcodonforthegenescanbefoundonthe5'3'strand of DNA

StatementII: Thestart/initiatorcodonforthegenescanbefoundonthe3'5'strand of DNA Choose the correct answer from the options given below

(a) Statement I is incorrect and statement II is correct

(b) Statement I is correct and statement II is incorrect

(c) Both the statements are correct

(d) Both the statements are incorrect

57. Somatic embryogenesis is a desired method of plant regeneration for plant transformation because

(a) It mainly results into chloroplast transformation

(b) It largely favours marker-free transgenics

(c) It produces confirmed transgenics

(d) It produces chimeric transgenics

58. Joining two DNA molecules using an adapter involves the following steps

A. Artificial synthesis of an adapter

B. Treating with polynucleotide kinase

C. Blunt end ligation to add an adapter to a blunt-ended DNA molecule

D. Joining two DNA molecules with sticky ends

What is the correct sequence in which these steps are performed?

(a) B, A, D, C (b) A, C, B, D

(c) D, B, C, A 1 (d) C, A, D, B

59. *Bt* brinjal was developed for resistance to

(a) Salinity (b) Fruit and shoot borer

(c) Damping off disease (d) Drought

60. Which of the following is also known as golden fibre?

(a) Cotton (b) Silk

(c) Jute (d) Flax

61. What types of DNA molecules are copied using rolling circle replication process?

(a) Bacterial chromosomes (b) Some bacteriophage genomes (such as λ)

(c) Mitochondrial genome (d) Yeast chromosomes

62. Plasmid vector will essentially have

(a) Origin of replication, a selectable marker and multiple cloning sites

(b) Promoter

(c) Selectable marker and multiple cloning sites

(d) Terminator

63. Genetic engineering involves the following steps

A. Gene manipulation and vector construction

B. Selection and confirmation of the transformants

C. Gene isolation

D. Genetic transformation

What is the correct sequence in which these steps are performed?

(a) B, A, D, C (b) A, D, C, B

(c) D, B, C, A (d) C, A, D, B

64. Alkaline phosphatase is used to

(a) Add the phosphate group onto free 5' terminus of a DNA molecule

(b) Add one or more deoxyribonucleotides onto the 5' terminus of a DNA molecule

(c) Remove the phosphate group at 5' terminus of a DNA molecule

(d) Change the conformation of covalently closed circular DNA

65. Which of the restriction enzyme does produce fragments with blunt end upon restriction digestion?

(a) *Bam*H (b) *Hind*III

(c) *Not*I (d) *Sma*I

66. Which of the following is NOT a millet crop?

(a) Jowar (b) Wheat

(c) Ragi (d) Bajra

67. A transgenic event is differentiated from other events by

(a) Copy number

(b) Site of integration

(c) Both copy number and site of integration

(d) Type of protein produced

68. Maximum number of types of alleles a recombinant inbred line population derived from crossing two genotypes of rice can generally show at an SSR locus is

(a) 4 (b) 3

(c) 2 (d) 1

69. Plasmid DNA isolation methods are primarily based on the

(a) Location (b) Sequence

(c) Size and conformation (d) Restriction sites

70. Tissue culture for plant transformation should be free from

(a) Somaclonal variation (b) *Agrobacterium* co-cultivation

(c) High rate of shoot regeneration (d) High rate of root regeneration

71. Function of ligase is

A) to repair single-stranded breaks that arise in double-stranded DNA

B) to join together two individual fragments of double stranded DNA

Choose the correct answer from the options given below:

(a) Both A and B are correct

(b) Both A and B are incorrect

(c) Only A is correct and B is incorrect

(d) Only B is correct and A is incorrect

72. Agar, the gelling agent used for tissue culture media, is obtained from

(a) Green algae (b) Red algae (Gelidium amansii)

(c) Blue-green algae (d) Brown algae

73. Suspension culture is generally used for

(a) Haploid production (b) Micropropagation

(c) *In vitro* pollination (d) Production of secondary metabolites

74. Which of the following is an example of a protein domain?

(a) β-sheet (b) Zinc finger

(c) Exon (d) Globin protein

75. What is an open reading frame (ORF)?

(a) All of the nucleotides of a gene that are transcribed into mRNA

(b) The nucleotides of a gene that make up the codons specifying amino acids

(c) The nucleotides of an mRNA molecule before the introns have been removed

(d) The amino acid sequence of a polypeptide

76. The isoelectric point of a protein is defined as

(a) The pH at which a protein has no net charge

(b) The pH at which a protein loses its activity

(c) The pH at which a protein has maximum activity

(d) The pH at which a protein's amino acids are all ionized

77. Hot-start PCR involves the following steps

A. Adding PCR mixtures (water, buffer, dNTPs, primers, template)

B. Adding polymerase

C. Attaining reaction temperature

D. Continuing PCR to amplify the specific amplicons

What is the correct sequence in which these steps are performed?

(a) B, A, D, C (b) A, C, B, D

(c) D, B, C, A (d) C, A, D, B

78. Match List I with List II

List I	List II
A. pBR322	I. Vector facilitating in vitro transcription
B. pUC8	II. Cloning vector based on bacteriophage
C. pGEM3Z	III. E. coli plasmid vector to clone DNA
D. M13mp1	IV. A lac selection plasmid
	V. F plasmid based vector for handling longer DNA inserts for genomic library construction

Choose the correct answer from the options given below:

(a) A-III, B-IV, C-I, D-II (b) A-I, B-III, C-II, D-IV

(c) A-II, B-I, C-III, D-IV (d) A-IV, B-II, C-III, D-V

79. Which of the following statement(s) are correct

Statement I: Bal31 exonuclease removes the nucleotides from both the strands of a double-stranded DNA

Statement II: E. coli exonuclease III degrades both the strands of a double-stranded DNA

Choose the correct answer from the options given below

(a) Only the statement I is correct and the statement II is incorrect

(b) Only the statement II is correct and the statement I is incorrect

(c) Both the statement I and statement II are correct

(d) Both the statement I and statement II are incorrect

80. Codon-anticodon interactions occur by

(a) Covalent bonds (b) Electrostatic interactions

(c) Hydrogen bonds (d) Hydrophobic interactions

81. The region of the soil around the root surface is called

(a) Rhizosphere (b) Endosphere

(c) Phyllosphere (d) Hydrosphere

82. Water is a liquid over the range of temperatures most compatible with life. The energy required to convert one mole of liquid water to one mole of water vapor is called

(a) Heat of vaporization (b) Dielectric constant

(c) Heat of fusion (d) Hydration shell

83. Exudation of sap from the cut surface of the stem of a well-watered herbaceous plant is due to the positive pressure in the xylem, which is termed as

(a) Root pressure (b) Transpiration pull

(c) Osmoregulation (d) Root hydraulic conductivity

84. **Which one of the following is NOT involved in photosynthetic electron transport process?**
(a) Plastoquinone(PQ) (b) Pheophytin (Pheo)
(c) ATP synthase (d) Photosystem I

85. **In metabolically active cells, the osmotic pressure is equal to**
(a) Pressure potential (b) Solute potential
(c) Matric potential (d) Water potential

86. **The membrane protein channels or pores controlling the selective movement of water are**
(a) Aquaporins (b) Channel proteins
(c) Carrier proteins (d) Transport proteins

87. **Match List I with List II**

List I	List II
A. F. W. Went	I. Abscisic acid
B. E. Kurosawa	II. Cytokinins
C. F. Skoog	III. Gibberellin
D. P. F. Waring	IV. Auxin
	V. Ethyelene

Choose the correct answer from the options given below:
(a) A -IV , B - III, C -II , D - I (b) A -IV , B -II , C -III , D -V
(c) A -I , B -II , C -IV , D -V (d) A -IV , B -V , C -II , D -III

88. **The regulation of plant development by light is termed as**
(a) Photomorphogenesis (b) Photosynthesis
(c) Differentiation (d) Photorespiration

89. **Natural rubber is a polymer of**
(a) Isopentenyl units (b) Amino acids
(c) Glucose units (d) Steroid glycosides

90. **The seeds requiring light for induction of germination process is termed as**
(a) Positively photoblastic (b) Long-day
(c) Negatively photoblastic (d) Short day

91. **The precursor for ethylene synthesis in plants is**
(a) Tryptophan (b) Methionine
(c) Indole (d) ent-Kaurene

92. **The principal function of the light-dependent reactions of photosynthesis is to generate**
(a) NADPH and ATP (b) NADH and ATP
(c) FADH2 and ATP (d) NADP and ATP

93. In plants, the tubular extensions of plasma-membrane that connect the cytoplasm of adjacent cells are called

(a) Apoplast (b) Symplast

(c) Plasmodesmata (d) Root hairs

94. The breakdown product of DNA (capable of inducing cell division) which was later identified as cytokinin is

(a) *trans*-Zeatin (b) Kinetin

(c) Benzyladenine (d) Thidiazuron

95. The amount of carbon dioxide assimilated by photosynthesis divided by amount of water transpired is known as

(a) Water use efficiency (b) Transpiration ratio

(c) Quantum efficiency (d) Quantum yield

96. Which one of the following is correctly matched

A. Sugarcane: Kranz anatomy **B. Pineapple: C_4 plant**

C. Rice: C_3 plant **D. Groundnut: C_3 plant**

Choose the *correct* answer from the options given below:

(a) A, C and D only (b) A, B and D only

(c) A and B only (d) B and D only

97. The organelles involved in the process of photorespiration are

(a) Chloroplast, Mitochondria, and Golgybody

(b) Chloroplast, Mitochondria, and Peroxisomes

(c) Chloroplast, Mitochondria, and Nucleus

(d) Chloroplast, Mitochondria, and Vacuole

98. In mitochondria, the enzymes of the Krebs Cycle are located

(a) On the membrane cristae (b) In the matrix

(c) In between the membranes (d) Smooth inner membrane

99. Pressure potential in the xylem of an actively transpiring plant is

(a) Zero (b) Negative

(c) Positive (d) ~ 0.1 MPa

100. The world's leading exporter of sunflower seed oil is

(a) Germany (b) Ukraine

(c) Cananda (d) India

101. Based on water use efficiency, identify the sequence of crops in ascending order:

(a) Pearl millet < redgram< maize

(b) Greengram < finger millet< pineapple

(c) Pineapple < rice< cotton

(d) Sorghum < wheat< brinjal

102. Identify the correct sequence of e flow leading to ATP synthesis in cyclic-photophosphorylation

(a) P 700 → PC → Cyt b_6f complex → Ao/A1 → FeS_X → Fd → P700

(b) P 700 → Cyt b_6f complex → Ao/A1 → FeSx → Fd → PC → P700

(c) P 700 → Fd → Cyt b_6f complex → PC → Ao/A1 → FeS_X → P700

(d) P 700 → Ao/A1 → FeSx → Fd → Cyt b_6f complex → PC → P700

103. Identify the sequence of nutrient elements in crops, capable of getting re-distributed under deficiency conditions in ascending order

(a) N < Fe< B (b) K <Mn< Ca

(c) P <Mg< B (d) B <Fe< N

104. An example of Short Day Plant

(a) Spinach (b) Barley

(c) Henbane (d) Cucumber

105. Abscisic acid synthesis in plants involves :

A. Isoprenoid patway **B. SAM pathway**

C. Carotenoid pathway **D. Indole-acetal doxim pathway**

Choose the *correct* answer from the options given below:

(a) A, B and D only (b) A and C only

(c) C and D only (d) B and D only

106. Given below are two statements

Statement I: Methyl viologen, also known as Paraquat, used as herbicide inhibits light reactions of photosynthesis.

Statement II: Paraquat blocks electron flow at plastoquinone acceptors of PS II by competing for binding sites of plastoquinone.

In light of the above statements, choose the *most appropriate* answer from the options given below

(a) Both Statement I and Statement II are correct

(b) Both Statement I and Statement II are incorrect

(c) Statement I is correct but Statement II is incorrect

(d) Statement I is incorrect but Statement II is correct

107. The essential co-factor in water-oxidizing process at PS II in the light reactions of photosynthesis

(a) Mg^{2+} (b) Mn^{2+}

(c) Fe^{2+} (d) Cu

108. In acid soils at pH of < 5, the element that becomes available in toxic amounts is 5, the element that becomes available in toxic amounts

(a) Nitrogen (b) Phosphorus

(c) Aluminium (d) Molybdenum

109. Phytochrome is localised within:

(a) Chloroplast (b) Mitochondria

(c) Plasma membrane (d) Cytosol

110. Bakane disease in rice is associated with discovery of this plant hormone

(a) Ethylene (b) Cytokinins

(c) Abscisic acid (d) Gibberellins

111. When tissue culture medium contains both auxin and kinetin in the ratio of 10:1, the pith cells of tobacco grow and form callus. Further,

Statement I: If the ratio of auxin to cytokinin is less in the medium, a number of roots are initiated from callus

Statement II: If the ratio is more (less cytokinin than auxin), a number of shoot buds are initiated. In light of the above statements

Choose the *most appropriate* answer from the options given below

(a) Both Statement I and Statement II are correct

(b) Both Statement I and Statement II are incorrect

(c) Statement I is correct but Statement II is incorrect

(d) Statement I is incorrect but Statement II is correct

112. Short term treatment for about 16-20 hours, given to cut flowers with high concentration of sucrose, AgNO Benzyl 3,Adenine etc to increase the shelf-life is called

(a) Grating (b) Holding

(c) Pulsing (d) Priming

113. Match List I with List II

List I	List II
(Nutrient Element)	**(Physiological function)**
A. Calcium	**I. Constituent of superoxide dismutase**
B. Zinc	**II. Activation of RuBP carboxylase**
C. Sulphur	**III. Synthesis of chlorophyll precursor**
D. Magnesium	**IV. Constituent of glutathione**
	V. Structural component of chromosomes

Choose the correct answer from the options given below:

(a) A - I, B - V, C - II, D - III (b) A - II, B - III, C - I, D - V

(c) A - III, B - II, C - V, D - I (d) A - V, B - I, C - IV, D - II

114. Match List I with List II

List I (Type of senescence)	List II (Example)
A. Overall senescence	I. Sugarbeet
B. Progressive senescence	II. Gulmohar
C. Top senescence	III. *Lemna*
D. Deciduous senescence	IV. Wheat
	V. Tobacco

Choose the correct answer from the options given below:

(a) A - V , B - IV , C - III, D - I (b) A - IV, B - V, C - I , D - II
(c) A - V, B - IV, C - II , D - I (d) A - III, B - IV, C - I , D - V

115. Most widely used chemicals for sugarcane ripening :

A. Ethrel **B. GA**
C. Salicylic acid **D. Glyphosate**

Choose the *correct* answer from the options given below:

(a) A and D only (b) B and C only
(c) A and C only (d) C and D only

116. Identify "Recalcitrant seeds":

A. Mungbean B. Rubber
C. Tea D. Wheat

Choose the *correct* answer from the options given below:

(a) A, B and D only (b) B, C and D only
(c) B and C only (d) A and D only

117. Examples of "Climacteric fruits":

A. Banana **B. Grape**
C. Tomato **D. Mango**

Choose the *correct* answer from the options given below:

(a) A, B and D only (b) B, C and D only
(c) A, B and C only (d) A, C and D only

118. The compound discovered as natural trigger for the metaboic explosion that raises the temperature of thermogenic inflorescence of *Arum lillie* is:

(a) Abscisic acid (b) Salicylic acid
(c) Brassino steroids (d) Ethylene

119. The following are found translocated abundantly in phloem

A. Aminoacids **B. Sucrose**
C. Phosphorus **D. Potassium**

Choose the *correct* answer from the options given below:

(a) A, B and D only (b) B, C and D only

(c) B and D only (d) A and D only

120. Identify the "Growth Retardants" used in crop plants

A. Maleic Hydrazide **B. Chloromequat Chloride**

C. Paclobutrazole **D. Triiodobenzoic acid**

Choose the *correct* answer from the options given below:

(a) A, C and D only (b) A, B and D only

(c) B, C and D only (d) B and C only

ANSWERS

1. (a)	2. (c)	3. (c)	4. (c)	5. (b)	6. (b)	7. (c)	8. (a)	9. (d)	10. (a)
11. (a)	12. (d)	13. (d)	14. (c)	15. (a)	16. (c)	17. (b)	18. (c)	19. (b)	20. (d)
21. (a)	22. (a)	23. (d)	24. (c)	25. (b)	26. (b)	27. (b)	28. (b)	29. (a)	30. (a)
31. (a)	32. (a)	33. (b)	34.(a)	35. (d)	36. (b)	37. (a)	38. (a)	39. (b)	40. (a)
41. (a)	42. (a)	43. (b)	44. (a)	45. (d)	46. (d)	47(b)	48. (a)	49. (c)	50. (a)
51. (b)	52. (c)	53. (d)	54. (d)	55. (b)	56. (b)	57. (c)	58. (b)	59. (b)	60. (c)
61.(c)	62. (a)	63. (b)	64. (c)	65. (c)	66. (b)	67. (c)	68. (c)	69.(c)	70. (b)
71. (d)	72. (b)	73. (d)	74. (b)	75. (b)	76. (a)	77. (a)	78. (a)	79. (c)	80. (c)
81. (a)	82.(a)	83.(a)	84.(c)	85. (b)	86. (a)	87. (b)	88. (a)	89. (a)	90. (a)
91. (b)	92. (a)	93. (c)	94. (b)	95. (a)	96. (b)	97. (b)	98. (b)	99. (b)	100.(b)
101. (a)	102. (d)	103. (a)	104.(d)	105. (c)	106. (c)	107. (b)	108. (c)	109. (d)	110. (d)
111. (c)	112. (c)	113. (c)	114. (c)	115. (c)	116. (b)	117. (a)	118. (c)	119. (c)	120. (a)

CHAPTER - 1

General Genetics

Genetics

The word genetics, coined by William Bateson (1906), is derived from the Latin word genesis, which means birth. Genetics is the study of heredity and variation. Heredity refers to the transmission of genetic information, more precisely genes, from parents to offspring. Variation refers to the difference among the individuals of a species for a character. It may be due to heredity or environment. Modern genetics is traditionally classified into three branches as transmission genetics, molecular genetics and population genetics.

Year	Scientists and Discovery
1869	Friedrich Miescher identified DNA
1901	Hugo de vries coined the term "mutation"
1902	Sutton & Boveri proposes Chromosomal theory of inheritance
1906	Bateson coined the term "genetics"
1909	-Johannsen coined the term "gene". -Archibald Garrod published the book "Inborn errors of metabolism"
1910	T.H. Morgan studied linkage and crossing over in Drosophila and coined the term " crossing over " (Nobel prize, 1933).
1911	Sturtevant constructed the first linkage map in Drosophila.
1917-1923	Bridges gave genic balance theory and non-disjunction theory. He described the different chromosomal abnormalities and aberrations.
1925	Bernestein proposed multiple allelic inheritance of ABO blood group
1927	H.J. Muller reported use of CIB technique to show X-rays are mutagenic (Nobel prize, 1946).
1928	Griffith discovered transformation in *Diplococcus pneumoniae.*
1930	R.A. Fisher, J.B.S. Haldane and Sewell Wright laid the foundation for population genetics.
1941	Beadle and Tatum proposed "one gene-one enzyme" in Neurospora (Nobel prize 1958).
1944	Avery, MacLeod and McCarty demonstrated the transforming principle as DNA, the genetic material
1950	Chargaff proved that in DNA, the amount of adenine is equal to amount of thymine; the amount of guanine is equal to amount of cytosine and the content of purine is equal to content of pyrimidine.

1952	Hershey (Nobel prize, 1969) and Chase revealed DNA as the genetic material of bacteriophage.
1953	J. D. Watson, F.H.C. Crick (Nobel prize, 1962) proposed double-helical structure model of DNA using X-ray diffraction data of Wilkins (Nobel prize, 1962) and base composition data of Chargaff.
1955	Benzer described fine structure of the phage T4rII locus
1956	Tjio and Levan resolved the normal diploid chromosome number of humans as 46
1957	Frankel Conrat and Singer established RNA as the genetic material of tobacco mosaic virus
1958	Meselson and Stahl identified semi-conservative replication of DNA
1961	- Nirenberg and Mathaei cracked the genetic code present on mRNA - Jacob and Monod proposed the "Operon model" for gene regulation (Nobel prize, 1978)
1965	Harris Watkins produced hybrid cell fusing somatic cells of mouse and man.
1970	Nathan and Smith isolated first restriction endonucleases (Nobel prize, 1968).
1972	Berg produces first recombinant in vitro (Nobel prize, 1980).
1977	Maxim, Gilbert and Singer published DNA sequencing techniques (Nobel prize, 1980)
1983	-Kary Mullis and others developed the polymerase chain reaction (PCR) for quick amplification of DNA. -Nobel prize for Barbera McClintock for her jumping genes concept.
1986	Nirenberg and Har Govind Khorana established the complete genetic code.
1990	-Gene therapy used to treat human genetic diseases in U.S.A. - Human Genome Project was launched.
1995	DNA of the bacterium *Haemophilus influenzae* was fully sequenced.
1997	Ian Wilmut cloned a sheep "Dolly"
1998	Rough draft of human genome map showing more than 30,000 genes was released. Development of Terminator gene technology by Delta and Pine Co and Bollguard cotton by Monsanto
2001	Craig Venter and Francis Collins announced first complete draft of human genome.
2003	The Human Genome Project was successfully completed.

Famous Indian Plant Breeders and their Contributions

- **T.S. Venkatraman** - An eminent sugarcane breeder, he transferred thick stem and high sugar contents from tropical noble cane to North Indian Canes. This process is known as noblization of sugarcane.
- **B.P. Pal** - An eminent Wheat breeder,developed superior disease resistant N.P. varieties of wheat.
- **M.S. Swaminathan-** Responsible for green revolution in India, developed highyielding varieties of Wheat and Rice

- **Pushkarnath** - Famous potato breeder
- **N.G.P. Rao** - An eminent sorghum breeder
- **K. Ramaiah** - A renowned rice breeder
- **Ram Dhan Singh** - Famous wheat breeder
- **D.S. Athwal** - Famous pearlmillet breeder
- **Bosisen** - An eminent maize breeder
- **Dharampal Singh** - An eminent oil-seed breeder
- **C.T. Patel** - Famous cotton breeder who developed world's first cotton hybrid in 1970
- **V. Santhanam** - Famous cotton breeder

Plant Genetics

Plant Genetics is the study of plant genes, heredity, and genetic variations in plants. It is the sub-field of botany. Plant Genetics helps us to understand the genetic makeup of plants. It also involves understanding genetic information that passes from one generation to another and the development and functioning of plants in various environments. Plant genetics research significantly impacts economics by modifying staple crops, increasing yields, providing disease and insect resistance, herbicide resistance, and improving nutritional value.

History of Plant Genetics

History of plant genetics is as follows:

- Plant Genetics was started with the George Mendels, who was born on 20th of July in 1822 in Austria Hungary.
- He is known as "Father of Genetics" because of his work on pea plant and showed the inheritence of particular traits and named as Mendel's law of Inheritence.
- W.E. Castle showed that seperation and selection of individual traits over time.
- G.H. Hardy and W. Weinberg, gave the concept of Hardy-Weinberg equilibrium in 1908.
- Nils and Heribert-Nilsson showed that crossing plants and creation of hybrids results in heterosis and hybrid vigor.

Model Plants for Plant Genetics

These are list of model plants on which genetic engeenering is performed. These are as follows:

Arabidopsis thaliana

Arabidopsis thaliana has following features:

1. It has been the model plant for the century in the field of plant genetics.
2. It is the first plant to get its whole genome sequenced in 2000 years.
3. It has a small genome, which makes its sequencing more possible.
4. Its genome size is 125 Mbp and have 25000 of genes.
5. It has its own database which is known as The Arabidopsis Information Resources (TAIR).
6. There are various researches which has been done or being in the process like genetic manipulation, biotic and abiotic resistance study.

Brachypodium distachyon

Brachypodium distachyon has following features:

1. It is model grass and many research have been performed on this due its excellence as temperate cereals.
2. Brachypodium is a diploid and have genome size of 355 Mbp.
3. It life cycle is small thus making it ideal for plant genetic study.

Nicotiana benthamiana

Nicotiana benthamiana has following features:

1. It is also an ideal plant model for genetic study related to plant- pathogen interactions and transgenic approaches.
2. It has a broad and larger leaf which makes Agrobacterium tumefaciens to infect it easily.
3. Nicotiana is used for expression of pathogens and new traits as well.

There are other organisms also who are actively involved in Plant Genetics, which includes

- Alga- Chlamydomonas reinhardtii
- Moss - Physcomitrella patens
- Clover - Medicago truncatula
- Snapdragon - Antirrhinum majus
- C4 Grass - Setaria viridis
- Maize

Genetically Modified Crops

GM crops are those crops which has been altered by inducing gene of interest using genetic engineering such as CRISPR tool. Genetic engineering allow to introduce gene of interest in selected crop and have better control over it.

Benefits of GM crop

Benefits of GM crop includes:

- GM crops can complete increasing demands of population.
- GM crops are resistance to pathogen attack and herbicides.

They also have genetically induced vitamin A such rice to erase its deficiency in poor people.

They have more nutritional values.

Examples of GM Crops

Examples of GM crops includes:

- Soybean: Resistance to herbicides and insects
- Potato: Resistance to insects, prevent browning
- Alfa Alfa: Resistance to herbicides
- Papaya: Resistance to ring spot virus disease
- GM cotton: Resistance to pests
- GM rice: Rich in Vitamin A

Define cell? And Write the Features of Cell Theory?

- Cell is the basic unit of organization or structure of all living matter.
- It was first discovered by Robert Hook in 1665 in cork tissue.
- Loewy and Sickevitz in 1963 defined a cell as "a unit of biological activity delimited by a semi-permeable membrane and capable of self-reproduction in a medium free of other living system".
- The cell has also been defined as "a unit of life that is the smallest unit which can carry on the activities indispensable to life to grow, to synthesi new living material and to produce new cells".

Features of Cell Theory:

- The cell theory or cell doctrine was formulated independently by M.J. Schleiden and Theodor Schwann in 1838-39.
- It states that the animals and plants differ from each other superficially. But they have same pattern of organization and construction.
- It further states that the bodies of both animals and plants are composed of cells and that each cell can not only act independently but also functions as an integral part of complete organism.
- Thus the cell is considered as morphological and physiological unit of living organisms or in words of Schwann and Schleiden as "functional and biological unit".

- Cells reproduce, assimilate, respire, respond to changes in the environment and absorb water and other materials from internal or external courses.
- R. Virchow formulated cell lineage theory i.e. cell is hereditary unit of an organism.

What is meant by Plastids? And explain its types?

- Plastids are the cytoplasmic organelles of the cells of plants and some protozoans such as Euglena.
- Whereas the cells of the bacteria, fungi and animals contain chromatophores instead of plastids.
- Plastids perform most important biological activities such as the synthesis of food and storage of carbohydrates, lipids and proteins.
- The term plastid is derived from the Greek word "plastikas" means formed or moulded and was used by A.P.W. Schimper in 1885.
- He classified the plastids into the following types based on their structure, pigments and function. 1. Chromoplasts (coloured) 2. Leucoplasts (colourless)

Chromoplasts: (Greek words, chroma = colour; plast = living)

- These are the coloured plastids of plant cells.
- They contain a variety of pigments and synthesizes the food through photosynthesis.
- Based on the type of pigment present in them, the chromoplasts of microorganisms and plant cells are as follows.

(*a*) **Chloroplasts:** (Greek words, chlor = green; plast = living)

- These are most widely occurring chromoplasts of the plants.
- They occur mostly in the green algae and higher plants.
- The chloroplasts contain the pigments chlorophyll A and chlorophyll B. They also contain DNA and RNA.

(*b*) **Phaeoplasts:** (Greek words, phaeo = dark brown; plast = living)

- These contain the pigment "Fucoxanthin", which absorbs the light.
- They occur in the diatoms, dinoflagellates and brown algae.

(*c*) **Rhodoplast:** (Greek words, rhodo = red; plast = living)

- The rhodoplast contains the pigment phycoerythrin which absorbs light.
- The rhodoplast occur in red algae.

Leucoplasts: (Greek words, leuco = white; plast = living)

- These are the colourless plastids which store the food material such as carbohydrates, lipids and proteins.
- The leucoplasts are rod like or spheroid in shape and occur in the embryonic cells, sex cells and meristematic cells.

- The most common leucoplasts of the plants cells are as follows
  (*a*) **Amyloplasts:** (Greek word, amyl = starch) these synthesize and store starch and occur in those cells which store starch.
  (*b*) **Elaioplasts:** These store lipids and occur in seeds of monocotyledons and dicotyledons.
  (*c*) **Proteinoplasts or proteoplasts:** These are the protein storing plastids which mostly occur in seed and contain few thylakoids.

Mitochondria and Chloroplasts are originated from prokaryotes? Substantiate your answer?

Mitochondria and chloroplast originate from prokaryotes. It is supported by so many similarities between these cell organelles and prokaryotes like

1. Nature of chromosomal DNA i.e. circular in nature.
2. Size of ribosomes i.e. 70S type.
3. Mode of replication of DNA i.e. by D – loop method.
4. Mitochondria attached to inner cellular membrane like mesosomes of bacteria i.e. both have respiratory function.
5. As like prokaryotes, there is no repetitive DNA in these organelles.
6. And also mitochondria is similar to invasive aerobic bacteria and chloroplast is similar to blue green algae in terms of length ((~3.4 - 44μ) and amount (~1 × 107 daltons) of DNA.
7. In prokaryotes, protein synthesis is inhibited by antibiotics like chloramphinical as like chloroplast and mitochondria. But nuclear protein synthesis inhibited by cycloheximide but not chloramphinical.

These evidences strongly support the origin of mitochondria and chloroplast from prokaryotes. That's why chloroplast and mitochondria were also referred to as prokaryotic cell within eukaryotic cell and also semi autonomous organelles.

Write the difference between euchromatin and heterochromatin? Explain how heterochromatin is advantageous over *euchromatin*?

S.No	Heterochromatin	Euchromatin
1	Represent darkly stained regions	Lightly stained regions
2	Contains few inactive genes	Contains lot of active genes
3	Covers small region of chromosome	Larger region of chromosome
4	Usually found near centromere and telomere	Found in the middle of chromosome between centromere and telomere
5	Two types – Constitutive and facultative	Only one type
6	Late replicating	Normal replicating
7	Usually no active part in transcription	Plays active role in transcription

Contd...

S.No	Heterochromatin	Euchromatin
8	30 nm fibre	3-8 nm fibre
9	Dark in interphase	Light in interphase
10	Late replicating sequence	Early replicating sequence
11	Rich in ATC	AT and GC are present
12	No condensation and decondensation	It undergoes condensation and decondensation
13	At particular places (centromere and telomere)	All over chromosome

Heterochromatin is further classified into two groups: a) Constitutive and b) Facultative

1. Constitutive heterochromatin: It is present in all cells at identical positions on both homologous chromosomes of a pair.
2. Facultative heterochromatin: It varies in state in different cell types, at different stages or sometimes, from one homologous chromosome to another. A well known example of facultative heterochromatin is the *Barr body*, an inactivated X chromosome in somatic cells of mammalian female(XX).

Heterochromatin is advantageous over euchromatin in terms of its composition of highly repetitive DNA i.e. tandem repeats. So, it is useful as molecular marker i.e. variable number of tandem repeats (VNTRs).

Explain the experimental evidences supporting the unineme Model of Chromosome structure?

Experimental evidences support the unineme models of chromosome structure are

(*i*) Lambrush chromosome establishment

(*ii*) Visco elastrometric evidence

(*iii*) Microspectro photometric evidence

(*iv*) Autoradiographic evidence.

Lambrush chromosome establishment:

- Lambrush chromosomes seen in diplotene stage of prophase I of primary oocytes of vertibrates and invertibrates
- Length of lambrush chromosomes is 800 μm in length
- When those chromosomes treated with DNAse, RNAse and protease

  Treatment of DNAse – loops are disintegrated

  Protease and RNAse – matrix will disintegrate
- After breaking, DNA strand is 20A° in diameter.
- It gives the evidence of single DNA per chromosome

Visco elastrometric evidence:

- Visco elastrometry means a technique of analysing viscosity of molecule present in the solution i.e. DNA molecule present in solution.
- In this, the cells are lysed in container to minimise the shearing and isolation of DNA.
- These DNA molecules are subjected to driving force by rotating cylinder and by stopping cylinder the DNA retains to original position.
- Here recoiling time is the function of size of DNA present in cell.
- If size of DNA increases, time to recoiling is also increases.
- In this study, the size of DNA molecule is 4.1×10^{10} daltons in drosophila.

Micro spectrophotometric method:

- In this light is used to study the size of DNA i.e. light is passed through the DNA
- Here the size of DNA studied is 4.3×10^{10} daltons in drosophila
- In both spectro photometric and visco elstrometric methods, the size of DNA is almost equal.

Autoradigraphy:

- In this, permit to lyse the cells in gently to minimise shearing and isolation of DNA
- After that these lysed DNA are diffused on to a radioactive membrane
- Exposed to β – particle emulsion
- Enlargement of DNA will be seen on film
- In this study, the contour length of DNA is 1.2 cm and amount is $2.4 – 3.2 \times 10^{10}$ daltons.

Write short notes on the following? All questions should be in Normal sentence case?

1. Salivary gland chromosomes
2. Lambrush chromosomes
3. B- chromosomes

Salivary gland chromosomes:

1. First found in Dipterian insects by E.G. Balbani.
2. It is somotically paired, so appear half (n), not (2n) chromosome.
3. Distinct pattern of transverse band (chromatic + a chromatic) is found.
4. Band form reversible puff, chromosomal puff, Balbiani ring associated with different gene activation.
5. Also called a polytene chromosome.
6. Repeated cycles of endo-replication (endomitosis) are found.
7. Five long and one short strand (form chromacentre at centre).
8. Long (X chromosome), 4 chromosome (II and III).
9. Male flies Y chromosome is in centre and not seen as separate.

10. Puffs are sites of RNA synthesis.
11. Ecdysome induces puffs in specific band.
12. Loss or addition of band is also affecting the phenotype. E.g. 16A region in X-chromosome of Drosophila.

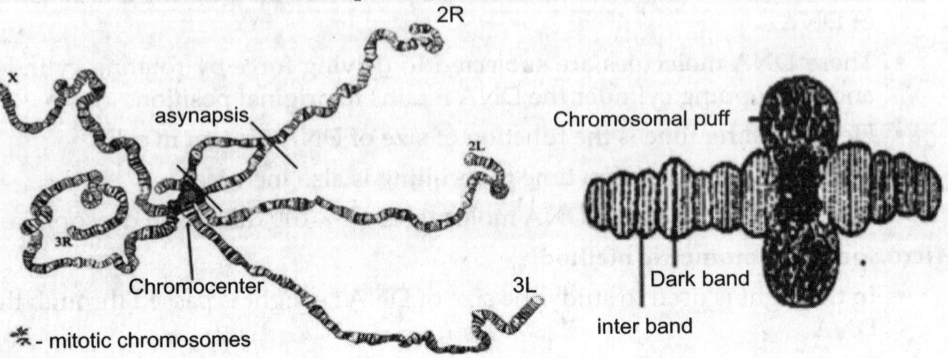

Lambrush chromosome: (Ruckert in 1882)

1. Found in oocytes of invertebrate and vertebrate, except mammal.
2. Distinctly observed in prolonged diplotene stage.
3. Pair of lateral loop from majority of chromomere is formed.
4. One to nine pair of loop from single chromomere is formed.
5. 800-1000μ in length.
6. Loop - 9.5 μ to 200 μ.
7. Loops are slowly withdrawn at metaphase.
8. Loop - single chromatid (fully extended single DNA double helix)
9. Site for RNA synthesis.
10. Function is to produce large quantities of protein and RNA stored in eggs.
11. Gene transcripts stored in inactive but stable form active only after eggs are fertilised.

B- chromosome: B-chromosomes are supernumerary chromosome. It may be missing or found as extra chromosome in many species, both plants and animals.

Features:

1. Dispensable not found in all individuals of a species may not be found in all cells.
2. Non-homologous with any of the basic chromosome.
3. Non-mendelian inheritance.
4. Smaller than A-chromosomes.
5. Unique pattern of heterochromatin distribution and telocentric in nature.
6. In general, genetically alert.
7. When present in high number; suppress vigour and fertility.

8. Mostly found in out breeds.
9. Sometimes restricted to aerial parts i.e. absent in roots. E.g. *Aegilops speltoids* and *Aegilops mutica.*
10. Effect on pollen fertility and seed set.
11. Flowering time delayed and other characters like plant height, tiller no adversely affected.

Explain in brief about nucleolar chromosomes?

Nucleolar Chromosomes:

The chromosome forming the nucleolus is called the nucleolar chromosome.

- Normally one or more chromosomes in the haploid set (genome) are nucleolar chromosomes.

  E.g. In maize (2n= 20), chromosome 6 forms the nucleolus, where as in barely (2n=14), two chromosomes, namely, 6 and 7 are nucleolar chromosomes. And in human, (2n=46), chromosomes of the D group (13, 14, 15) and G group (21, 22) are nucleolar chromosomes.
- The region of the chromosome which is active in nucleolus formation is called nucleolar organiser region or nucleolar organising region
- It contains ribosomal DNA (r DNA) or r RNA cistrons.
- The observable number of the nucleoli does not correspond to the number of nucleolar organiser chromosomes.
- The nucleolar organiser region is negatively heteropycnotic and is called the secondary constriction.
- The portion of the chromosomal segment separated by the secondary constriction from the main body of the chromosome is called satellite. And the chromosome bearing a satellite is termed as SAT- Chromosome or satellite chromosome.
- The nucleolar organiser region can be translocated from one chromosome to the other.
- When the nucleolar organizer is lost, a large number of small nucleoli are formed in all the chromosomes.
- This indicates that the main nucleolar organiser functions to form nucleolus and suppresses other loci.
- Nucleolar suppression may be observed in allopolyploids where genomes of two different species are brought into a single nucleus. .

With a focus on chromosomes, what are the key events during interphase and M –phase in eukaryotic cell cycle?

A. **Interphase:** Nucleus is not in a state of division.

   G1 phase – Lot of RNA and proteins are synthesized

S phase – Chromosomes are duplicated

G2 phase – RNA and protein synthesis continued

B. M – Phase

1. **Prophase:**
   - Initiation of spindle formation and condensation of duplicated chromosomes
   - Formation of spindle is accompanied by fragmentation of endoplasmic reticulum and golgi complex
   - Nucleolus disappears
   - Nuclear membrane breaks up in to many small vesicles and microtubules.
2. **Metaphase:**
   - Attachment of spindle microtubule to the kinetochores indicates that cell enters into metaphase.
   - Through the operation of spindle apparatus, the duplicated chromosome comes to lie in a single plane in the middle of the cell. This equatorial plane is called the metaphase plate.
3. **Anaphase:**
   - The sister chromatids of duplicated chromosomes are separated from each other during anaphase of meiosis.
   - This separation is accomplished by shortening the microtubules attached to the kinetochores and by degrading the materials that hold sister chromatids hold together.
4. **Telophase:**
   - Once separation is achieved, the chromosomes decondensed into network of chromatin fibres
   - Nucleolus and nuclear membrane will be reformed.

Arrange the following in a correct sequential order in relation to meiotic cell division?

(*a*) Disjunction of chromosomes

(*b*) Disjunction of chromatids

(*c*) Terminalisation of chromosomes

(*d*) Formation of cell plate

(*e*) Pairing of homologous chromosomes

(*f*) Co-orientation of chromosomes

(*g*) Auto orientation of chromosomes

(*h*) Exchange of segments between non sister chromatids of homologous chromosomes

(*i*) Reappearance of nucleolus and nuclear membrane

Correct order is e (Zygotene), h (Pachytene), c (Diplotene), f (Metaphase I), a (Anaphase I), g (Metaphase II), b (Anaphase II), i (Telophase II) and d (Cytokinesis)

What are the differences between mitosis and meiosis from genetics point of view? Whether meiosis is boon or bane for the organisms? Explain.

S. No	Feature	Mitosis	Meiosis
1	Occurrence	Both haploid and diploid cells	Only in diploid cells
2	Crossing over and synapsis	Does not occur	Occurs
3	Characteristic feature of metaphase plate	At metaphase plate chromosomes belong to either maternal or paternal parents are arranged. So there is no variation	At metaphase plate chromosomes belong to both maternal and paternal parents are arranged. So it gives lot variation
4	Division of centromere	Occurs at Anaphase	Occurs at Anaphase II only. At Anaphase I only disjunction of chromosomes takes place.
5	Other features	It is not the basis for explanation of mendelian principle of independent assortment	It is the basis for explanation of mendelian principle of independent assortment

- Meiosis boon or advantage for the organisms because in meiosis at Metaphase I, there will be random arrangement of maternal and paternal chromosomes i.e homologous chromosomes. There by it facilitates the separation of homologous chromosomes at Anaphase I.
- By this pairing, in drosophila we can observe 2^4 combinations of possibility of variability in the form of gametes. And also in human beings 2^{23} is possibility of gametic variability. In meiosis, there will be a chance of occurrence of crossing over, so we can't imagine the recombination percentage in meiosis.
- In sexually reproducing species, meiosis is essential for the continuity of generation. Because meiosis results in the formation of male and female gametes and union of such gametes leads to the development of zygotes and there by new individual.

What is meant by linkage? And write types and characteristic features of linkage?

The tendency of genes to remain together in their original combination during inheritance is called linkage.

- Mendel's law of independent assortment is applicable only when the genes are located in different chromosomes while linkage refers to the genes located in the same chromosome.
- The phenomenon of linkage was first reported by Bateson and Punnet in 1906. They studied flower colour and pollen shape in sweet pea involving two varieties / races.

Characteristic features of Linkage:

1. Linkage involves two or more genes which are located in same chromosome in a linear fashion.
2. Linkage reduces variability.
3. Linkage may involve either dominant or recessive alleles (coupling phase) or some dominant and some recessive alleles (repulsion phase).
4. It may involve either all desirable traits or all undesirable traits or some desirable and some undesirable traits.
5. It is observed for oligo-genic traits as well as polygenic traits.
6. Linkage usually involves those genes which are located close to each other.
7. The strength of linkage depends on the distance between the linked genes. Lesser the distance, higher the strength and vice versa.
8. Presence of linkage leads to higher frequency of parental types than recombinants in test cross. When two genes are linked the segregation ratio of dihybrid test cross progeny deviates significantly from 1:1:1:1 ratio.
9. Linkage can be determined from test cross progeny data.
10. If crossing over does not occur, all genes located on one chromosome are expected to be inherited together. Thus the maximum number of linkage groups possible in an organism is equal to the haploid chromosome number.
11. Linkage can be broken by repeated intermating of randomly selected plants in segregating population for several generations or by mutagenic treatment.
12. Besides pleiotropy, linkage is an important cause of genetic correlation between various plant characters.

Explain the method of detection of linkage along with significance of linkage in crop improvement?

Detection of linkage:

Test cross is the most common method of detecting the linkage. In this method, the

F_1 heterozygous at two loci (AB/ab) is crossed to a double recessive parent (ab/ab) and the phenotypic ratio of test cross progeny is examined. If the phenotypic ratio of test cross progeny shows 1:1:1:1 ratio of parental and recombinant genotypes, it indicates absence of linkage. If the frequency of parental types and recombinant types deviate significantly from the normal dihybrid test cross ratio of 1:1:1:1, it reveals presence of linkage between two genes under study.

Another way to detect the presence or absence of linkage is to self pollinate the individual heterozygous at two loci. If there is complete dominance at each locus and no epistasis, the segregation ratio of the progeny will be 9:3:3:1. Presence of linkage either in coupling or repulsion phase will lead to significant deviation from 9:3:3:1 ratio. The deviation of observed values from the expected ratio is tested with the help of χ^2 test.

Significance of Linkage in Plant Breeding:

1. Effect on selection
2. Effect on genetic variances
3. Effect on genetic correlation

Effect on selection: Linkage between two (or) more loci controlling different desirable characters is advantageous. Due to linkages occur together frequently in segregating population. For this, breeder has to grow large populations than with unlinked loci.

Effect on genetic variance: Estimates of genetic variance for quantitative characters are greatly influenced.

- Linkage disequilibria influence the gene action by two way of causing bias in estimates of genetic variances.
- Repulsion phase linkage caused upward bias in dominance variance and downward bias in additive variance.

Effect on genetic correlation: Linked characters generally show high value of genetic correlation and also of co heritability.

- If linked-two desirable characters can be improved simultaneously.
- If not breeder has to combine from two different sources.

Why the organisms should have linkage groups? How many linkage groups are possible in an organism? Explain with examples of human beings and drosophila?

- Group of genes present on same chromosome are said to be linked to gather.
- Organisms should have linkage groups because these are represented by haploid (n) chromosome number i.e. two homologous chromosomes are similar that's why we represent with only one chromosomal set.
- Thus the maximum number of linkage groups possible in an organism is equal to the haploid chromosome number (n).

E.g. Onion 2n = 16 n = 8 maximum linkage groups possible = 8

Maize 2n = 20 n = 10 maximum linkage groups possible = 10

- But in case of heterogametic sex chromosomes (i.e. it represents non homologous chromosomes), number of linkage groups equal to one more than haploid chromosome number (n+1). i.e. in this case linkage groups mainly depends on both autosomes and sex chromosomes.

  Number of linkage groups = Number of autosome pairs + Number of sex chromosome pairs.

  Linkage groups in human beings (XX-XY):

  In female (XX), number of linkage groups = n = 23

  But in males (XY), number of linkage groups = n + 1 = 24

  Linkage groups in drosophila (XX-XY):

  In female (XX), number of linkage groups = n = 4

  But in males (XY), number of linkage groups = n + 1 = 5

What is meant by crossing over? And explain the cytological proof of crossing over?

The term crossing over was first used by Morgan and Cattell in 1912. The exchange of precisely homologous segments between non-sister chromatids of homologous chromosomes is called crossing over.

- First cytological proof for genetic crossing over was proposed by Curt Stern in Drosophila in 1931
- He used cytological markers in his studies, female fly in which one X-chromosome was broken in to two segments. Out of these two segments, one behaved as X-chromosome and the other had small portion of Y-chromosome attached to its one end.
- In female fly, the broken X- chromosome had one mutant allele (carnation) for eye colour and another dominant allele (B) for bar eye shape. The other X- chromosome with attached portion of Y chromosome had alleles for normal eye colour (red eye) and normal eye shape. Thus the phenotype of female was barred.
- A cross of females was made with carnation male (car +). As a result of crossing over female flies produce four types of gametes, *viz.*, two parental types or non crossover types (car B and ++) and two recombinant types or cross over types (car+ and Bar+). The male flies produce only two types of gametes (car+ and Y), because of no crossing over in drosophila male.
- **Conclusion:** After crossing, stern observed that in red bar flies, one X-chromosome was normal and other was fragmented. The fragmented X chromosome also had attached portion part of Y chromosome. Such combinations in red bar are possible only through exchange of segments between non sister chromatids of homologous chromosomes.
- This proved that genetic crossing over is the result of cytological crossing over.

Write the Factors affecting the Crossing Over along with the significance of Crossing Over in Crop Improvement?

Factors effecting crossing over:

1. Sex: individuals with heterogametic sex have less crossing over than homogametic sex. E.g. Drosophila male.
2. Maternal age: Bridges showed that maternal age increase, decreasing crossing over.
3. Temperature: 22°C is medium temperature .both cool and high temperature increases crossing over.
4. Cytoplasmic effect: Thoday, Lawrence and Boom proved that cytoplasmic factors reduce crossing over. E.g. Tifton male sterility cytoplasm in bajra.
5. Nutritional, chemical and radiation effect: concentration of metallic ions like Ca and Mg.(R.P. Levine)
   - Suzuki found that crossing over frequency is increased when infected with antibiotics like actinomycin – D and myteromycin – C.
   - X – ray radiation increase the frequency of crossing over in drosophila .
6. Genotype effect: crossing over between two genes may vary due to difference in strain genotype.
7. Chromosome structure: Inversions are crossing over suppressors.
8. Centromere effect: Nearer to centromere the genes have less frequency of crossing over.

Significance of crossing over in crop improvement:

1. It increases variability
2. It helps to break linkages
3. It makes possible to construct chromosome maps.

Explain the relationship between crossing over and chiasma formation?

There are two theories to explain the relationship between crossing over and chiasma formation are classical theory and chiasma type theory.

Classical theory:

- Pressure on non homologous chromosomes leads to chiasma formation
- It says that chiasma preceeds crossing over
- According to this theory, crossing over occurs after diplotene stage
- There is no one to one relation between chiasma and crossing over by this theory.

Chiasma type theory:

- Proposed by Janssens and later elaborated by Darlington.
- Crossing over precceds chiasma formation
- Chiasma formed after pachytene stage
- Explains the relation of one to one relation between chiasma and crossing over
- Most accepted theory.

Explain in brief about coincidence and interference?

Coincidence: It refers to the occurrence of two or more distinct crossing overs at the same time in the same region of a pair of homologous chromosomes and as a result, a double cross over product is obtained.

Coefficient of coincidence is estimated by using the formula:

$$\text{Coefficient of coincidence} = \frac{\text{Observed frequency of double cross over}}{\text{Expected frequency of double cross over}}$$

(The ratio between the observed and the expected frequencies of double crossovers is called coefficient of coincidence)

Interference: The occurrence of crossing over in one region of a chromosome interferes with its occurrence in the neighbouring segment. This is known as interference. The term interference was coined by Muller.

- It may also be defined as the tendency of one crossing over to prevent another crossing over from occurring in its vicinity. This is called positive interference. And in this coefficient of coincidence is less than one.
- Sometimes, one crossing over enhances the chance of another crossing over in the adjacent region. This is termed as negative interference. And in this coefficient of coincidence is always more than one. Eg: *Aspergillus,* bacteriophages.
- The effect of interference reduces as the distance from the first crossing over increases.
- The intensity of interference may be estimated as coefficient of interference.

  Coefficient of interference = 1 - coefficient of coincidence
- the value of the coefficient of coincidence is less than one for positive interference, greater than one for negative interference, one for absence of interference and zero for complete or absolute interference.

From the three point test cross mapping experiment, the following genotypic frequencies were obtained.

Xyz	Xyz	xYZ	Xyz	XYZ	xyZ	xYZ	Xyz
2	380	96	109	364	24	3	22

Explain the following

1. What is the correct gene order?
2. From the data provided, draw a linkage map showing the distances between the genes?
3. What is the coefficient of coincidence in this cross?

1. For gene order, we have to compare the double cross over recombinants with parental types. Here change occur in 'x' gene. So 'x' is in centre. And order is YXZ.

2.

Class	1	2	3	4	5	6	7	8
Genotypes	xyz	XYZ	xYZ	Xyz	xyZ	XYz	Xyz	xYZ
Frequency	380	364	96	109	24	22	2	3
Recombinant type	Parental types		Single cross over between x and y		Single cross over between x and z		Double cross over types	

Distance between x and y = (class3 + class 4 + class 7 + class 8)/ Total

= 210/1000 = 0.21 = 21cM

Distance between x and z = (class5 + class 6 + class 7 + class 8)/ Total

= 51/1000 = 0.051 = 5.1cM

Total = 21 cM + 5.1 cM = 26.1 cM

We can also obtain this by =

= 0×(744/1000) + 1×(251/1000) + 2×(5/1000)

= 0.261 = 26.1 cM

Linkage map: Y ——— 21 cM ——— X ——— 5.1 cM ——— Z

3. Coefficient of coincidence (c)

= observed double cross over frequency/ expected double cross over frequency

Observed double cross over frequency = 5/1000 = 0.005

Expected double cross over frequency = 0.21 × 0.051 = 0.01

Therefore, c = 0.005/ 0.01 = 0.5

Coefficient of interference = 1 – coefficient of coincidence

= 1 – 0.5 = 0.5

Here, coefficient of coincidence is 0.5. it indicates the 50% of expected double cross over takes place.

Explain briefly about genetic, cytogenetic and physical maps? And also mention the correlation between these maps?

Genetic maps/ Linkage maps: It refers to schematic representation of places where genes reside on chromosomes in linear specific order analogous to beads on strings.

- These are based on recombination frequency between markers and frequency of recombinants directly propotional to distance between the genes.
- These percent frequencies are used as map units for preparing linkage maps.
- A map unit is also called, a centi- Morghan (cM), after the name of the scientist Morghan, who first constructed the linkage map in drosophila.
- Linkage map also constructed by using molecular markers such as RFLP and SSR etc.

Cytogenetic maps: It refers to representation of location of various genes in chromosome relative to microscopically visible land marks in chromosome with respect to banding pattern.

- These are constructed mainly using FISH, Somatic cell hybridisation and analysis of changes in polytene chromosomes in drosophila.
- Cytogenetic map is only accurate within limits of -5Mbs (1 Mb = 10^6 bp) along the chromosome; the resolution is much better in species having polytene chromosomes.

Physical maps: It refers to genes are depicted in same order as they occurred in chromosome and the distances between them are shown in basepairs or Kbs or Mbs.

- These are constructed mainly based on Restriction mapping and STS mapping etc.

Correlation between linkage, cytogenetic and physical maps:

1. Insitu hybridization is a common method in which YAC or cosmid clones can be hybridized to intact chromosomes to determine their location. This approach allows correlation of a cytological map with physical map.
2. In case of those genes that have already been cloned, linkage map can be correlated with a cytogenetic map by using the cloned gene as a probe in insitu hybridisation.
3. A linkage map can be correlated with cytogenetic map under genetically mapped gene can be localised to particular chromosomal region in human somatic cell hybrids.
4. Polymorphic STS markers can be used in both physical and linkage mapping; they provide a powerful approach to correlate location on linkage map with those of physical map.

5. E.g. For linkage map, distance between Sc (scute, abnormal bristle formation) and w (white eye) of drosophila is located at 1.5 map units.

   For cytogenetic map, gene Sc located on 1B3 on X – chromosome and w is located on 3C2.

   For physical map, genes sc and w were separated by 1.5×10^6 bp.

   And also 1 cM = 1000 kb in human beings, 1600 kb in mice and 1500 kb in maize. These differences arise due to differences in frequencies of recombination in different species or chromosomal regions.

How somatic cell hybridisation technique is used to assign genes on human chromosomes? Explain.

Techniques mainly used to assign genes on human chromosomes are

1. Chromosomal banding
2. Somatic cell hybridisation

Somatic cell hybridisation: In this one cell is hybridised with other cell.

E.g. Human cell is hybridised with mouse cell.

- Littlefield explained the technique, how we can identify the hybridised cells, so that in mammalian cells one protein called aminopterin prevents the enzymes required for DNA replication.

  These two enzymes are HPRT (Hypoxanthin phosphoribosyl transferase) and TK (Thymidine kinase)

  $HPRT^-/TK^-$ - are unable to synthesize DNA.

  $HPRT^+/TK^+$ - are able to synthesize DNA.

- Here, he mixed the mouse cell with $HPRT^+/TK^-$ and human cell with $HPRT^-/TK^+$. And then these cells kept in HAT medium (Hypoxanthine, Aminopterin and Thymidine).
- So by the presence of aminopterine, $HPRT^-/TK^-$ cells were not survive and only $HPRT^+/TK^+$ will survive.
- This is not only useful for identification of hybrid cells but also for mapping of TK gene on chromosome 17.

  Synteny test: In this, two loci present on one chromosome was identified.

  E.g. B – locus link with lactate dehydrogenase also link with B- locus of peptidase.

  Assignment test: In this, loci present on particular chromosome are identified.

  E.g. Blood clotting, glycoprotein tissue factor - III, present on chromosome-1 was identified.

Explain in detail about the Mendels laws of inheritance along with its exceptions?

Mendelism / Laws of Mendel: Mendel's cross breeding experiments with garden peas showed certain numerical relations among the progeny. The relationship between the character pairs and the kinds and the ratio of the progeny is known as Mendelism. Mendelism can be summarised in the following laws which are called Mendel's laws of inheritance.

1. **The law of segregation or the law of purity of gametes:** It states that when a pair of alleles / allelomorphs is brought together in a hybrid (F_1) they remain together without contaminating each other and they separate or segregate from each other into a gamete in a complete and pure form during the formation of gametes.
   - The law is universal in its application and it has been found to occur in plants as well as animals.
2. **Law of independent assortment:** The factors in an allelomorphic pair separate independently to the separation of factors in the other allelomorphic pair.
3. **Law of Dominance:** States that in a hybrid one factor of the allelomorphic pair expresses itself completely over the other.
4. **Law of unit characters:** State that each factor (gene) controls the inheritance of single character. These factors occur in pairs in each diploid organism.

Explanation of Laws of Mendel:

Monohybrid ratio: The Mendel's first law i.e. Law of segregation or purity of gametes can be explained by considering the monohybrid ratio i.e. by studying inheritance of only one character. For example: In pea, round seed shape is dominant over wrinkled seed shape.

- Phenotype ratio: 3 round : 1 wrinkled
- Genotypic ratio: 1 RR : 2R : 1rr

Parents Red flower (RR) × White flower (rr)

↓

F_1 Red (Rr)

↓

F_2

	R	R
R	RR (Red)	Rr (Red)
r	Rr (Red)	rr (White)

Two different alleles of the same gene i.e. 'R' and 'r' were brought together in the hybrid (F_1). Even though the hybrid was round seeded in the next generation (F_2) it produced both round and wrinkled seeded progeny. Thus both the alleles for round shape (R) and wrinkled shape (r) remained together in the hybrid without contaminating each other. In F_2 generation (selfing of (F_1) hybrid), the different phenotypes could be recovered because the two alleles in F_1 remained pure and did not contaminate each other thus producing two types of gametes from F_1 i.e. (R) and (r). The separation of homologous chromosomes during anaphase I of meiosis may be regarded as the reason for segregation of the two alleles of a gene. This is because the alleles of a gene are located in an identical position in the two homologous chromosomes.

Dihybrid Ratio: Mendel's second law i.e. Law of independent assortment can be explained by studying the inheritance of two characters at a time, simultaneously.

- Independent segregation for two genes can be explained by assuming that the two genes are located in two different chromosomes.
- The two alleles of a gene will be located in the two homologues of the concerned chromosome.
- Independent separation of these two pairs of chromosomes at anaphase I of meiosis will lead to the independent segregation of the genes located in them.
- Thus, any allele of one gene is equally likely to combine with any allele of the other gene and pass into the same gamete.
- Independent segregation of two genes produces four different types of gametes in equal proportion. A random union among these gametes gives rise to sixteen possible zygotes. These zygotes yield a 9:3:3:1 phenotypic ratio, which is known as the typical dihybrid ratio.
- When two pairs of independent alleles enter into F_1 combination, both of them have their independent dominant effect. These alleles segregate when gametes are formed. But the assortment occurs independently at random and quite freely.
- E.g. When plants of garden pea with yellow round seeds (YYRR) were crossed with plants having green wrinkled seeds (yyrr), yellow round seed plants (YyRr) were obtained in F_1. Thus yellow colour of seed exhibited dominance over green and round seed shape over wrinkled seed shape independently. The F_1 produces four types of gametes YR, Yr, yR and yr. Selfing of F_1 gives rise to yellow round, yellow wrinkled, green round and green wrinkled individuals in 9:3:3:1 ratio. This is possible only when the alleles of two genes controlling the two characters assort independently to one another.

Parents Yellow Round (YYRR) × Green Wrinkled (yyrr)

↓

F_1 Yellow Round (YyRr)

↓

F_2

	YR	Yr	yR	Yr
YR	YYRR	YYRr	YyRR	YyRr
Yr	YYRr	YYrr	YyRr	Yyrr
yR	Yyrr	YyRr	YyRR	yyRr
yr	YyRr	Yyrr	yyRr	Yyrr

- Phenotypic ratio: 9 Yellow round : 3 Yellow wrinkled : 3 green round : 1 green wrinkled
- Genotypic ratio: 1 YYRR : 2 YYRr : 1 YYrr : 2 YyRR : 4 YyRr : 2 Yyrr : 1 yyRR : 2 yyRr : 1 yyrr

What are the exceptions to Mendelian law of independent assortment? What is the physical basis and evident for recombination? How the distance between genes and gene order are typically determined in eukaryotes?

Exceptions to Mendel's law of independent assortment are linkage and crossing over.

- During meiosis, in the absence of linkage leads to formation of parental and recombinants in 50:50 ratio.
- But in the presence of linkage leads to the more than 50% of parental forms than recombinants.
- Sometimes crossing over coupled with linkage also leads to deviation of law of independent assortment.

	Chromosomic representation	Metaphase configuration
Absence of linkage (Genes present on different chromosomes)	A B — — a b	A B — — a b (or) A B — — a b 50:50 parental forms and recombinant forms
Presence of linkage (Genes present on same chromosomes)	A B —— —— a b	A B —— —— >50% parental forms a b A b —— —— <50% recombinant forms a b

Genetic distance in eukaryotes:

- In diploid individuals, two point test cross is used to know the distance between the genes. i.e. by using fomula = number of recombinant progeny/ total number of progeny.
- Even three point test cross also used for testing genetic distance i.e. using single crosses as like two point test cross.
- In case of haploid individuals,
  (*i*) For unordered spores (E.g. Yeast)
  Genetic distance = (½ Tetra types + Non parental ditypes)/ Total number × 100
  (*ii*) For ordered spores (E.g. Neurospora)
  Genetic distance = (1/2 Second division segregants)/ Total number × 100

Gene order:

- In diploid individuals, gene order is determined by using three point test cross i.e. by comparing the parental forms with double cross over types.
  E.g. In drosophila, the arrangement of three genes (pr- purple eyes, c-curved wings and b- body character) as

  b pr c
  ______
  ______ Parental types
  \+ + +

  b + c
  ______
  ______ Double cross over types
  \+ pr +

  Change in sequence at 'pr' gene, so the order is b – pr- c.
- In haploid individuals, gene order can be determined by using map distance for ordered and unordered spores.

What is meant by epistasis? How do you differentiate complementary and duplicate gene action? How do you explain biochemically?

When expression of one gene depends on the presence or absence of another gene in an individual, it is known as gene interaction. The interaction of genes at different loci that affect the same character is called epistasis.

1. **Complementary epistasis/ Duplicate recessive epistasis (9:7):** When recessive allele at either of two loci can mask the expression of dominant alleles at the two loci is known as recessive epistasis.
   - E.g. Flower colour in sweat pea. In this purple colour is governed by two dominant genes P and C. When these genes are in two separate individuals (PPcc or ppCC) or recessive (ppcc) produce white flowers.

- Here, recessive allele p epistatic to C/c alleles and mask the expression of these alleles and other recessive allele c is epistatic to P/p alleles and mask their expression.
- Thus, the normal dihybrid ratio of 9:3:3:1 is modified to 9:7 ratio in F_2 generation.

Parents Purple flower (PPCC) × white flower (ppcc)

↓

F_1 Purple flower (PpCc)

F_2

	PC	Pc	pC	pc
PC	PPCC (P)	PPCc (P)	PpCC (P)	PpCc (P)
Pc	PPCc (P)	PPcc (W)	PpCc (P)	Ppcc (W)
pC	PpCC (P)	PpCc (P)	ppCC (W)	ppCc (W)
pc	PpCc (P)	Ppcc(W)	ppCc(W)	ppcc (W)

Biochemical analysis:

Here both dominant genes are necessary for expression of phenotype.

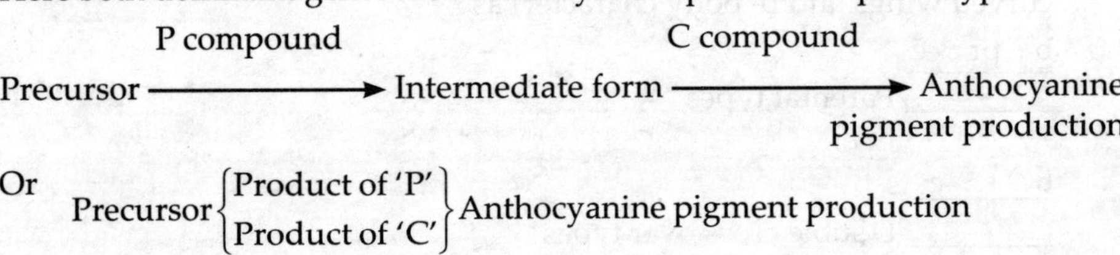

2. **Duplicate gene action/ Duplicate dominant epistasis (15:1):** When a dominant allele at either of two loci can mask the expression of recessive alleles at the two loci is known as duplicate dominant epistasis.
   - E.g. Awn character in rice. In this development of awn is controlled by two dominant genes A and B. And awnless condition develops only when both genes were in homozygous recessive state (aabb).
   - Here, allele A is epistatic to B/b alleles and other allele B is epistatic to A/a alleles.
   - Thus, the normal dihybrid ratio of 9:3:3:1 is modified to 15:1 ratio in F_2 generation.
   - Similar gene interaction found in nodulation in peanut and nonfloating character in rice

Parents Awned rice (AABB) × Awnless rice (aabb)

↓

F_1 Awned rice (AaBb)

↓

F_2

	AB	Ab	aB	Ab
AB	AABB (A)	AABb (A)	AaBB (A)	AaBb (A)
Ab	AABb (A)	AAbb (A)	AaBb (A)	Aabb (A)
aB	AaBB (A)	AaBb (A)	AaBB (A)	aaBb (A)
ab	AaBb (A)	Aabb (A)	aaBb (A)	aabb (a)

Biochemical analysis:

Here presence of any dominant gene leads to expression of phenotype.

A or B gene

Precursor ⟶ Awned character in rice

Write short notes on the following?

(*i*) Multiple alleles

(*ii*) Pseudo alleles and Iso alleles

(*iii*) Penetrance and expressivity

Multiple alleles: Generally a gene has two alternative forms called alleles. Usually one of them is dominant over the other. The two alleles of a gene determine the two contrasting forms of a single character. E.g. Tall (T) and dwarf (t) plant height in garden pea.

But in many cases, several alleles of a single gene are known to exist and each one of them governs a distinct form of the concerned character or trait. Such a situation is known as multiple allelism and all the alleles of a single gene are called multiple alleles. Many genes in both animals and plants exhibit multiple alleles. E.g. Blood group in human beings, fur colour/coat colour in rabbits and self-incompatibility alleles in plants.

Characteristic features of multiple alleles:

1. Multiple alleles always belong to the same locus in a chromosome.
2. One allele is present at a locus at a time in a chromosome.
3. Multiple alleles always control the same character of an individual.
4. However, the phenotypic expression of the character will differ depending on the alleles present.
5. There is no crossing over in a multiple allelic series.
6. In a multiple allelic series, wild type is almost always dominant over the mutant type.
7. A cross between two strains homozygous for mutant alleles will always produce a mutant phenotype and never a wild phenotype. In other words, multiple alleles do not show complementation.

8. Further, F_2 generations from such crosses show typical monohybrid ratio for the concerned trait.

Pseudo allele and Iso allele:

Pseudo alleles: It refers to closely linked and functionally related genes.

- These govern different expressions of the same character.
- These are considered to occupy a complex locus which is divided into sub loci. Thus, they occupy different positions, but on the same complex locus.
- They exhibit low frequency of genetic recombination by crossing over, in other words, crossing over occurs between pseudo alleles, but at a very low frequency.
- They exhibit Cis/Trans position effect. In tran's heterozygotes such mutants produce mutant phenotype, but in cis-heterozygote they produce a wild phenotype.
- E.g. Lozenge eye in Drosophila and Star asteroid eye in *Drosophila*

Isoallele: An allele which is similar in its phenotypic expression to that of other independently occurring allele is called isoallele. Or isoalleles are those alleles which act within the same phenotypic range of each other.

These are of two types

1. Mutant isoalleles: Alleles act within the phenotypic range of a mutant character.
2. Normal isoalleles: Alleles act within the phenotypic range of wild character.

Penetrance and expressivity:

Penetrance; It is the ability of a gene to express itself in all the individuals, which carry it in the appropriate genotype. (or) The frequency with which a gene produces a phenotypic or visible effect in the individuals which carry it in the appropriate genotype is known as penetrance.

- It refers to the proportion of individuals which exhibit phenotypic effect of a specific gene carried by them.
- Penetrance of some genes is limited to one sex only. For example: Milk production in cattle or human beings.
- It may be affected by environmental conditions.
- Penetrance is expressed in percentage.

It is of two types

1. **Complete penetrance:**
    - When all the individuals that carry a particular gene exhibit its phenotypic effect, it is known as complete penetrance.
    - In this case, all homozygous dominant individuals will exhibit one phenotype, while all homozygous recessives will exhibit another phenotype. Recessive alleles have no or zero penetrance in heterozygous condition.

2. **Incomplete penetrance:**
   - When specific gene does not express their effect in all the individuals which carry them in appropriate genotype, it is known as incomplete penetrance.
   - For example: The recessive gene producing partial chlorophyll deficiency in the cotyledonary leaves of lima bean shows incomplete penetrance as it expresses itself only in 10% of the individuals.
   - Almost all the genes showing incomplete penetrance exhibit variable expressivity as well

Expressivity: The degree of phenotypic expression of a gene in different individuals is called expressivity. It is also influenced by environmental conditions in some cases. Expressivity is also of two types.

1. **Uniform expressivity:**
   - When the phenotypic expression of a gene is identical or similar in all the individuals, which carry such a gene, it is known as uniform expressivity.
   - Most of the qualitative characters exhibit uniform expressivity. E.g. Seed shape in pea rr genotypes have wrinkled seed shape, while RR or Rr genotypes exhibit round seed shape.
2. **Variable expressivity:**
   - When the phenotypic expression differs in different carriers of a gene, it is known as variable expressivity.
   - E.g. Recessive gene producing partial chlorophyll deficiency in cotyledonary leaves of lima beans.

Explain why some genes have incomplete penetrance and variable expressivity? Explain with one example for each?

Incomplete penetrance: Incomplete penetrance means gene is present in organism but it is not completely expressed in phenotype.

Variable expressivity: Genotypes which results in the variable phenotypic expression.

Incomplete penetrance and variable expressivity are mainly due to

(*i*) Any modification in biochemical pathways for controlling that trait.

(*ii*) Physical environmental factors like temperature, light, humidity and nutrition influence incomplete penetrance and variable expressivity.

(*iii*) Biological environment factors like gender and genetic background will influence the incomplete penetrance and variable expressivity.

Example for incomplete penetrance:

- Polydactyly in human beings
- It is due to one extra finger and toes in human beings

- It is not completely expressed
- It is due to autosomal dominant genes

Example for variable expressivity:

- In drosophila, a dominant mutant causes the lobe eye shape.
- But in heterozygous condition, in some individuals it shows lobe eye shape, in some individuals it shows compound eyes i.e. it exhibits lot of variation.

Enlist the different mechanisms of sex determination found in plants and animals? And explain genic balance system of sex determination.

There are three important mechanisms of sex determination viz., chromosomal sex determination, monogenic sex determination and environmental sex determination.

S. No	Mechanism of sex determination	Animals	Plants
1 a i	Chromosomal system Allosomal system XX-XY	Drosophila and man	Melandrium, cannabis, coccinia, humulus, saxinia and bryonea
ii iii iv b c	XX-XO XO-XX ZW-ZZ Diploid-Haploid system Genic balance system	Grasshoppers and many insects hemiptera and orthoptera Fumea Birds, butter flies and moths Honey bee Drosophila	Dioscorea and Valisnaria Fragaria
2	Monogenic system	Drosophila, man, goats, pigs and dogs	Asparagus, maize, spinach and papaya
3	Environmental effect	Bonelia and Dinophilus	Equisetum, cucumber, muskmelon and cannabis

Genic balance system:

- This system was proposed by Bridges in 1922 working with drosophila.
- It states that sex determining genes are present on both X chromosomes as well as autosomes.
- Here sex expression is determined by the balance of genes on autosomes and X chromosome i.e. the ratio of X chromosomes to that of autosomes (X/A ratio).
- Here, Y chromosome has no role in determination of sex but it is essential for male fertility; as a result, XO flies are phenotypically males but they are sterile.

Ratio of chromosome and autosome (X:A)	Sex index (X/A)	Expression of sex
2X:2A	1.00	Female
1X:2A	0.50	Male
2X:3A	0.67	Inter sex
1X:3A	0.33	Super male
3X:2A	1.50	Super female

What is compound chromosomal system of sex determination? Give an example and explain how this complex chromosomal compliment arranged during gamete formation?

Compound chromosomal system mainly represents the presence of heterogametic sex chromosomes in male. There by it form some different configurations at meiotic metaphase I

E.g. In Ascaris incurve

In female, 26 autosomes + 16 X – chromosomes

In male, 26 autosomes + 8 X – chromosomes + 1 Y – chromosome

- During meiosis metaphase I, these all chromosomes arranged end to end and form a single chromosome like structure.
- In this during gamete formation, male produces two gametes of one X and Y chromosome.

What are dosage compensation mechanisms that operate in mammals? How they help to maintain gene balance?

Dosage compensation mechanisms in mammals:

1. Hyperactivation of X – chromosome present in male. E.g. Drosophila.
   - In drosophila, hyperactivation results due to complex proteins, these are male specific lethal proteins.
   - Five genes code for these proteins are Msl_1, Msl_2, Msl_3, Mle (maleless) and Mof (male absent).
   - Here RNA also involved and coded by genes rox1 and rox2
2. Inactivation of one X – chromosome present in female. E.g. Cat and mice.
   - It was first observed by Mary Lyon in embryonic cells of cat and mice.
   - It affects the fur colour of cat and mice and it is controlled by two alleles i.e. B (black pigmentation) and b (orange pigmentation).
   - Here inactivation is random process, so equal chance of inactivation of two chromosomes. So it shows mosaicism. This type of phenotype is tortoise shell phenotype.
   - Murraya bar also observed inactive body called bar body in females only not in males.

- Molecular analysis of bar body shows that methyl groups attached to this inactivated X- chromosome.
- Inactivation of X – chromosome will started at any point in chromosome is called X – chromosome inactivation centre (XIC). It is due to binding of X – chromosome inactivation specific transcript (XIST) to that site.

3. Hyperactivation i.e. partial repression of X – linked genes in the somatic cells of hermaphrodites. E.g. Nematode - *Caenorabdities elegens*

Under which circumstances the trait is transmitted only through male parent? What are pseudo autosomal regions and what is their significance?

Some characters like

- Testis determining factor (male fertility)
- Hairs on ear lobes
- H – Y antigen synthesis

These are governed by genes present on Y – chromosome only. So these characters are transmitted only through male parent.

Genes present on Y – chromosome is called holandric genes.

Pseudo autosomal regions:

(*i*) These are the regions that present at the ends of the chromosomes, so that X and Y chromosomes pairs during meiosis.

(*ii*) These regions do not follow sex- linked inheritance.

(*iii*) Instead, they follow autosomal inheritance, so these are called as pseudo autosomal regions.

Significance:

- By presence of these regions only, X and Y chromosomes pair during meiosis
- Within these regions, there is no inactivation of active genes during dosage compensation.
- E.g. Steroid sulphate synthase gene – Red cell antigen Xga in human beings.

Explain in brief about the chromosomal theory of inheritance?

It was proposed by Sutton and Boveri and their main description is parallelism between chromosomes and genes.

Mendel said that

- Factors present in pair
- Factors segregate during gamete formation
- Factors maintain individuality

According to chromosomal theory of inheritance, it is well established that many genes are located in each chromosome in a linear fashion. And also

- Like factors, chromosomes present in pair
- Like factors, chromosomes also segregate during gamete formation
- Like factors, chromosomes are present in single dose in gametic cells and double dose in somatic cells
- Like factors, chromosomes maintain individuality.

First time, Morghan in 1910 presented a more direct evidence supporting the chromosomal theory of evidence i.e. gene was located in a specific chromosome of an organism i.e. pattern of transmission of white gene was identical with that of X – chromosome of drosophila.

Explain how non – disjunction of chromosomes was used as proof of the chromosomal theory of inheritance?

Calvin bridges proposed the non – disjunction of X – chromosomes as proof of chromosomal basis of heredity.

X^wX^w (White eye) × X^{WY} (Red eye)

↓

X^WX^w (red) and X^wY (White)

But exceptions that in the females, some of them are white eyed and in males some of them are red eyed flies.

- Bridges explained these results by proposing that exceptional F_1 flies were result of abnormal X- chromosome behaviour during meiosis in the females of the parental generation.
- Ordinarily, the X – chromosomes in these females should disjoin or separate from each other during meiosis.
- Occasionally however, they might fail to separate and producing an egg with two X – chromosomes or an egg with no X – chromosome at all.
- Fertilisation of such abnormal eggs by normal sperm would produce zygotes with an abnormal number of sex chromosomes.

X^wX^w (White eye) × X^WY (Red eye)

↓

Gametes	X^w	X^wX^w	Null X
X^W	X^WX^w	$X^WX^wX^w$	X^W
	Red eye female	Meta female dies	Red eye male Indicates the X is essential chromosome to develop into female
Y	X^wY White eye male	$X^w X^w Y$ Exceptional white eye female	Y Dies

This non-disjunction of chromosomes mainly occurs due to the aberrations in meiosis i.e. during anaphase there is irregularities in separation. So it leads to formation of trivalents or univalents.

What is meant by cytoplasmic inheritance? And write its types and characteristic features of cytoplasmic inheritance?

Inheritance due to genes located in cytoplasm (plasmagenes) is called cytoplasmic inheritance. Since genes govering traits showing cytoplasmic inheritance are located outside the nucleus and in the cytoplasm, they are refered to as plasmagenes.

The cytoplasmic inheritance is of two types: 1) Plastid inheritance and 2) mitochondrial inheritance.

1. **Plastidial or chloroplast inheritance:** Plastids self duplicated and have some amount of DNA and plays an important role in cytoplasmic inheritance.
   - Plastids have green pigments called chloroplasts.
   - Chloroplasts contain a unique circular DNA (cp DNA) in the stroma that is completely different from the nuclear genome.
   - E.g. Leaf varigation in *Mirabilis jalapa,* Lozap in maize
2. **Mitochondrial inheritance:** The inheritance of some characters, such as cytoplasmic male sterility in plants, pokyness in *Neurospora* etc., is governed by mitochondrial DNA (mtDNA).

Characteristic features of cytoplasmic inheritance

1. **Reciprocal difference:** Reciprocal crosses show marked differences for characters governed by plasmagenes. In most cases, plasmagenes from only female parent are transmitted and hence this phenomenon is also called uniparental inheritance.
2. **Lack of segregation:** In general, F_1, F_2, F_3 and subsequent generations do not show segregation for a cytoplasmically inherited trait, as F_1 individuals receive plasmagenes from female parent only.
3. **Somatic segregation:** Plasmagenes generally show the features in somatic tissues such as leaf variegation features which is of rare occurrence in case of nuclear genes.
4. **Association with organelle DNA:** Several plasmagenes have been shown to be associated either with chloroplast or mitochondrial DNA. For example: Cytoplasmic Male Sterility (CMS) in sorghum and maize is associated with mitochondrial DNA.
5. **Nuclear transplantation:** Nuclear transplantation means nucleus of a cell is removed and replaced by nucleus of another genotype from a different cell.
6. **Mutagenesis:** Some mutagens are highly specific mutagens which act only on the plasmagenes and do not affect nuclear genes E.g. Ethidium bromide,

Induction of mutations by such agents or chemicals in a gene clearly indicates that it is a plasmagene.

7. **Lack of chromosomal location:** In many organisms extensive linkage maps of nuclear genes are available. If a gene is shown to be located in one of these linkage groups, obviously it cannot be a plasmagene.
8. **Transfer of nuclear genome through back crosses:** Nucleus of a variety or species may be transferred into cytoplasm of another variety or species through repeated back crossing with former, which is used as recurrent male parent.
   - Lines produced in this way are called alloplasmic lines, since they have cytoplasm and nucleus from different species.
9. **Lack of association with a parasite or symbiont or virus:** Only those cytoplasmically inherited traits which are not associated with parasites, symbionts or viruses can be regarded to be governed by plasmagenes.

Define mutations? And write its historical land marks in the use of mutations?

Mutation can be defined as a sudden heritable change in the character of an organism which is not due to either segregation or recombination.

- The term mutation was first used by Hugo de Vries to describe the sudden phenotypic changes which were heritable, while working with *Oenothera lamarckiana*.
- But the earliest record of mutations dates back to 1791 when Seth Wright noticed a male lamb with unusually short legs in his flock of sheep. This lamb served as a source of short leg trait for the development of Ancon breed of sheep.
- However, the systematic studies on mutations were started in 1910 by T.H. Morgan who used *Drosophila melanogaster* for his studies.
- In 1927, H.J. Muller demonstrated for the first time the artificial induction of mutations by using x-rays in *Drosophila*.
- Similarly in 1928, L.J. Stadler demonstrated an increase in the rate of mutations due to x-rays in barley and maize.
- Induction of mutations by chemicals in fungus *Aspergillus* was demonstrated by R.A. Steinberg in 1939.
- C. Auerbach and J.N. Robinson in 1946 used chemicals to induce mutations in *Drosophila*.
- The first plant breeding programme using mutations (mutation breeding) was initiated in 1929 in Sweden, Germany and Russia.
- In India it was initiated in early 1930s.

Explain in Detail about the Molecular Basis of Mutations?

Molecular basis of mutations:

The term mutation is presently used to cover only those changes which alter the chemical structure of the gene at molecular level. Such changes are commonly referred to as "point mutations".

- Point mutations involve a change in the base sequence of a gene which results in the production of a mutant phenotype.
- Point mutations can be subdivided into the following three classes on the basis of molecular change associated with them.
- 1. Base substitution 2. Base deletion and 3. Base addition

1. **Base substitution:** When a single base in a DNA molecule is replaced by another base it is known as base substitution. This can be of two types.
   (*a*) **Transition:** Replacement of a purine by another purine or a pyrimidine by another pyrimidine. (or) The substitution of a purine by another purine or of a pyrimidine by another pyrimidine base in DNA or RNA is known as transition. (A G or C T).
   (*b*) **Transversion:** Replacement of a purine by a pyrimidine and vice versa. (or) The substitution of a purine by a pyrimidine or of a pyrimidine by a purine in DNA or RNA is known as transversion. (A or G C or T or U).
2. **Base deletion :** In base deletion, one or more bases are altogether deleted.
3. **Base addition:** There is insertion of one or more bases.
   - If the number of bases added or deleted is not a multiple of three, a frameshift mutation is obtained, as the reading frame in such case is shifted from the point of addition or deletion onwards.
   - Hence, in a frameshift mutation, all the aminoacids of a polypeptide chain located beyond the site of mutation are substituted / altered.

   Frameshift mutations: The mutations which arise due to addition or deletion of nucleotides in mRNA are known as frameshift mutations, because the reading frame of base triplets (codons) beyond the point of addition or deletion is altered as a consequence of such mutations.

Write short notes on chimerism?

Chimera is defined as a mixture of genetically diverse tissues in the same shoot.

- These tissues frequently form mosaic pattern.
- The most common and easily observed chimeras are leaf variegations observed in horticultural plants like Codium and Acalypha.
- These chimeras can be of three types.

1. **Sectorial:** The tissues possess the characteristics which closely resemble the parents from which these tissues are originally derived.
2. **Periclinal:** The core is of one plant and the epidermal region is of another plant. As a result, the leaves may be of one plant type and the flowers and fruits are of another plant.
3. **Chromosomal chimera:** The most frequent being the ploid-chimera. It has been a very common experience to meet with cells especially in the root tips, with higher chromosomal number in an otherwise diploid tissue. Plants under experimental control have been sometimes observed to give rise suddenly to sectors or whole branches differing in chromosome number.

Write the detection of mutations in plants along with significance of mutations in crop improvement?

Detection of mutations in plants:

1. Generally the seeds of a variety or strain are treated with the mutagens and grown to obtain M_1 plants.
2. The M_1 plants are selfed to avoid out crossing due to partial male sterility in M_1 plants.
3. The seeds thus obtained represent the M_2 generation. The M_2 plants are grown and the plants having mutant features are counted.
4. Then the frequency of a given mutation is estimated as percent ratio between the number of plants exhibiting a mutant phenotype in M_2 and the total number of plants in M_2.

$$\text{Mutation frequency (\%)} = \frac{\text{Number of plants exhibiting mutant phenotype in } M_2}{\text{Total number of plants in } M_2}$$

Significance of mutations in Plant Breeding:

1. When a variety is exceptionally good except for one or few characters.
2. When recessive characters is desirable and transfer of that character from wild species is difficult.
3. When a desirable character is linked with an undesirable character.
4. If there is no known source of resistance gene in the available germplasm.
5. To create variability.
6. To develop male sterile lines.
7. To create variations in vegetatively propagated plants.

Explain the methods used for detection of morphological mutations in living organisms?

Detection of mutations depends on their types.

- Morphological mutations are detected either by change in the phenotype of an individual or by change in segregation ratio in a cross between normal (with marker) and irradiated individuals.
- The molecular mutations are detected by a change in the nucleotide, and
- The biochemical mutation can be detected by alteration in a biochemical reaction.

Methods of detection of morphological mutants have been developed mainly with *Drosophila*. Four methods *viz.*,

1. CIB method (used for detection of sex linked recessive mutations in *Drosophila*)
2. Muller's 5 method (used for detection of sex linked recessive mutations in *Drosophila*)
3. Attached X- chromosome method (used for detection of sex linked visible mutations in *Drosophila*)
4. Curly lobe plum method (used for detection of visible autosomal mutations)

For CIB and Curly lobe plum method refer the cytogenetics (translocations) chapter.

Muller's 5 method: This method is an improved version of CIB method, this method differs from CIB in two important aspects.

1. First, this method utilizes apricot recessive gene in place of recessive lethal in CIB method.
2. Second, the female is homozygous for bar apricot genes, whereas it is heterozygous for IB genes in CIB method.

   In this method, the mutation is detected by the absence of wild males in F_2 progeny.

Steps:

1. A homozygous bar apricot female is crossed with mutagen treated male. In F_1 we get two types of progeny, *viz.*, and heterozygote bar females and bar apricot males.
2. These F_1 are intermated. This produces four types of individuals. Half of the females are homozygous bar apricot, and half are bar heterozygous. Among the males, half are bar apricot and half should be normal.

   If lethal mutation is induced, the normal male will be absent in the progeny.

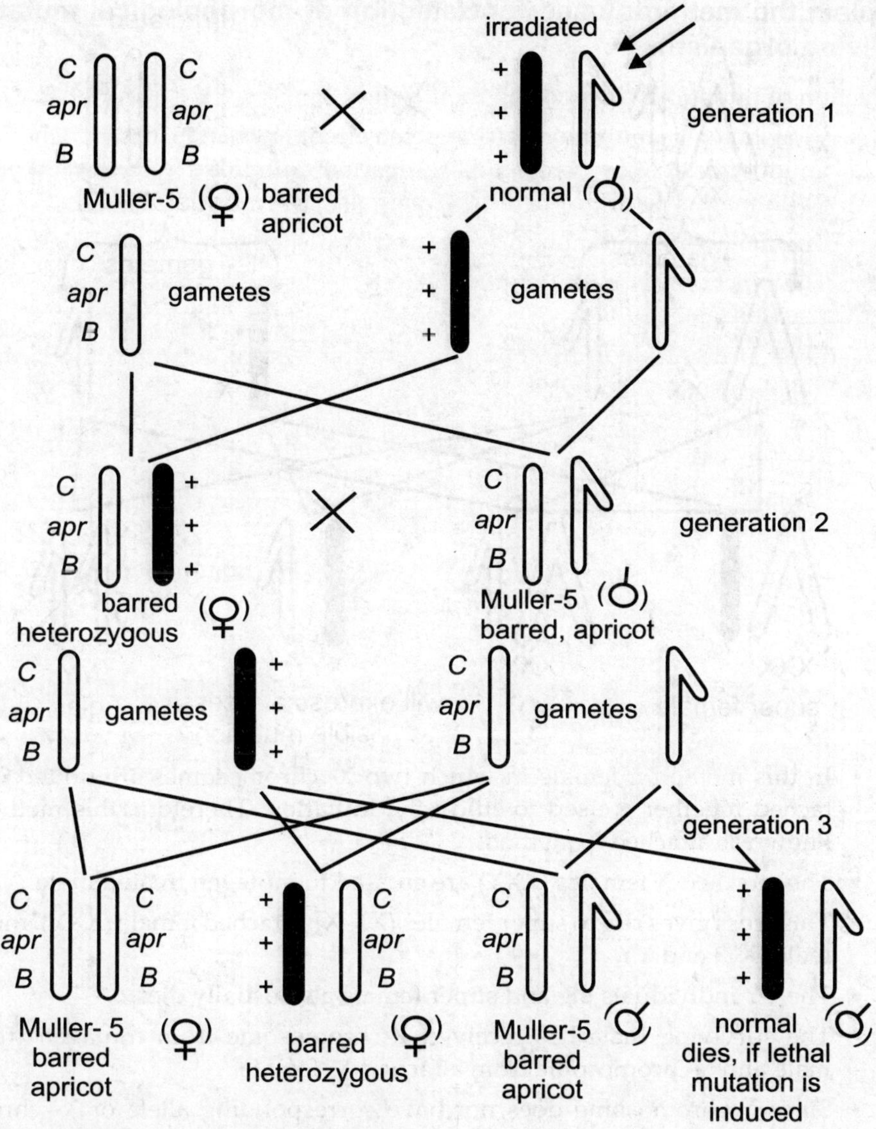

Attached X- chromosome method:

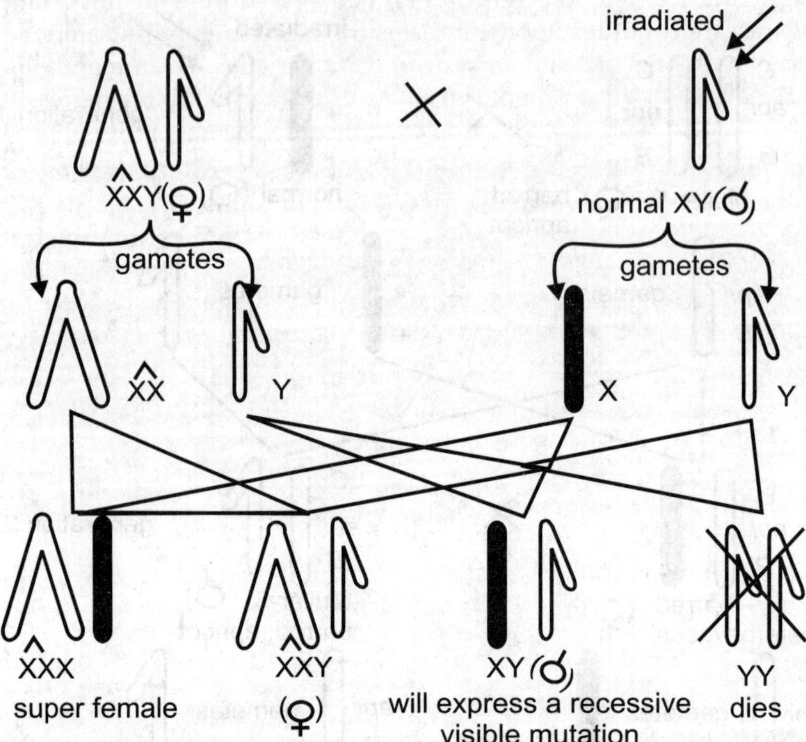

- In this method a female in which two X- chromosomes are united or attached together is used to study the mutation. Therefore, this method is known as attached X-method.
- The attached X females (XXY) are crossed to mutagen treated male.
- This cross gives rise to super females (XX-X), attached female (XXY), mutant male (XY) and YY.
- The YY individuals die and super female also usually dies.
- The surviving male has received X-chromosome from mutagen treated male and Y-chromosome from attached X- female.
- Since Y chromosome does not have corresponding allele of X- chromosome, even recessive mutation will express in such male which can be easily detected.

Advancements in Plant Breeding and Genetics

Plant breeding and genetics have been instrumental in revolutionizing agriculture, contributing significantly to food security, nutrition, and sustainable development. Over the years, significant advancements in technology and research have propelled the field forward, ushering in an era of precision agriculture and sustainable crop. the remarkable successes achieved through plant breeding and genetics, as well as the potential benefits that these technologies hold for the future of agriculture and food production. It will also address the importance of responsible research and development to ensure the sustainability, safety, and equitable distribution of genetically improved plant varieties. The integration of plant breeding and genetics has ushered in a new era of agricultural progress, shaping the future of global food security and environmental sustainability.

- **Traditional plant breeding:** Traditional plant breeding has been practiced for thousands of years, where farmers and horticulturists selectively crossed plants with desired characteristics to produce offspring with preferred traits. This process involved careful observation and the manual transfer of pollen between plants. While effective, traditional breeding methods were time-consuming and limited in the scope of traits that could be targeted.
- **Genetic discoveries:** The 20th century brought about groundbreaking discoveries in genetics, including Gregor Mendel's foundational work on the principles of inheritance. These discoveries laid the groundwork for modern plant breeding practices. In the 1950s, the discovery of the DNA structure by Watson and Crick unlocked a new era in genetics, allowing scientists to delve into the molecular basis of heredity.
- **Hybridization and hybrid vigor:** The concept of hybridization revolutionized plant breeding. Hybrid varieties, produced by crossing genetically diverse parents, demonstrated a phenomenon known as hybrid Vigor or heterosis. This characteristic resulted in offspring that exhibited superior traits, such as higher yields, increased disease resistance, and improved stress tolerance.
- **Biotechnology and genetic engineering:** The advent of biotechnology and genetic engineering marked a transformative stage in plant breeding. Scientists gained the ability to manipulate genes directly, enabling the introduction of specific desirable traits into plants. Genetically modified organisms (GMOs) were developed with traits like pest resistance, herbicide tolerance, and increased nutritional content. While GMOs sparked debates about safety and environmental concerns, they undeniably provided tools to enhance crop performance.
- **Marker-assisted selection (MAS):** Marker-Assisted Selection (MAS) emerged as a crucial technique that allows breeders to select plants with desired traits more efficiently. Instead of relying solely on phenotypic characteristics, which can take years to manifest, MAS employs genetic markers linked to target traits. This expedites the breeding process, resulting in the development of improved varieties in a shorter timeframe.
- **Genome editing techniques:** More recently, genome editing technologies like CRISPR-Cas9 have transformed plant breeding. CRISPR allows precise modifications of specific genes, without introducing foreign DNA as in traditional genetic engineering. This technique has vast potential for creating crops with enhanced traits, such as increased nutritional content, disease resistance, and improved environmental adaptation.

- **Data-driven breeding and AI:** The integration of data-driven approaches and artificial intelligence (AI) has revolutionized plant breeding. Sophisticated algorithms analyze vast datasets, including genomic information, environmental factors, and historical crop performance data. AI helps predict the most promising crosses, shortening the breeding cycle and increasing the chances of developing high performing varieties.

Plant breeding and genetics have come a long way from traditional selection methods to sophisticated genetic engineering and AI driven predictions. These advancements have allowed breeders to develop crops that can thrive in challenging environments, resist diseases and pests, and meet the nutritional demands of a growing global population. As technology continues to evolve, plant breeders are better equipped to address future agricultural challenges, ensuring a greener and more sustainable future for all.

CHAPTER - 2

Genome Organization of Crop Plants

What is meant by complementation test? Write its characteristic features and limitations?

Complementation test (Cis – trans complementation):

1. Complementation or trans test is an operational definition of genes.
2. Complementation or lack of it assured by the phenotype (wild or mutant) of each trans heterozygote.
3. Each cis - heterozygote contain one wild type chromosome, should have the wild type phenotype whether the mutations are in the same gene or different gene.
4. Indeed, the cis – heterozygote must have the wild type for the results of trans test to be valid.
5. If cis – heterozygote has mutant phenotype trans test cannot be used to determine the whether two mutations are in the same gene.
6. In trans test, if F_1 mutant phenotype are produced than these are allelic. If F_1 is wild, these are non allelic.
7. This test is called "complementation test" means non allelic results in production of wild type phenotype.

Limitations:

1. Trans test cannot be used for dominant mutations because cis heterozygote contain one chromosome with wild type copies of all relevant genes, these should specify functional products and wild type phenotype only.
2. Cis test are not informative in studies of dominant or codominant or in cases where intra genic complementation occurs.
3. Cis – test are uninformative because of epistatic interactions.
4. Polar mutations fail to complement mutations in genes subject to polar effect.

Explain in brief about evolutionary concept of gene?

Classical definition of gene: Gene is the unit of function (one gene specifies one character), recombination and mutation.

Modern definition of gene: Gene may be defined as the unit of function, i.e. the sequence of DNA, that specifies one polypeptide or one RNA molecule; this unit includes the 5′- (upstream) and 3′- (downstream) noncoding regulatory sequences as well as introns of the gene. Such a gene can be identified by complementation test.

Summary of the evolutionary concept of gene		
Year	**Scientist**	**Concept of a gene**
	Functional definition of gene	
1866	G.J. Mendel	A 'unit factor' that controls one specific phenotypic trait (one gene-one character)
1902	Sir A.E. Garrod	One mutant gene-one mutant block
1940	G.W. Beadle and E.L. Tatum	One gene specifies one enzyme (one gene-one enzyme hypothesis)
1957	U.M. Ingram	One gene encodes one polypeptide (one gene-one polypeptide hypothesis)
1960s	C. Yanofski, and co-workers, and A. Sarabhai and coworkers	Gene is a continuous sequence of nucleotides specifying a collinear sequence of amino acids in a polypeptide. (Difficulty: Overlapping genes, gene-within gene, interrupted genes, antibody gene segemnts)
	Recombination test	
Before 1940		Recombination occurs between, but not within genes (gene is the unit of recombination)
1940	C.Oliver	Intragenic recombination shown to occur in the *lozenge* gene of *Drosophila*
	Complementation test (operational definition of gene)	
Early 1940s	E.B. Lewis	Complementation test for functional allelism in *Drosophila:* later used by S. Benzer to analyse the r^{II} locus of T_4 bacteriophage

Write the silent features of DNA? And explain the significance of complementary and antiparallel nature of DNA?

The salient features of double helix structure of DNA are:

1. The DNA molecule consists of two polynucleotide chains wound around each other in a right-handed double helix.
2. The two strands of a DNA molecule are oriented anti-parallel to each other i.e. the 5′ end of one strand is located with the 3′ end of the other strand at the same end of a DNA molecule.
3. Each polydeoxyribonucleotide strand is composed of many deoxyribonucleotides joined together by phosphodiester linkage between their sugar and phosphate residues.

4. The half steps of one strand extend to meet half steps of the other strand and the base pairs are called complementary base pairs. The adenine present in one stand of a DNA molecule is linked by two hydrogen bonds with the thymine located opposite to it in the second strand, and vice-versa.
5. Similarly, guanine located in one strand forms three hydrogen bonds with the cytosine present opposite to it in the second strand, and vice-versa i.e. complementary base pairing and these base pairs are called complementary base pairs or Watson and crick base pairs.
6. The pairing of one purine and one pyrimidine maintains the constant width of the DNA double helix. The bases are connected by hydrogen bonds.
7. Although the hydrogen bonds are weaker, the fact that so many of them occur along the length of DNA double helix provides a high degree of stability and rigidity to the molecule.
8. The diameter of this helix is 20°A, while its pitch (the length of helix required to complete one turn) is 34°A.
9. In each DNA strand, the bases occur at a regular interval of 3.4°A so that about 10 base pairs are present in one pitch of a DNA double helix.
10. The helix has two external grooves, a deep wide one, called major groove and a shallow narrow one, called minor groove. Both these groves are large enough to allow protein molecules to come in contact with the bases.
11. This DNA structure offers a ready explanation of how a molecule could form perfect copies of itself.
12. During replication, the two strands of a DNA molecule unwind and the unpaired bases in the single-stranded regions of the two strands by hydrogen bonds with their complementary bases present in the cytoplasm as free nucleotides. These nucleotides become joined by phospho-diester linkages generating complementary strands of the old ones with the help of appropriate enzymes.

Significance of complementary base pairing:

1. Structural convenience of which DNA has a constant diameter of 20°A and also stability of structure.
2. For replication to occur DNA must have complementary units
3. It is also fundamental for transcription i.e. store house of genetic information.

Significance of antiparallel nature:

1. It provides chance for nucleotides to pair, without antiparallel nature two nucleotides cannot pair with each other. It needs hydrogen bonds to occur in between each other. So for matching of the nucleotides for and for occurrence of hydrogen bonds there must be a antiparallel nature.
2. It provides polarity to a DNA. Therefore transcription and translation always occur in one direction.
3. It also gives identity to each of nucleotides based on their strand in which they are present.

Explain in brief about the structure and types of RNA?

Structure of RNA:

- RNA like DNA is a polynucleotide. RNA nucleotides have ribose sugar, which participate in the formation of sugar phosphate backbone of RNA.
- Thymine is absent and is replaced by Uracil.
- Usually RNA is a single stranded structure. Single stranded RNA is the genetic material in most plant viruses E.g. TMV.
- Double stranded RNA is also found to be the genetic material in some organisms. E.g. Plant wound and tumour viruses.
- RNA performs nongenetic function.

There are three main types or forms of RNA.

1. messenger RNA (m-RNA):

- It constitutes about 5-10% of the total cellular RNA.
- It is a single stranded base for base complimentary copy of one of the DNA strands of a gene.
- It provides the information for the amino acid sequence of the polypeptide specified by that gene.
- Generally a single prokaryotic mRNA molecule codes for more than one polypeptide; such a mRNA is known as polycistronic mRNA. All eukaryotic mRNAs are monocistronic i.e. coding for a single polypeptide specified by a single cistron.

2. ribosomal RNA (r – RNA):

- Occur in association with proteins and is organized into special bodies of about 20°·A diameter called ribosomes.
- The size of ribosomes is expressed in terms of 'S' units, based on the rate of sedimentation in an ultracentrifuge.
- It constitutes about 80% of the total cellular RNA.
- The function of rRNA is binding of mRNA and tRNA to ribosomes.

3. transfer RNA (t – RNA):

- It is also known as soluble RNA (sRNA).
- It constitutes about 10-15% of total RNA of the cell.
- It is a class of RNA which is of small size of 3S type and generally have 70 – 80 nucleotides. The longest t – RNA has 87 nucleotides.
- It's main function is to carry various types of amino acids and attach them to mRNA template for synthesis of protein.
- Each t – RNA species has a specific anticodon which base pairs with the appropriate m – RNA codon.
- The nucleotide sequence of the first tRNA (yeast alanyl tRNA) was determined by Robert Holley (1965), who proposed the clover leaf model of secondary structure of tRNA.

- Each tRNA is specific for each aminoacid.
- tRNA molecule contains the sequence of CCA at the 3′end, which is called amino acid attachment site. At the CCA end, it is joined to the single amino acid molecule for which that tRNA is specific. For example, a tRNA molecule specific for lysine cannot bind to the arginine.
- tRNA consists of three loops; (a) DHU-loop or D-loop (aminoacyl recognition region), (b) antic odon loop and (c) thymine loop (ribosome attachment region). Anticodon loop contains a short sequence of bases, which permits temporary complementary pairing with the codons of mRNA.

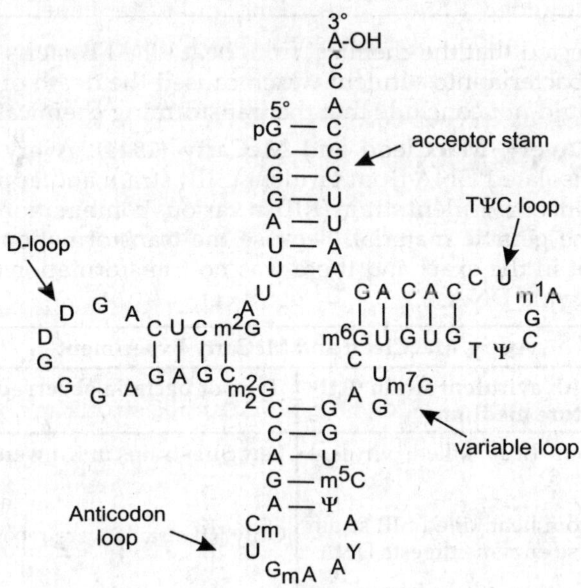

Explain with experimental proof for the following?

1. DNA is a genetic material
2. RNA is a genetic material
3. Semi conservative method of DNA replication

DNA is genetic material:

Experiments of Grrifith (1928): Griffith's systematic experiments with *Diplococcus pneumonia* gave a clue that a chemical substance was responsible for genetic transformation.

- There are two types of strains of *Diplococcus pneumonia viz.*, virulent (SIII) with smooth surface and avirulent (RII) with rough surface.
- Bacteria with smooth surface have coating of polysaccharide. This coating is absent in bacteria with rough surface. The virulent strain causes pneumonia in mice when injected resulting in death of mice.

Effects of two strains of *Diplococcus pneumonia* on mice		
Strain of ***Diplococcus pneumonia*** injected into mice	**Effect on mice**	**Whether SIII was recovered in mice**
Culture of virulent strain	Mice died	SIII recovered
Culture of avirulent strain RII	Mice survive	SIII not recovered
Heat killed culture of virulent strain SIII	Mice survive	SIII not recovered
Mixture of a avirulent RII and heat killed virulent SIII	Mice died	SIII recovered

- This suggested that the chemical from heat killed bacteris transformed the avirulent bacteria into virulent which caused the death of mice. However, Griffith could not conclude that the transforming chemical is DNA.

Experiments of Avery, MacCleod and McCarty (1944): Avery, MacCleod and McCarty in 1944 isolated DNA from virulent (SIII) strain and applied the same in the culture medium of avirulent strain (RII) in various combinations and concluded that DNA was the genetic material. Because the transformation occurred when DNA was present in the exact and there was no transformation when DNA was digested with enzyme DNAse.

Avery, MacCleod and McCarty Experiment	
Type of extract with avirulent strain (RII) in culture medium	**Type of bacteria observed after few hours**
DNA extract from heat killed virulent strain (SIII)	Few SIII strains in RII were observed
Extract of DNA from heat killed SIII strain treated with DNAse enzyme digests DNA	Only RII strains were recovered
DNA extract from heat killed SIII strain treated with RNAse enzyme digests RNA	Some SIII strains with RII were recovered
DNA extract from heat killed SIII strain treated with proteases enzyme which digest proteins	Few SIII colonies with RII were recovered

Experiments of Hershey and Chase (1951):

- Hershey and Chase conducted experiments with T_2 phage of E.coli.
- In one sample, the DNA of T_2 phage was labelled with radioactive phosphorous (^{32}P) because DNA contains phosphorous but not sulphur and in other sample protein portion of T_2 phage was labelled with radioactive sulphur (^{35}S) because protein contains sulphur but not phosphorous.
- These two labelled phages were used to infect two different samples of normal bacteria.
- Result: It was observed that the phages which were labelled with radioactive phosphorous transmitted this character to their daughter phages but radioactive sulphur not transmitted and remained outside the bacterial cell
- Conclusion: Provided strong evidence that DNA is the genetic material.

Summary of Hershey and Chase experiment	
T_2 phage labelled with radioactive phosphorous ^{32}P (DNA labelled)	Recovered daughter phages with radioactive phosphorous ^{32}P
T_2 phage labelled with radioactive sulphur ^{35}S (Protein labelled)	Recovered daughter phages without radioactive sulphur ^{35}S

RNA is genetic material:

- Frankel – Conrat and Singer demonstrated in 1957 that RNA functions as the genetic material by using reassociation studies in TMV (Tobacco Mosaic Virus).
- In this, proteins and RNA of TMV can be separated chemically, when they are remixed under appropriate conditions, they reassociate to produce active TMV particles.

Strain used	Reassociation		Symptoms on infecting the tobacco leaves	Isolated TMV particles
	RNA from strain	Protein from strain		
Strain A TMV particles	A	A	Strain A	Strain A TMV particles
Strain B TMV particles	B	B	Strain B	Strain B TMV particles
Strain A + Strain B	A	B	Strain A	Strain A TMV particles
Strain A + Strain B	B	A	Strain B	Strain B TMV particles

- They observed that only RNA of TMV has the capacity to prevent the disease, and that the type of proteins present in the virus particles is determined by the RNA.
- I.e. only RNA fraction of TMV is capable of producing the disease, and hence appears to be the genetic material.

Evidence for Semi conservative replication: The experiment conducted by Matthew Meselson and Franklin Stahl in 1958 on E. coli provided a conclusive proof that replication of DNA is by semi conservative model.

1. Meselson and Stahl labeled DNA of *E. coli* bacteria with heavy nitrogen i.e. ^{15}N by growing them on a medium containing ^{15}N for many generations to replace the normal nitrogen (^{14}N).
2. The density of normal and heavy nitrogen differs. The ^{14}N is lighter (1.710 g/cm) than ^{15}N (1.724 g/cm).
3. It is possible to detect such minute differences in density through density gradient centrifugation. Distinct bands are formed in the centrifuge tube for different density DNA.
4. DNA extracts of *E. coli* with ^{15}N gave a characteristic heavy band at one end

of a tube that had been centrifuged at a high speed in an ultra-centrifuge.

5. These labeled cells were then grown on a normal unlabelled media containing ^{14}N for one generation.
6. DNA was again extracted and processed and it was found to consist of a hybrid DNA containing both ^{14}N and ^{15}N at the same time.
7. This indicated that the DNA had not replicated in two separate labelled and unlabelled forms.
8. The next generation of growth on unlabelled DNA was found to be in amounts equal to the partially labelled hybrid DNA.
9. Additional generation of growth on unlabelled media gave a relative increase in the amount of unlabelled DNA.
10. After two generations half the DNA was with intermediate density and half with light bands which further confirm semiconservative mode of DNA replication.
11. After third generation, ¾ DNA was found with ^{14}N and ¼ with hybrid nitrogen (^{14}N+^{15}N).
12. When the hybrid DNA was denatured by heating upto 100°C it was found to produce two separate single strands and in the ultracentrifuge density gradient, it was observed to form two separate bands; one band containing ^{15}N and the other ^{14}N. Thus it was concluded that DNA replication was by semi-conservative mode.

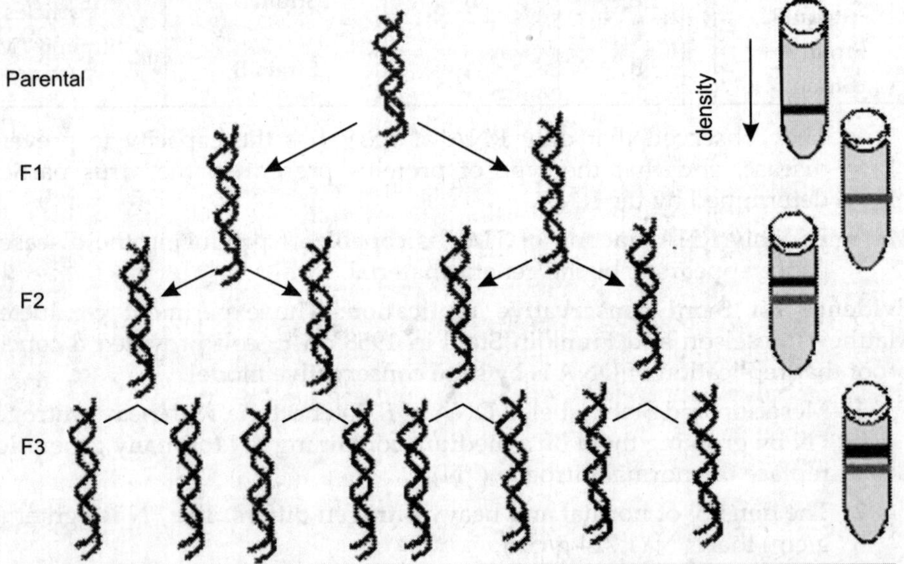

The major objection put forth for the semi-conservative replication was that the DNA molecule must unwind a number of times (1/10 of the total number of nucleotides) which cannot be accomplished without breaking within a short span of time say two minutes.

The presence of replication forks in E. coli chromosomes first shown by J. Cairns in 1963 using the technique of autoradiography. He grew E. coli cells in the presence of ^3H- thymidine (tritiated thymidine) since thymine is ordinarily found in DNA only. The E. coli chromosome is a double stranded circular chromosome. It was shown that the two circular component strands separate during replication with each strand duplicating individually producing a θ shaped structures during replication. This indicates that unwinding and replication proceed simultaneously.

What is meant by replication of DNA? And explain in brief about the semiconservative replication of DNA?

The process by which a DNA molecule makes its identical copies is called DNA replication.

Modes of DNA Replication:

There are three possible modes of DNA replication.

1. **Dispersive:** In dispersive mode of replication, the old DNA molecule would break into several pieces, each fragment would replicate and the old and new segments would recombine randomly to yield the progeny DNA molecule. Each progeny molecule would have both old and new segments along its length.
2. **Conservative:** According to conservative scheme, the two newly synthesized strands following the replication of a DNA molecule would associate to form one double helix, while the other two old strands would remain together as one double helix.
3. **Semi conservative:** In this model of DNA replication, each newly synthesized strand of DNA would remain associated with old strand against which it was synthesized. Thus, each progeny DNA molecule would consist of one old and one newly synthesized strand.

Method of semi-conservative replication:

1. DNA replication was found to begin at various initiation points, called origin of replication, and proceed bi-directionally.
2. Two enzymes, DNA gyrase and DNA helicase induce unwinding of complementary strands of DNA.
3. Single-strand DNA binding (SSB) proteins bind to the single-stranded DNA, stabilizing it and preventing it from reanneaing.
4. An enzyme, primase, initiates replication by synthesizing the primer.
5. DNA polymerases synthesize the complementary strand by progressively adding deoxyribonucleotides.
6. The DNA replication always proceeds in $5' \rightarrow 3'$ direction.
7. During replication one strand of DNA can replicate continuously and the other strand discontinuously or in pieces. The continuously replicating

strand is known as leading strand and the discontinuously replicating strand is known as lagging strand.

8. The replication of lagging strand generates small polynucleotide fragments called 'Okazaki fragments' which are later joined together by the enzyme, DNA ligase.

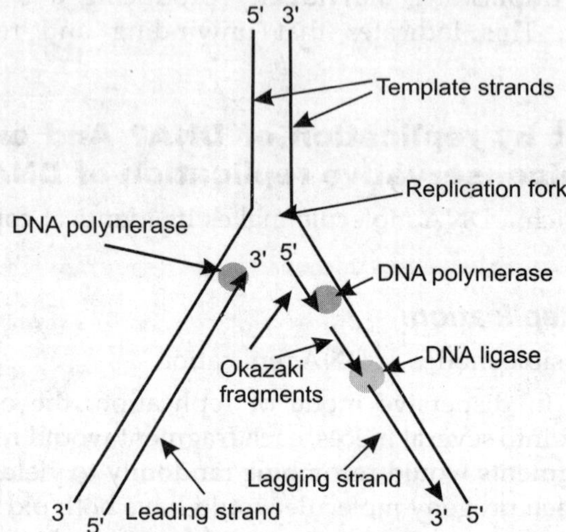

Substantiate your answer for the following questions?

(*a*) The rate of prokaryotic DNA replication is faster than that of eukaryotic organism.

(*b*) The nature of replication of telomeric DNA is different from that the remaining DNA of chromosomes.

(a) The rate of prokaryotic DNA replication is faster than that of eukaryotic organism because

(*i*) In prokaryotes only DNA will be synthesized, but not histone proteins.

(*ii*) But in eukaryotes along with DNA, histone proteins also synthesized, and also formation of nucleosomes is occur. That's why; it takes more time for replication.

(*iii*) For this compensation, in eukaryotes there will be many replicons present. Each replicon starts the replication at same time to complete the replication faster.

(b) Nature of replication of telomeric DNA is different from that of remaining DNA of chromosomes because.

(*i*) Telomeres are end portions of chromosomes that contain the repeated DNA sequences.

(*ii*) DNA sequences of telomeres is generally 5′TTAGGG3′

(*iii*) For the replication of telomeres a specific enzyme required i.e. telomerase.

(*iv*) Telomerase is a ribonucleotide sequence containing RNA template

(*v*) Telomere replication occurs in three steps.

(*a*) Binding of RNA template to DNA strand

5′TTAGGGTTAGGGTTAGGGTTA3′ ⟶ DNA template
3′AAUCCC5′ ⟶ RNA template

(*b*) Polymerisation

5′TTAGGGTTAGGGTTAGGGTTAGGG3′
3′AAUCCC5′

Here, nucleotide bases are added to DNA strand which are complementary to RNA template.

(*c*) Translation

5′TTAGGGTTAGGGTTAGGGTTAGGGTTA3′
3′AAUCCC5′

Telomeric sequence is single strand and it forms a hair pin like structure and 3′ end more longer than 5′ end to over come the loss of nucleotide pairs.

Write the reasons to say that RNA evolved earlier to DNA (RNA is progenitor of life forms)?

RNA world was given by Gilbert

1. Uracil is progenitor of thymine and uracil is found only in RNA i.e all deoxy ribonucletides comes from ribonucletides
2. RNA has self catalytic activity
3. RNA is required for replication and translation
4. Discovery of reverse transcription
5. Discovery of ribozymes
6. Discovery of telomerases
7. Discovery of peptidyl transferase i.e. rRNA
8. Phosphodiester bond formation is mediated by the rna
9. Discovery of smart ribozymes (invitro synthesis by Charles wilson)
10. Many evidences like epigenetics, gene silencing and recent gene regulations showed that RNA was the genetic material that is progenitor of life and also expected to be genetic material of life

Write short notes on the following?

1. Levels of organization of DNA
2. C- value paradox

3. Chromosomal DNA and evolution
4. DNA content and adaptability

Levels of organization of DNA:

Nucleosome – solenoid model of DNA organization was proposed by Kornberg and Thomas in 1974 and is the most accepted. According to this model, the chromatin is composed of a repeating unit called nucleosome. One complete nucleosome consists of

(*a*) Nucleosome core (consists of two molecules of each of histones H_2a, H_2b, H_3 and H_4).

(*b*) Linker DNA

(*c*) An average of one molecule of histone H_1 and

(*d*) Other associated chromosomal proteins.

The three levels of organisation of DNA are

1. The first level of condensation involves packaging DNA as a negative super coil in to nucleosome, to produce the 10 nm diameter interphase chromatin fibre. This clearly involves an octamer of histone molecules, two each of histones H_2a, H_2b, H_3 and H_4. Each nucleosome has 6 nm height and 11 nm diameter.
2. This second level of condensation involves the additional folding of this 10 nm nucleosome fibre to produce 30 nm chromatin fibre characteristic mitotic and meiotic chromosomes. Histone H_1 is involved in this super coiling of 10 nm nucleosome fibre to produce 30 nm chromatin fibres. This process is called as solenoid formation.
3. Finally, non histone chromosomal proteins forms scaffold that is involved in condensing the 30 nm chromatin fibre into highly packed metaphase chromosome of 84 nm diameter.

C-value paradox:

- Refers to the lack of correlation between genome size and genetic complexity.
- The total amount of DNA in the haploid genome is a characteristic of each living species and is known as 'C' value.
- 'C' values range from < 106 bp (mycoplasma) to > 1011 bp (some plants and amphibians).
- E.g. Toad xenopus and man have genomes of essentially the same size though we assume that man is more complex in terms of genetic development.
- It is not understood why natural selection allows this variation and whether it has evolutionary significance.
- There are two main reasons for 'C' value variations.

(*a*) Organisation of eukaryotic genes: In eukaryotes the actual coding sequence of a gene is much smaller than the average size of genes.

- Due to the presence of 'introns'-intervenal sequences interrupting coding sequences 'exons'.
- Such genes are called 'split' genes of interrupted genes.
- The size of eukaryotic in RNA is about 1/5th of the size of the gene coding for it.

(*b*) Repetitive DNA:

- The actual number of genes per genome is much lower than the number estimated on the basis of haploid DNA content.
- This is mainly due to the presence of 'repetitive DNA'.

Chromosomal DNA & evolution:

- Evolution in several plant and animal genera has proceeded with change in chromosomal DNA and chromosome length (E.g. in *Lolium, Festuca, Lathyrus, Vicia,* etc.; Furuta and Nishikawa, 1991).
- In a comparison, annuals have been found to have a lower DNA content than the perennials (E.g. *Crepis, Helianthus*)
- And inbreeders were found to have lower DNA content than outbreeders (E.g. *Lathyrus*).
- This suggests that evolutionary trend is towards decrease in DNA content, accompanying evolution of annuals from perennials and that of inbreeders from out breeders,
- Although there may be exceptions to this general trend. There are also indications of qualitative rather than quantitative change in DNA during evolution in some cases, suggesting that evolution may also proceed without any change in the nuclear DNA content.

DNA content (genome size) & adaptability:

- In several recent studies correlations of nuclear DNA content with either the altitude or the latitude have been worked out.
- For instance adaption to higher altitude has been shown to be correlated with higher DNA contents in *Secale*.
- Similarly, a negative correlation of nuclear DNA content with latitudes (lower DNA content at higher latitudes) has been shown in maize.
- The 4C DNA content in maize populations collected from different altitudes (100ft. to 5300ft.) varied from 11.1 pg to 13.5 pg.
- Similarly, 4C DNA content in northern (higher latitude) flint maize was 9.9 pg, while that in Mexican lines was 11.9 pg.
- The 4C DNA content in maize thus has been shown to vary from 9.9 pg to 13.5 pg, which is a variation of about 40% and is substantial at the intraspecific level.

- In both these cases of adaptation of altitude and latitude, lower genome size seems to be favored by shorter growing season - seems to be correlated with early maturity.
- However, in a number of hybrid maize populations examined, no correlation of nuclear DNA content with relative maturity or with adaptation to latitudes was observed
- The lack of correlations in hybrid maize has been attributed to the breeding structure of these populations,

What is meant by split genes? And write its features?

Split genes of eukaryotes:

- The coding regions (*exons*: from expressed sequences) of most eukaryotic genes are interrupted by several noncoding sequences called (*introns*: from intervening sequences) - genes are called split genes or interrupted genes.
- Some genes of eukaryotes *e.g.* 96% of yeast, 17% in Drosophila and 6% in mammals are not split.
- Split genes have been found in an Archebacterium and a phage of E. coli.
- First split gene was described in 1977 by Chambdon and coworkers. It was the ovalbumin gene of chicken.
- Hybridization of mature ovalbumin mRNA with single stranded preparation of DNA - seven loops of varying sizes.
- These loops represent regions of ovalbumin gene not present in the mature RNA that is introns.
- When heterogeneous RNA used in hybridization (DNA - RNA), loops corresponding to introns are not detected.

RNA splicing: This clearly indicates that both introns and exons are transcribed to yield hnRNA, which is then processed to remove introns, and produce functional RNA.

Features of Split Genes:

Each interrupted gene begins with an exon and ends with an exon.

1. The same interrupted gene organization is present in all the tissues of an organism.
2. Exons occur in the same precise order in the mRNA in which they occur in the gene.
3. Most introns of nuclear genes are blocked in all the reading frames, i.e., termination codons occur frequently in their reading frames.
4. Genes related by evolution have related organizations - variation in the lengths of genes is mainly due the differences in the lengths of their introns.
5. Exons tend to be fairly small relative to the sizes of the genes - code for less than 100 amino acids (less than 50 in vertebrates).

6. The switch uninterrupted to largely interrupted organization of genes occurs in lower eukaryotes.
7. The size of interrupted genes increases within higher eukaryotes and genes become significantly larger in insects mainly due to very long introns.
8. During evolution, introns have diverged both in base sequence (due to base substitutions) and in size (due to deletions and insertions).
9. Since mutations in introns are not subject to selection, while those in exons are, exons are conserved, while introns are variable.
10. Since exons are conserved, they can be used as probes for the isolation of same genes from other organisms or related genes, and also for fishing out genes with unknown functions.

Explain in brief about the evolutionary pattern of Interrupted genes?

Evolution of Interrupted Genes:

The manner in which interrupted genes evolved is not clear.

1. **Introns early model:** The current proteins have evolved by combining over a period of time, together ancestral proteins that were originally separate and encoded by different genes.
   - These genes were brought together in a sequence, in a manner that the sequences flanking the genes were retained as introns, while the sequences encoding the modules formed to the exons.
   - Primitive state and the source of non – interrupted genes.
2. **Introns late model:** The non – interrupted state is primitive and introns were subsequently inserted to create the split gene.
   - The significance of split organization of eukaryotic genes is not clear.
   - Introns are relics of evolutionary processes that brought together different ancestral genes to form new larger genes.
   - It is also possible that at least some introns have been introduced within certain exons during the evolution.
   - Introns may also provide for increased recombination rates between exons of a gene and thus be of some significance in the creation of genetic variation.
   - Some introns may code for enzymes involved in the processing of hnRNA.
   - Thus different introns may play different role; but some introns may have no known biological function.

Give a description of transposable elements? And explain the Ac – Ds system in maize?

- The process by which a segment of DNA can move from one position to another in a genome is called transposition.

- These movable segments are called transposable elements or transposons.
- The movement of transposable elements to a new site in the genome is sponsored by the enzyme transposase.
- Transposition may due to
  (*i*) Conservative process, which involves the excision of the sequence from its original position followed by its reinsertion resulting in the changing its position in the genome with increasing its copy number. E.g. IS elements, Composite elements, Ac-Ds and P- elements.
  (*ii*) Replicative process, which results in an increase in copy number because during this process the original element remains in place while a copy is inserted at the new position. E.g. Tn3 elements.

Types of transposons:

I. RNA transposons (Transposition via RNA intermediate i.e. Retrotransposition).

(*a*) RNA transposons with long terminal sequences (LTRs)

E.g. Ty1/ Copia retroelements or Ty3/ gypsy retroelements in yeast.

(*b*) RNA transposons that lack LTRs

E.g. LINES (Long interspersed nuclear elements) and SINES (Short interspersed nuclear elements).

II. DNA transposons (Transposons donot require RNA intermediate)

(*a*) DNA transposons in prokaryotes

E.g. Insertion sequences (IS1, IS186), Composite transposons (Tn10, Tn5 and Tn903), Tn3 type transposons and transposable phages.

(*b*) DNA transposons in eukaryotes

E.g. Ac/Ds elements in maize and P- elements in drosophila

Ac – Ds system in maize:

- In 1950, Barbara Mc Clintock discovered the Ac – Ds system of transposable elements in maize.
- Mc Clintock called these elements as controlling elements.
- These controlling elements can be classified in to two groups
  (*i*) Autonomous elements have the ability of transposition.
  E.g. Ac (activator)
  (*ii*) Non autonomous elements do not have the ability to transposition.
  E.g. Ds (dissociator)
- Non autonomous elements transpose only when an autonomous member (Ac) of the same family is present in the same nucleus.

- These Ac – Ds elements responsible for striping and spotting in maize kernels.

What is meant by Genetic code? And write the properties of genetic code?

Genetic code: The genetic code is the set of rules by which information encoded in genetic material (DNA or RNA sequences) is transled into proteins (amino acid sequences) by living cells.

Properties of Genetic code; There are several important features or properties of a genetic code.

1. **The code is triplet:** The triplet code was first suggested by Gammow in 1954. In a triplet code, three RNA bases code for one amino acid.
2. **The code is universal:** The same genetic code is applicable to all forms of organisms from microbes to human beings.

   However, some exceptions to the universality of genetic code occur in ciliates (unicellular protozoa) and in mitochondria. E.g. In some ciliates, codons UAA and UGA code for glutamine instead of acting as termination codons. Similarly in mitochondria, UGA codes for tryptophan instead of signalling termination.
3. **The code is comma less:** The genetic code is without a comma or break. The codons are continuous and there are no breaks between the codons. A change or deletion of a single base in the code will alter the entire sequence of aminoacid to be synthesized.
4. **The code is non-overlapping:** Three nucleotides or bases code for one aminoacid and six bases will code for two amino acids. In a non – overlapping code, one base or letter is read only once.
5. **The code is non – ambiguous:** Out of the 64 codons, 61 code for 20 different aminoacid, while 3 are nonsense codons. None of the codons code for more than one aminoacid.
6. **The code is degenerate:** In most cases, several codons code for the same aminoacid. Only two amino acids, viz., tryptophan and methionine are coded by one codon each. Nine amino acids are coded by two codons each, one amino acid (isoleucine) by three codons each, five amino acids by four codons each and three amino acids by six codons each. This type of redundancy of genetic code is called degeneracy of genetic code. Such a system provides protection to the organisms against many harmful mutations. If one base of codon is mutated, there are other codons, which will code for the same amino acid and thus there will be no alteration in polypeptide chain.
7. **The code has polarity:** The code has a definite direction for reading of message, which is referred to as polarity. Reading of code in opposite

direction will naturally specify for another amino acid example: GUC codes for valine, if reversed, CUG codes for Leucine.

Write short notes on the following?

1. Wobble hypothesis
2. Central dogma

Wobble hypothesis:

Crick postulated the wobble hypothesis in 1966 to account for the degeneracy of genetic code.

- Degeneracy of genetic code arises because of
  - (*i*) Presence of more than one variety of tRNA specific to one amino acid, and
  - (*ii*) The ability of an anticodon (on tRNA) to base pair with more than one codon (on mRNA).
  - (*iii*) The first two bases on the codon are usually the important ones; the third base may be in some cases one of the four bases and the triplet would still code for the same amino acid as long as the first two bases are the same and have the same sequence. This concept is known as the Wobble hypothesis.
  - (*iv*) But base pairing at the third base of codons with the first base of anticodon is not precise.

Base in the first position of the anticodon	Base recognizes in the third position of the codon
U	A or G
C	G only
A	U only
G	C or U
I (Inosine)	U, C or U
Q (Quenosine)	U or C
2- Thiouridine	Only A

Central dogma:

In 1958 Crick proposed that the information present in DNA (in the form of base sequences) is transferred to RNA and then from RNA it is transferred to proteins in the form of aminoacid sequence and that this information does not flow in the reverse direction i.e. from protein to RNA to DNA. This relationship between DNA, RNA and protein is known as central dogma. DNA molecules however provide the information for their own replication.

Transcription Translation

DNA ⟶ RNA ⟶ Protein

However in cells infected by certain RNA viruses E.g. TMV, the viral RNA produces new copies of itself with the help of RNA replicase. In some viruses, the DNA is synthesized from RNA in the presence of an enzyme, reverse transcriptase. This process is known as reverse transcription. This was first reported by Temin and Baltimore in 1970.

Explain in detail about the different steps involved in protein synthesis?

The process of protein synthesis where amino acids are brought by the tRNA and sequentially arranged based on the codons of the mRNA is known as translation (i. e. the nucleotide sequence is translated into the amino acid sequence).

- The translation process requires mRNA, rRNA, ribosomes, 20 kinds of aminoacids and their specific tRNAs and many translation factors.
- The process of translation (protein synthesis) consists of five major steps *viz.*,
  1. Activation of aminoacids
  2. Transfer of aminoacids to tRNA
  3. Chain initiation
  4. Chain elongation and
  5. Chain termination. Each step is governed by specific enzymes and cofactors.

Steps:

1. In the cytoplasm, the amino acids are activated in the presence of ATP and linked to their respective tRNAs by a process called charging of tRNA in the presence of an enzyme aminoacyl synthetase.

$$\text{Aminoacid} + \text{ATP} \xrightarrow[\text{Synthetase}]{\text{Aminoacyl} - \text{tRNA}} \text{Aminoacyl - AMP} + \text{Pyrophosphate}$$

2. Thus a number of tRNA molecules, pick up aminoacids freely floating in the cytoplasm and forms aminoacyl-tRNAs.

$$\text{Aminoacyl - AMP} + \text{tRNA} \xrightarrow[\text{Synthetase}]{\text{Aminoacyl} - \text{tRNA}} \text{Aminoacyl - tRNA} + \text{AMP}$$

3. The processed mRNA enters the cytoplasm and binds to ribosomes, which serve as work benches for protein synthesis.
4. The ribosome consists of rRNAs and different proteins. Ribosome contains two subunits; the large subunit and the small subunit.
5. The process of translation starts when an initiating aminoacylated tRNA base pairs with an initation codon of an mRNA molecule that has been located by the small subunit of ribosome.

6. Then the larger subunit joins. Two separate and distinct sites are available in the ribosome to which the tRNAs can bind; A (acceptor or aminoacyl attachment) site and P (peptidyl) site.
7. An aminoacyl-tRNA first attaches to site A (acceptor site or aminoacyl attachment site) the kind of aminoacyl t-RNA being determined by the sequence of mRNA (codon) attached to site A.
8. The peptide bonds are formed between the aminoacids which is catalysed by the enzyme peptidyl transferase.
9. The peptidyl tRNA along with the mRNA codon moves to the P (peptidyl) site making the A site available for the attachment of a new aminoacyl-tRNA.
10. Thus the translation proceeds and at the end a releasing factor binds to the stop codon terminating the translation.
11. The ribosome releases the polypeptide and mRNA and subsequently dissociates into two subunits.
12. Further processing of polypeptide chain into proteins and enzymes is done in the cytoplasm itself and depends upon the bonding properties of the amino acids joined in them.

Write short notes on the following?

1. One gene - one enzyme hypothesis and
2. One gene – one polypeptide hypothesis

One gene - one enzyme hypothesis: In 1941, Beadle and Tatum proposed the one-gene-one-enzyme hypothesis by their work on *Neurospora* biochemical mutants and concluded that each gene controlled a single specific biochemical reaction.

Research findings of Beadle and Tatum:

- They isolated mutant strains of Neurospora that were unable to synthesize pyridoxine (*pdx*), thiazole half of thiamine (*thi*) or p-aminobenzoic acid (*pab*) or with some biochemical defect.
- Wild type *Neurospora* grows on a minimal medium containing nitrate, sulphate, other inorganic salts, sugar and biotin.
- Asexual spores of wild type *Neurospora* were irradiated with UV or X- rays and mated with the wild type *Neurospora* of the opposite mating type.
- In this way, Beadle and Tatum were able to demonstrate that in each case the mutant culture was unable to synthesize a single biochemical and that in each such case this inability was governed by a single gene.
- Finally they concluded that each gene controls a single and specific biochemical reaction, most likely by specifying a single enzyme protein (one-gene-one-enzyme hypothesis).

Conidia of *Neurospora*

↓ Irradiated with UV or X-rays

Conidia used for mating with wild type *Neurospora*

↓ Grown on minimal medium

Individual ascospores isolated

↓ Cultured on complete medium

A large number of cultures derived from single ascospores

↓ Each culture tested on minimal medium

Cultures unable to grow on minimal medium identified; these are biochemical mutants, Each mutant culture tested as follows: one mutant showed these results

↓

Complete medium (Growth)

Minimal medium (No growth)

Minimal + vitamins (Growth)

Minimal + amino acids (No growth)

This particular mutant is unable to synthesize a vitamin

↓

Mutant tested on minimal medium + individual vitamins. This mutant grew on minimal medium + pyridoxine; it is unable to synthesize pyridoxine

Normal growth would occur in the culture tube containing the vitamin, which is not synthesized by the culture in question.

One-gene-one-polypeptide hypothesis:

- Many enzymes (and protein molecules) are composed of a single polypeptide and are termed as monomers, while others made up of two or more polypeptides are dimmers and tetramers etc.
- Each polypeptide is controlled by single specific gene E.g. Tryptophan synthatase in E.coli.
- Thus, one-gene-one-enzyme hypothesis is valid for only those enzymes that are composed of a single polypeptide.
- In addition, many genes specify nonenzyme protein molecules E.g. Haemoglobin, structural proteins.
- Therefore, one-gene-one-polypeptide hypothesis proposed by *Ingram* is more appropriate and accounts for monomeric as well as multimeric enzymes and for nonenzyme proteins.

- But, many genes do not produce any polypeptides E.g. Genes specifying tRNA and rRNA etc.
- Therefore, one-gene-one polypeptide and one-gene-one enzyme hypothesis would be regarded as applicable to only those genes, which specify polypeptide sequences.

Explain in brief about regulation of prokaryotic gene expression?

- Gene expression in prokaryotes is regulated at several different levels: transcription, mRNA processing, mRNA turnover, translation and post translation. However, the regulatory mechanisms with the largest effects on phenotype at the level of transcription.
- Based on regulation of transcription, the various regulatory mechanisms seem to fit into two categories:
    1. Mechanisms that involve the rapid turn-on and turn-off of gene expression in response to environmental change.
    2. Mechanisms referred to as programmed circuits or cascades of gene expression. In these cases, the sequential expression of genes is genetically pre-programmed, and the genes cannot usually be turned on out of sequence. E.g. When a lytic bacteriophage infects a bacterium, the viral genes are expressed in a predetermined sequence, and this sequence is directly correlated with the temporal sequence of gene-product involvement in the reproduction and morphogenesis of the virus.
- Constitutive, inducible and repressible gene expression:
    1. In prokaryotes, genes that specify housekeeping functions such as rRNAs, tRNAs, and ribosomal proteins are expressed constitutively. Other genes usually are expressed only when their products are needed.
    2. Genes that encode enzymes involved in catabolic pathways often expressed only in the presence of the substrate of the enzymes i.e. their expression is inducible.
    3. Genes that encode enzymes involved in anabolic pathways are turned off in the presence of the end product of the pathway i.e. their expression is repressible.
- Positive and negative control of gene expression:
    1. In positive control mechanism, the product of regulatory gene, an activator, is required to turn on the expression of structural genes.
    2. In negative control mechanisms, the product of a regulator gene, a repressor, is necessary to turn off the expression of the structural genes.
    3. Activators and repressors regulate gene expression by binding to sites

(operator) adjacent to the promoter of structural genes.

4. Whether or not the regulator proteins can bind to their binding sites depends on the presence or absence of small effector molecules that form complexes with the regulator proteins.
5. The effector molecules are called inducers in inducible systems and co-repressors in repressible system.

- Operons: co-ordinately regulated units of gene expression
  1. In bacteria, genes with related functions frequently occur in co-ordinately regulated units called operons.
  2. Each operon contains a set of contiguous structural genes, a promoter (the binding site for RNA polymerase) and an operator (the binding site for a regulatory protein called a repressor).
  3. When a repressor is bound to the operator, RNA polymerase cannot transcribe the structural genes in the operon. When the operator is free of repressor, RNA polymerase can transcribe the operon.
- Lactose operon in *E. coli*: Induction and catabolic repression.
  1. The E. *coli* lac operon is a negative inducible and catabolic repressible system.
  2. The three structural genes in the lac operon are transcribed at high levels only in the presence of lactose and the absence of glucose.
  3. In the absence of lactose, the lac repressor binds to the lac operator and prevents RNA polymerase from initiating transcription of the operon.
  4. Catabolic repression keeps operons such as lac encoding enzymes involve in catabolism from being induced in the presence of glucose, the preferred energy source.
- Tryptophan operon in E. *coli*: Repression and attenuation.
  1. The E. *coli* trp operon is a negative repressible system, transcription of five structural genes in the *trp* operon is repressed in the presence of significant concentrations of tryptophan.
  2. Operons such as *trp* that encode enzymes involved in amino acid biosynthetic pathways often are controlled by a second regulatory mechanism called attenuation.
  3. Attenuation occurs by the premature termination of transcription at a site in the mRNA leader sequence (the sequence 5′ to the coding region) when tryptophan is prevalent in the environment in which the bacteria are growing.

Write short notes on the following?

1. RNA interference (RNAi) mechanism

2. Riboswitch mechanism

1. **RNA$_i$ mechansim:** RNA molecules involved here are two types of 1. Short interfering RNAs (Si RNAs) 2. Micro RNAs (mi RNAs). Steps involved in RNA interference mechanism are
    - A large double stranded RNA molecule is diced in to small, double stranded interfering 21-28 bp long fragments by dicer enzyme.
    - The small interfering RNAs and proteins assemble into ribonucleoprotein particles
    - The small interfering RNA in ribonucleoprotein particle is unwound to produce an RNA- induced silencing complex (RISC)
    - RISC targets a sequence in a messenger RNA that is complementary to the interfering RNA
    - The RISC's interfering RNA basepairs with its target in the mRNA
    - If perfectly base paired, mRNA is cleaved. It leads to degradation of mRNA. The RISC associated with this RNA interference is called Short interfering RNAs.
    - If imperfectly base paired, translation of the mRNA is arrested i.e. polypeptide synthesis from the mRNA is repressed. The RISC associated with this RNA interference is called Micro RNAs.

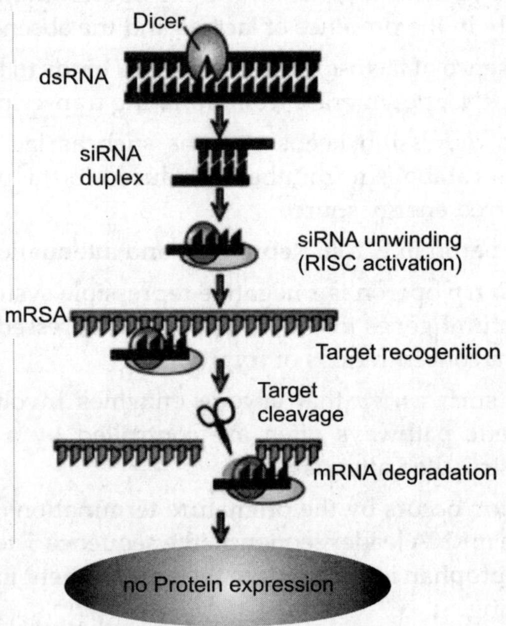

2. **Riboswitch mechanism:** It refers to natural RNA sensors that affect gene control via their capacity to bind small molecules.
    - An RNA leader sequence able to switch between an antiterminator

confirmation and a terminator confirmation.

- Transcription termination is triggered by direct binding of a small molecule to a 5′ untranslated leader mRNA
- The 5′ leader is able to adopt either of two confirmation Antiterminator confirmation: Transcription of gene continues past the leader and their remaining part of the gene.

  Terminator confirmation: Transcription is terminated
- Riboswitches exert their functions without the need for protein cofactors
- They act in feedback regulation mechanisms
- Once the level of an end product (TPP – Thiamine pyrophosphate) is a metabolic pathway rises, riboswitch binding occurs, triggering a repression of gene expression in the same pathway.
- Substrate specificity of riboswitches is extremely high.

CHAPTER - 3

Cytogenetics of Crop Plants

Explain in detail about the role of cytogenetics in the era of molecular genetics.

1. Cytogentics remain important in the age of molecular genetics, because they are is inseperable from overall genome analysis.
2. Century of cytogenetic research:
   - 1900 to 1949 – age of classical genetics
   - 1950 to present – age of molecular genetics
3. From 1990s there are many advances in cytogenetic and molecular cytogenetic techniques such as different chromosomal banding techniques; *in situ* hybridization, microscopy and DNA manipulation are developed.
   - FISH (Fluorescent *Insitu* Hybridization) was originally derived from ISH, using FISH, DNA sequences such as repetitive DNA sequences, multi-copy gen families and low or single copy gens can be physically localized directly on chromosomes.
   - In the 1990s, Fibre-FISH, greatly improved the sensitivity and resolution of the FISH technique.
   - GISH (Genomic Insitu Hybridization), a special technique of FISH that uses genomic DNA of a donor species as a probe in combination with excess amount of unlabelled blocking DNA, provides powerful technique to monitor chromatin introgression during interspecific hybridization.
4. Wheat genetic stocks developed using cytogenetic techniques:
   - E.R. Sears developed valuable cytogenetic stocks include nullisomic, tetrasomic, monosomic, ditelosomic and doubleditelosomic lines. These monosomic and telosomic lines allowed researchers to locate genes and DNA markers to individual chromosomes.
   - Deletion line stocks in wheat are powerful source for wheat genome project such as EST mapping.
5. Study of chromosomal land marks using cytogenetic techniques
   - Cytogenetic techniques are excellent tools to study structure and function of chromosomal land marks such as heterochromatin, rRNA genes, centromere and telomere.

6. Role of cytogenetics in breeding for disease resistance: In wheat to date, 23 leaf rust, 19 stem rust and 6 stripe rust resistant genes have been transferred from alien species to wheat.
7. Conclusions:
   - Large genome size and high percentage of repetitive DNA sequences make molecular genetic studies difficult. However, these characteristics make amenable to cytogenetic studies E.g. wheat.
   - Role of cytogenetics in identifying new disease resistant sources for durable resistance to different diseases cannot be replaced by other technique in cereals (especially wheat). While genetic engineering offers opportunity for future, the problems of gene identification, gene cloning and social acceptance of engineered derivatives are still to be solved.

Erlist and explain brief about different molecular cytogenetic techniques applied in plant breeding?

For understanding genome relationship and affinities-

1. Different banding techniques are used:-
   (*a*) Q-banding-with quinacrine mustard (Caspersson used in Vicia faba)
   (*b*) C-banding-Geimsa staning
   (*c*) N-banding-NOR region (Nucleolus organizing region)
   (*d*) G-banding-diluted trypsin + giemsa
   (*e*) R-banding-reverse to G banding
   (*f*) T-banding-for telomere region trificale has been beneficial most from banding analysis.
2. Analysis and quantification of DNA Content by
   (a) Feulgen micro spectrophotometry can detect DNA content in each stage of cell division.
   (*b*) Flow cytometry-using flourescent dye, with intense beam of light on nuclei, perform flow based chromosome sorting and can identify addition line. It is also an important tool to monitor environmental quality.
   (*c*) Quantitative fluorescent spectrophotometry.
   (*d*) Quantitative image analysis.
3. *In situ* hybridization to understand ancestry, synteny and phylogeny and to assign genome, linkage group and to identify genomic region of interest (GISH and FISH).
4. Chromosomal micro-dissection and cloning: it dissects out a specific chromosome and clone. Mostly applied to mammals, it has been used in wheat, rice, Vicia faba etc.
5. Three dimensional analysis of chromosome using evolution microscopy-

strategy of each character and 3-D image is visualized. It helps in understanding functional role of heterochromatin and also different chromosomal arrangements.

Banding patterns of chromosomes:

- Pardue and Gall (1970) – Banding patterns for identification of chromosomes.
- A band is the area of a chromosome that stains more deeply than the neighboring parts.
- Mitotic metaphase chromosomes are used for the study of banding pattern.

Explain in detail about techniques used for chromosome banding?

Techniques Used for Chromosome Banding:

C-band:

- Pardue and Gall - constitutive heterochromatin can be stained specifically by Giemsa-solution.
- Constitutive heterochromatin is located near the centromere, at telomeres and in the nucleolar organizer regions.
- It is composed of highly repetitive DNA.
- Banding is due to the differential staining reactions of the DNA of heterochromatin and euchromatin.
- Several treatments with acid, alkali or increased temperature.
- Denaturation of DNA is caused by these treatments.
- DNA renaturation occurs in treatments with sodium citrate at 60°C.
- Repetitive DNA renatures but low repetitive and unique DNAs do not renature.
- Results in differential staining of the specific chromosome regions.
- Used for various plant and animal species including human.
- The Y chromosome of mammals is mostly heterochromatic.

In Barley Linde-Laursen in 1978 divided the C-bands:

- Centromeric bands situated at one or both sides of the centromere,
- Intercalary bands, Telomeric bands and Bands beside the secondary constriction.
- Also used to identify the extra chromosomes of trisomics and telotrisomics.

G- bands:

- Giemsa staining following pretreatment with weak trypsin solution, urea or protease.
- It provides greater detail than C-banding.

- First used for human chromosomes by Sumner *et al.* in 1971.
- Not suitable for plant chromosomes.

Q-bands:

- Caspersson *et al.* in 1968
- Stained with Quinacrine mustard show bright and dark zones under UV light.
- For Human and mice chromosomes.

N-bands:

- Matsui and Sasaki in 1973.
- Air-dried chromosome slides are stained for 90 minutes with Giemsa.
- Following the extraction with 5% trichloroacetic acid at 95°C for 30 minutes.
- And then 0.1 NHCl at 60°C for 30minutes.
- N bands located at the secondary constriction, satellites, centromeres, telomeres and heterochromatin segments.
- N-bands represent certain structural non-histone proteins specifically linked to the nucleolar organizer region of the eukaryotic chromosomes.
- N-banding patterns have been used for the location of nucleolar organizer regions in the different organisms – plants and animals.
- A modified Giemsa-N-banding - Singh and Tsuchiya in 1982, combination of acetocarmine staining and Giemsa-N-banding.
- Centromeric region looks like a "diamond-shaped" structure.
- Early metaphase or prometaphase.

Other techniques:

- R-banding (Reverse Giemsa-banding)
- H-banding
- T-banding (Terminal banding)

Banding techniques for the identification of individual chromosomes of an organism but also to establish evolutionary relationships between different species.

Explain in brief about FISH Technique?

FISH (Fluorescent in-situ hybridization):

The FISH technique involves 1. A sample of dividing cells is dried in to microscopic slide and treating with formamide so that the chromosomes become denatured but do not lose their characteristic metaphase morphologies. 2. Labelling of probe with reporter molecule (biotin, dioxin). 3. Hybridization with the probe and finally incubation in immuno-fluorescent reagent before taking photographs with fluorescent microscope.

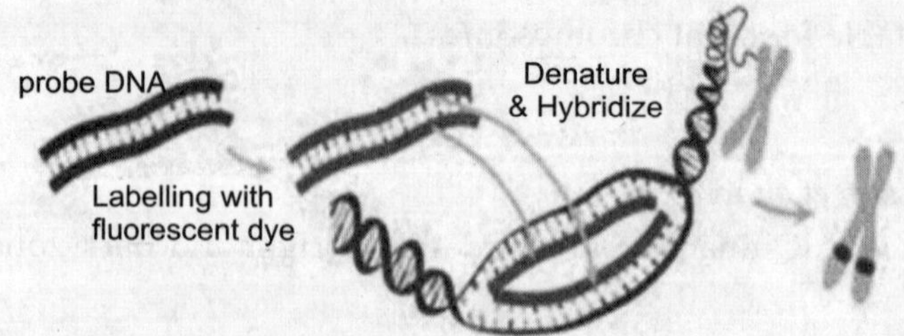

- The probes used in FISH can be total genomic DNA or chromosome specific DNA or single copy sequences. For, this reason FISH can be used for gene mapping and study of structural and numerical aberration of chromosome.
- Since the probes binds only to the specific sequence of a chromosome, fluorescence generates a colourful glowing band on the chromosome against a colourless background; as a result, the slightest amount of colour is readily detectable.
- FISH uses probes of 1Kb or greater size, usually, probes are 40 Kb or longer.
- Fluorescent signal obtained by FISH is mapped by measuring its position relative to the end of the short arm of the chromosome (The FLpter value).
- Fibre- FISH, makes use of DNA prepared by gel stretching or molecular combing and can distinguish markers less than 10 Kb apart as FISH gives resolution of 1 Mb only.
- This technique has some advantages over simple *in situ* hybridization, which are as follows:
  - Higher resolution, sensitivity and speed.
  - Simultaneous detection of two or more sequences in the same nucleus.
  - Hybridization of repetitive sequence can be suppressed.
  - Entire genome or single copy sequence can be highlighted.

Chromosomal aberrations: It is of two types.

1. Structural changes in chromosomes
2. Numerical changes in chromosomes

Explain in brief about the structural chromosomal aberrations?

Structural changes in chromosomes: Structural chromosomal aberrations alter the chromosome structure i.e. the number, the sequence or the kind of genes present in chromosomes.

There are four common types of structural chromosomal aberrations.

1. Deletion or deficiency – Loss of a segment of a chromosome. It may be terminal or intercalary deletion.
2. Duplication (Repeat) – A chromosome segment present in more than two copies in the same nucleus. It may be tandem, reverse tandem, displaced and transposed duplication.

3. Inversion – A chromosome segment contains genes in a sequence, which is the reverse of the normal. It may be paracentric or pericentric inversion.
4. Translocation – Chromosome segment integrated into non homologous chromosomes. It may be simple or reciprocal translocation.

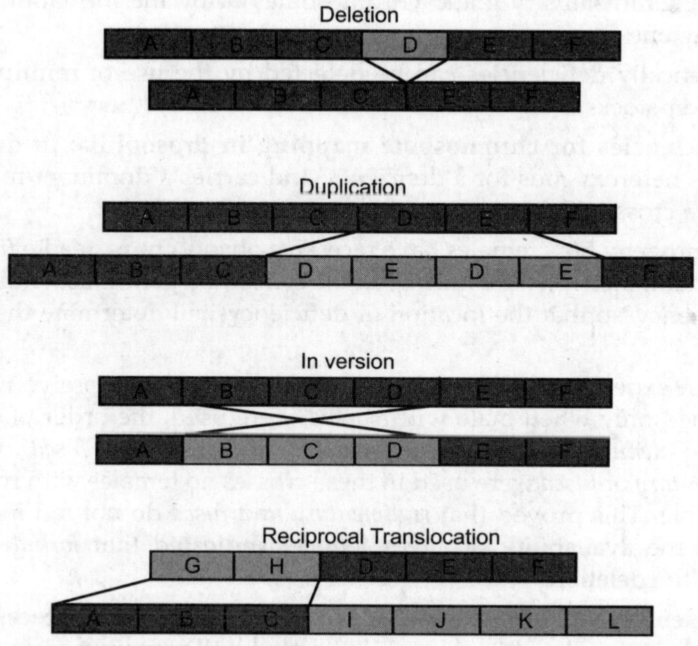

What is meant by deficiency? Explain how deficiencies are used for mapping the genes on the chromosomes?

Deficiency: Deficiency is due to loss of a part of a chromosome. Smaller deficiencies in heterozygote are tolerated. Such individuals at meiosis will form a loop in a bivalent that can be observed at pachytene stage.

Due to defiency, recessive allele will behave like a dominant allele (Pseudodominance). This phenomenon was utilized for location of genes on specific chromosome in drosophila, maize and other organisms.

E.g. Salivary gland chromosomes of drosophila where even small deficiency can be detected. And also in humans deficiency on chromosome 5 is associated with cat like cry in Childs.

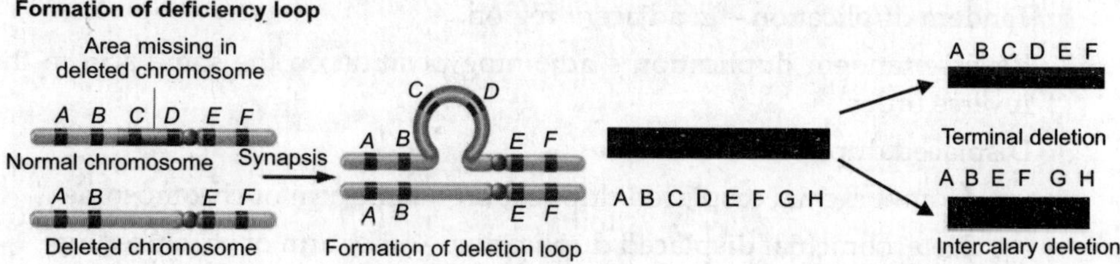

Use of deficiencies for mapping the gens on the chromosomes: Deletions can be detected and used for chromosome mapping either by cytological methods or genetical methods.

- Cytologically these can be detected with the help of missing bands or buklings in salivary gland chromosomes or by the formation of loops in pachytene chromosomes
- Genetically deficiencies can be detected by the use of multiple recessive marker stocks.

Use of deficiencies for chromosome mapping in drosophila: In drosophila, if female fly is heterozygous for a deficiency and carries a dominant marker, it can be used for a cross with a male carrying recessive marker.

If in the progeny, 50% females have recessive phenotype instead of the expected 100% dominant, then it would mean that the concerned gene falls within the region of the deficiency, so that the location of deficiency will determine the location of gene also.

E.g. In an experiment (Slizynoka, 1938), females with recessive trait (forked) were obtained only when male with '*forked*' were used, the order of genes being *w-r-f-B-fu (w - white; r = rudimentary; f = forked; B= Bar and fu = fused).* When males with *rudimentary* or *fused* were used in these crosses no females with recessive trait were available. This proved that *rudimentary* and *fused* do not fall in the deleted region and the availability of *forked* females indicated that *forked* character is included in the deletion.

Use of deficiencies for chromosome mapping in plants: In plants recessive marker stocks may be crossed as female with irradiated pollen from a male carrying the dominant allele or alternatively, the F_1 heterozygous seed may be irradiated. If the male gamete carries a deficiency or a deficiency is induced in the seed, the marker gene of the female will express itself. This phenomenon is called *psuedodominance* and can be used as a criterion for detection of deficiencies.

- The technique of *psuedodominance* was used for location of genes in the deleted segments cytologically.
- Very extensive studies on the use of deficiencies for chromosome mapping in tomato have been conducted by G.S. Khus in association with C.M. Rick.

What is meant by duplications? And explain its types?

Duplication: Duplication is the condition where addition of a chromosome part occurs. It is of different types.

1. Tandem duplication – in adjacent region
2. Reverse tandem duplication – adjoining position on the same arm in the reverse order
3. Displaced duplication
   - Homobrachial displaced duplication – same arm of chromosome
   - Heterobrachial displaced duplication – other arm of chromosome

4. Transposed duplication - duplication on different chromosome (non homologous chromosome).

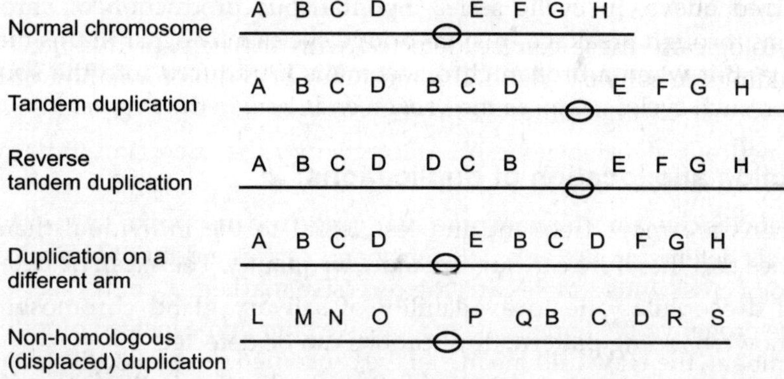

Among these a revrse tandem repeat may lead to bridge breakage events due to crossing over between non sister chromatids and sister chromatids. When crossing over between non sister chromatids, bridge may produce at anaphase I and crossing over between sister chromatids, bridge may produce at anaphase II repectively.

Write short note on the following?

1. Bridge breakage fusion cycle
2. Detection and location of duplications
3. Phenotypic effects of locations
4. Role of duplications in evolution

1. Bridge breakage fusion cycle:

In the gametophyte and endosperm of corn, ends of chromosomes that have recently been broken behave as though they were "sticky," as is shown by their tendency to adhere to one another. Extensive studies of broken chromosomes in corn have been made by Barbara Mc Clintock. She found that following reduplication of a broken chromosome the two sister chromatids may adhere at the point of previous breakage.

- The fused sister chromatids would be unable to separate readily.
- In effect, they constitute a single chromatid with two centromeres, a dicentric chromatid.
- As the centromeres move to opposite poles at anaphase, the dicentric chromatid stretches out, forming a chromatin bridge from one pole towards the other.
- This bridge eventually breaks, but the break does not always occur at the point of previous fusion.

When a chromosome bridge breaks, perhaps as a result of tension caused by the movement of two centromeres of the dicentric chromatid two new broken ends are formed. Each of these has the same qualities of adhesiveness that gave rise to the original fusion. This situation permits repetition of more similar events as described above, in cyclic series. Spontaneous production of chromosome aberrations through breakage -fusion -bridge cycles may occur in this manner for some time. But when a broken chromosome is introduced into the sporophytic generation such cycles cease, as the broken ends heal in the zygote.

2. Detection and location of duplications:

Duplications are more frequent and less lethal to the individual than are the deficiencies and therefore can survive more frequently. These can be detected by

- In drosophila, due to availability of salivary gland chromosomes with known banding pattern, duplications can be detected.
- In plants, pachytene pairing will show duplication in the form of loops.
- In haploid individual plants, the regions exhibiting pairing will indicate duplications, as shown in maize and pearl millet.
- Genetically, duplications can be detected through 15:1 or 9:7 ratios.

3. Phenotypic effects of duplications:

There are a number of examples particularly in drosophila, where phenotypic effects of duplication heterozygosity have been studied.

Organism	Gene duplication	Location on chromosome	Phenotypic effect
Drosophila melanogaster	Eyeless (ey)	Chromosome 4	Eyelessness
	Pale	Chromosome 3	Diluter of eosin eye colour
	Bar eye (B)	X-chromosome	Reduction in number of facets in eye
	Confluence (co)	Chromosome 3	Irregularly thickened and fused veins
	Vermillion (v)	X-chromosome	Vermillion eye

- The most popular example of duplication is the dominant trait, Bar eye (B) in drosophila. This character is located on the X-chromosome and is responsible for the reduction in the number of facets in the eye.

Trait	**Number of facets on the eye**
Wild	779
Duplication heterozygote	358
Duplication homozygote	68
Heterozygote double bar	45
Homozygote double bar	25

Ac-Ds elements in maize: These are also called as transposable elements. Here, Ac changes position by enzyme 'transposase'. So because of duplication it leads to postion and dosage effect in maize.

4. Role of duplications in evolution:

- In crops like bajra and maize duplications have played an important role in evolution.
- The detection of duplicate loci in some diploid plant and animal species also suggested a role of duplications in evolution.
- A comparison of chromosomes of different species within the genera *Allium* and *Lolium,* suggests that the evolution within the genus in both cases proceeded at least partly due to duplications.
- Presence of bivalents and end-to-end associations in these haploids suggest that segmental homologies exist among chromosomes of these species.
- Preferential intragenomic chromosome pairing observed in some diploid *Secale-Hordeum* hybrids also suggested segmental homologies among chromosomes of the same species.

What is meant by translocations? And explain the cytology of translocation heterozygotes?

Translocations: It includes unilateral or bilateral transfer of chromosome segment from one chromosome to another. An important class of translocation having evolutionary significance is known as reciprocal translocation which involves mutual exchange of non homologous chromosomes.

Cytology of translocation heterozygotes: If a translocation is present in one of the two sets of chromosomes, this will be a translocation heterozygote. In such plants normal pairing into bivalent is not possible among chromosomes involved in translocation. Due to pairing between homologous segment of chromosomes, a cross shaped (+) figure involving four chromosomes will be observed at pachytene. This ring of four chromosomes at Metaphse I can have one of the following three orientation.

- Alternate segreagation: The centromeres lying alternate to each other in the cross shaped figure move to the same pole. In other words, the adjacent chromosomes will orient towards opposite poles. As a result, the two normal chromosomes move to one pole, while the two translocated chromosomes move to the opposite pole.
- Adjacent I segregation: Adjacent chromosomes having non homologous centromeres will orient towards the same pole. In other words, the chromosomes having homologous centromeres will orient towards the opposite poles.

- Adjacent II segregation: Adjacent chromosomes having homologous centromeres will orient towards the same pole.

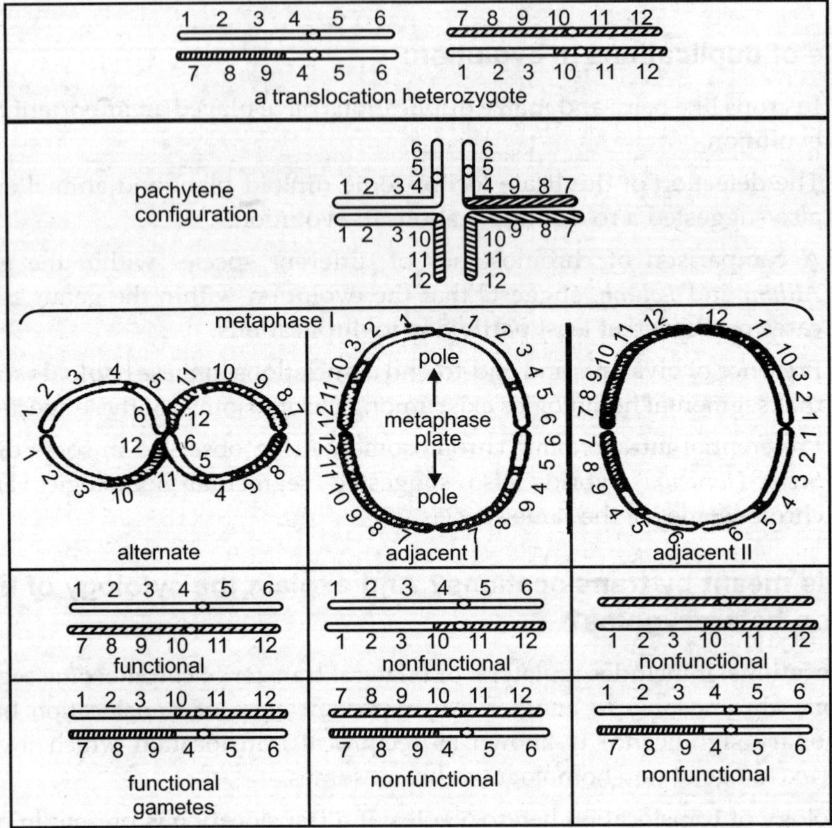

Explain in brief about the breeding behaviour of translocation heterozygotes.

- Presence of translocation heterozygote is detected by semi sterility and low seed set.
- This can be confirmed at meiosis by quadrivalent formation.
- This results from alternate disjunction.
- Functional gametes obtained from alternate disjunction are of two types, those with interchanged chromosomes and others with normal chromosomes.
- The functional gametes will give rise to three kinds of progeny namely normal, translocation heterozygote and translocation homozygotes in 1:2:1 ratio.

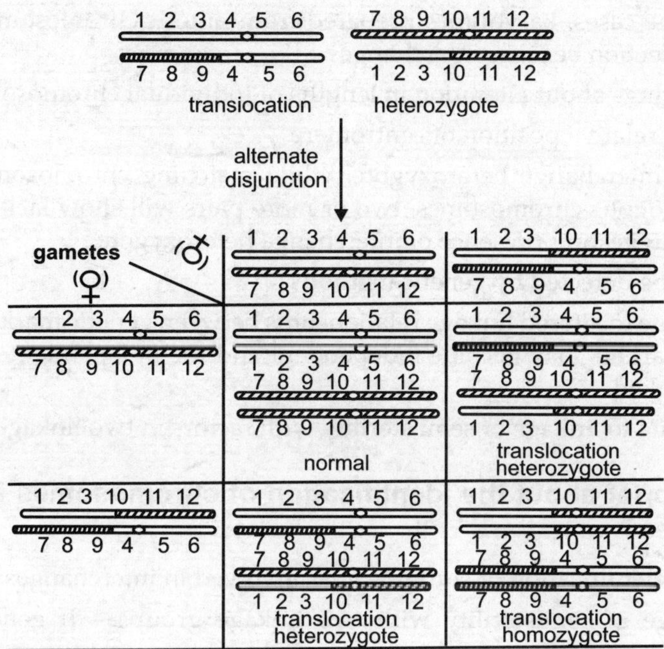

Translocation heterozygotes are always associated with semi-sterility. Justify your answer?

- Cytology of an interchange heterozygote suggested that degree of sterility will depend upon total frequency of adjacent 1 and adjacent 2 disjunctions and on frequency of crossing over in interstitial region.
- Fertile gametes will mainly result due to alternate disjunction subject to the condition that no crossing over in interstitial region takes place.
- Consequently interchange heterozygotes are partially sterile.
- Even when predominant alternate disjunction is present, but pollen abortion is still observed. This is attributed to crossing over in interstitial region.

Explain in brief about the detection methods of interchanges?

Methods of detection of interchanges:

- Interchange heterozygotes is usually associated with pollen abortion and reduced seed set, so we can detect interchanges mainly through pollen sterility and reduced seed set.
- The plants with 50% or higher sterility or seed set can subjected to meiotic study, where interchange heterozygosity can confirmed by the presence of
  - (*a*) Cross-shaped figure at pachytene
  - (*b*) Quadrivalent or higher chromosome association at Diakinesis or Metaphase I

- In some cases, karyotypes prepared from mitotic chromosomes also used for detection because interchanges
  (*a*) Brings about alteration in lengths of individual chromosomes and
  (*b*) In relative position of centromere
- In an interchange heterozygote, while matching chromosomes, pairs of homologous chromosomes, two or more pairs will show lack of matching thus suggesting presence of interchange heterozygote.
- It can be detected by genetic methods
  (*a*) Due to altered linkage relationships between genetic markers i.e linked markers may become independent and the independent may become linked.
  (*b*) Due to linkage of semi sterility with factors in two linkage groups.

Explain in brief about the identification of chromosomes involved in interchanges?

Methods for identification of chromosomes involved in interchanges:

1. Linkage of semisterility with two linkage groups – If genetic markers belonging to different groups are available in homozygous condition, these can be used for detection of chromosomes involved in interchange.
2. Linkage relations in interchange stocks *vs* normal stocks – Linkage relations become altered in interchange stocks (homozygotes). In interchange stock, while linked markers may become independent and the independent may become linked.
3. Pachytene analysis in interchange heterozygotes – Morphology of pachytene chromosomes in cross shaped figure can be used to identify chromosomes involved in interchange heterozygote.
4. Karyotypes in interchange stocks - Alteration in lengths of individual chromosomes.
5. Use of trisomic sets – When interchange homozygote crossed with all trisomics of a set and F_1 with 2x+1 chromosome may be analyzed for meiotic configuration. Observation of pentavalent in two of trisomics and these chromosomes involved in trisomic condition are also involved in interchanges being testd.
6. Intercross method – When we cross all translocations in a diallele fashion or to cross unidentified translocations with known translocation stocks.
   - Absence of quadrivalent or presence of single quadrivalent mean that the two interchanges involve same chromosomes.
   - Presence of two quadrivalents mean that two interchanges involve entirely different chromosomes
   - Presence of hexavalent means that one interchange chromosome is common in two interchanges.

7. Use of monosomics and double monotelodisomics – When monosomics were used used, two chromosomes involved in translocation were identified by trivalents instead of quadrivalents, because of univalent formation. Sometimes bivalent and univalent will form but not quadrivalent.

Cytogenetic localization of genes using interchanges:

Methods involved here are

- Linkage between marker gene and semisterility
- Overlapping translocation procedure – It makes the use of interstitial trisomic region obtained from two overlapping translocations. If the gene outside the interstitial region, then none of the progenies will give trisomic ratio.
- Use of B-A translocations – Interchange between normal A type chromosome and B chromosome can result from breakage and reunion have been reported in maize.
- Use of compound B-A translocations.

Define balanced lethal? Explain the role of balanced lethal in maintenance of permanent heterozygosity?

It was first discovered by Muller in Drosophila where if dominant alleles of two traits each associated with a recessive lethal effect, are present in heterozygous repulsion phase, homozygotes will not survive and permanent hybridity will be maintained.

- A similar balanced lethal system operates in Euonotheras, so that the heterozygosity of the different races is maintained in progeny over generations.
- In *Oenothera lamarckiana,* the lethality is zygotic as in the balanced lethal systems of Drosophila and would therefore, lead to 50% ovule abortion or seed set.
- It is therefore a wasteful mechanism.

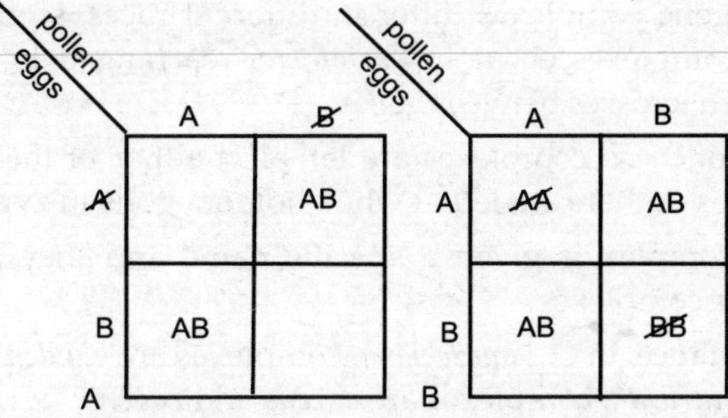

Thus, there are two balanced lethal mechanisms, one involving zygotic lethality and the other involving gametic lethality.

Gametic lethals:

- Gametic lethals are megaspore lethal or pollen lethal.
- In such cases, sperm do not fertilize eggs of the same genotypes due to the slow growth of pollen tubes or the slow development of embryosacs.
- Therefore only gametes of differing genotypes are compatable, producing only heterozygous plants.
- In *Oenothera lamarkiana* one gamete complex is called velans and the other is called gaudens.

Zygotic lethality:

- This occurs at zygotic level.
- The genotype which is homozygous for lethal genes will lose or die. Only heterozygotes will survive.
- Homozygous velans or gaudens are lethal in the zygotes and do not survive; thus only heterozygous velans / gaudens survives.

Explain in brief about Renner Complex System in *Oenothera*.

Renner complexes: Renner identified these complexes in *Oenothera lamarckiana*. Transmit only one of the two complexes present in the heterozygote through the pollen, and the other only through the egg. This has been interpreted as evidence of gametic lethals.

- Most races in O*enothera* are permanent hybrids, forming rings and undergoing alternate disjunction, thus forming only two types of functional gametes.
- Most of these races would be true breeding, and seed set will be perfect except cases like *O. lamarckiana,* where half of the seed did not germinate.
- These results were explained - assuming presence of two gene complexes in each race.
- Although, in a single race or species, only two gene complexes are found, these gene complexes differ in different races, so that a large number of such complexes occur in O*enothera,* rearranged in various combinations through crosses between races.
- Some of these complexes are lethal in either of the (Alpha) and (Beta) gametes - others are lethal when homozygous in zygotes.
- Each complex is given a specific name and they are collectively called Renner complexes.
- For instance, in *O. lamarckiana,* complexes are *gaudens and velans* and in case of *O. muricata,* complexes are *curens* and *rugens*.

Permanent heterozygosity maintained in Oenothera lamarkiana:

Translocation heterozygosity maintained, when chromosomes of concerned species having following features.

1. Centromeres must be median or submedian in position, this will allow chiasma formation on both arms and consequently results in ring formation.
2. Heterozygotes for translocations involve acrocentric or telocentric chromosome will form complex configuration at meiotic metaphase which are not ring shaped and in which adjacent chromosomes often pass to same pole. Since combinations of two adjacent chromosomes always include deletion and duplications for chromosomal segments, so a high proportion of sterility results from such rings.
3. Chiasma formation should confine to chromosome ends only. So that ring configurations are flexible. Which would result in a alternate disjunction
4. The disjunction of chiasmata may be promoted by the existence near the centromere, of large heterochromatic regions, in which chismata not formed.

Explain in detail about the utility of balanced lethal system.

Utility of balanced lethals:

1. **Maintenance of the complete permanent translocation heterozygous condition**

   E.g. *Oenothers biennis* strain suaveolens Grado

2. **For detection of mutations:**

   Mutations induced due to radiations or due to chemicals could be broadly of two types - (*i*) lethal mutations, (*ii*) visible mutations. Both these types could be located either on sex chromosomes or on autosomes (autosomes are chromosomes other than sex chromosomes). Different methods for detection of sex linked lethals, sex linked visibles and autosomal mutations are available.

Detection of sex linked lethal:

The CIB method:

- This method was proposed by H. J. Muller.
- This method involves use of a CIB stock which carries
- An inversion in heterozygous state to work as crossover suppressor (C),
- A recessive lethal on X-chromosome in heterozygous state, and
- A dominant marker, Barred (B) for the barred eye (narrow eye). One of the two X chromosomes in a female fly carried all these three features and the other X chromosome was normal. Male flies irradiated for induction of mutations were crossed to CIB females. Male progeny receiving CIB X-chromosome will die.

The CIB female flies obtained in progeny can be detected by barred phenotype. These are crossed to normal males. In the next generation, 50% of males receiving CIB X-chromosome will die. The other 50% males will receive X-chromosome, which may or may not carry the induced mutation. In case, lethal mutation was induced, no males will be observed. On the other hand, if no lethal mutation was induced, 50% males will survive. Thus, the CIB method was the most efficient method for detecting sex linked lethal mutations. This has since been improved upon and therefore, has been replaced by Muller-5 method.

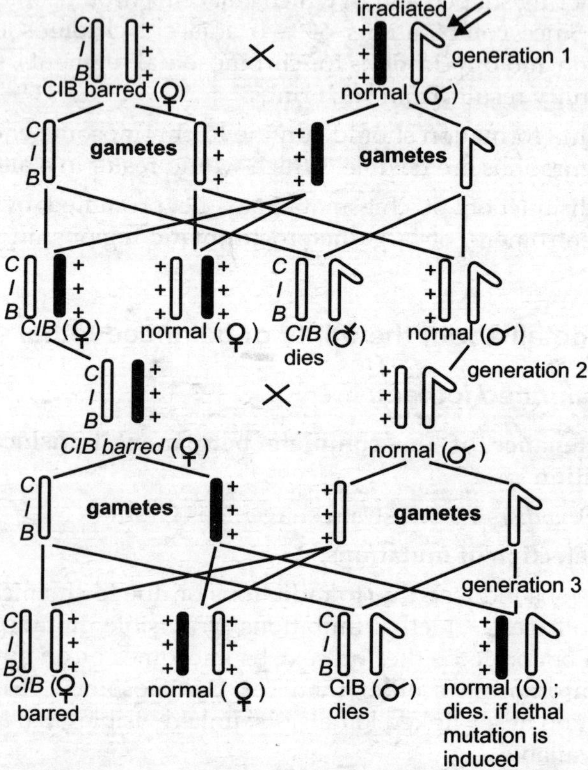

Detection of autosomal mutations (lethal or visible): Detection of mutations on autosomes (chromosomes other than sex chromosomes) makes use of a balanced lethal stock.

- Balanced lethal stock is one which carries recessive lethal alleles on two specific homologous chromosomes, their normal allele being present on its homologue; consequently homozygous individuals will not survive.
- The concept of balanced lethal would be better understood using a specific example.

For detection of visible mutations in Drosophila, a stock carried the dominant genes Cy (curly wing) and L(lobed eye)on one chromosome and Pm (plum eye i.e. brownish eye) on the other homologue so that the organism can be designated as Cy *L*/*Pm*. Phenotype of the stock which was heterozygous for Cy, L and *Pm,* was curly, lobe and plum. If such individuals are crossed among themselves, progeny is always heterozygous, because each of the two homologous chromosomes carried recessive lethal genes. Through the presence of inversion, crossing over was also suppressed, so that Cy and/or L may not be transferred to the chromosome carrying *Pm* and vice versa.

For detection of an autosomal mutation, Cy *L*/*Pm* stock is crossed to irradiated fly. From F_1, Cy L flies were again backcrossed to Cy *L*/*Pm* in order to obtain curly, lobe (heterozygotes) male and female flies carrying the same autosome to be tested. Such Cy L heterozygotes when crossed among themselves would give 1 CyL/CyL: 2 CyL/++: 1 ++/++. Since *CyL* homozygotes are lethal, phenotypic ratio will be 2: 1. However, if lethal mutation was induced, only curly will appear. If a recessive visible mutation was induced, this will appear in homozygous state in one-third of the progeny.

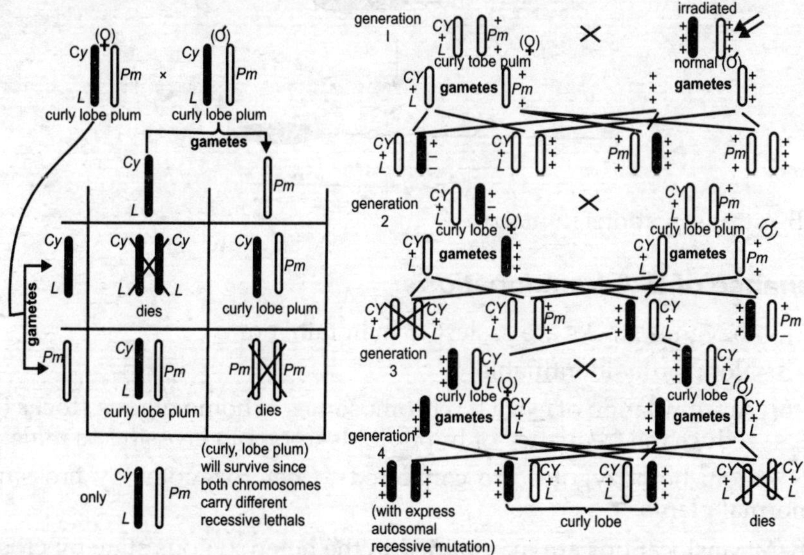

Benefits:

- Preservation of heterozygosity in *Oenothera* and other species.
- Ring formation associated with the balanced lethal system will reduces the loss of potential fertility in *Oenothera lamarkiana* and other sps.
- Useful for detection of autosomal mutations.
- Useful for detection of sex linked recessive mutations using stocks carrying balanced lethals.
- This system was useful in various crossing schemes in order to reveal the effects of different mutations and to remove the deleterious mutations.

What is meant by B-A Translocations? Explain in brief about maintenance of B-A translocations?

- Interchanges between normal A type chromosomes and B chromosomes can result from breakage and reunion and have been reported in maize, rye, *Lolium* and pearl millet.

- However, detailed breeding and cytological behaviour has been studied only in maize.
  1. The interchange heterozygote will have A, A^B, B and B^A forming a quadrivalent.
  2. The functional gametes due to alternate disjunction will have either A^B and B^A or A and B.
  3. Since B does not undergo disjunction at the second pollen mitosis, in the pollen with A^B and B^A, the two sperm nuclei will be hyperploids with A^B, B^A and B^A and hypoploid with A^B. On the female side, however, other duplication gametes like A^B-A and B^A-A resulting due to adjacent disjunction will also function.

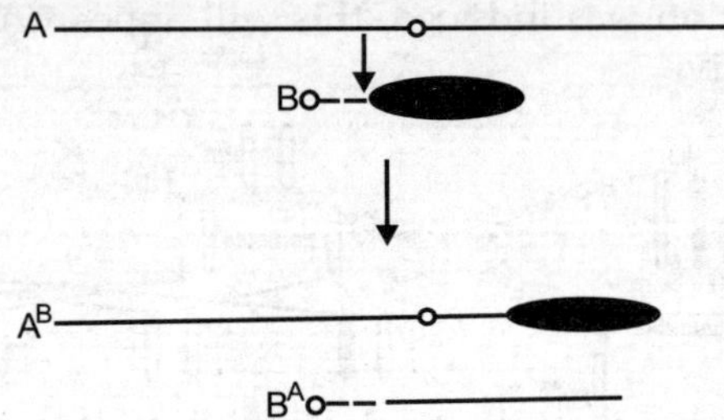

E.g. B-A translocations in maize.

Maintenance of B-A translocations:

- Homozygous stocks are easiest to maintain, but
  - Not readily identifiable
  - Accumulation of extra B^A chromosomes in homozygous stocks tends to reduce the frequency of hypoploids upon out crossing as male.
- Solution: homozygotes are converted to heterozygotes by crossing with normal plants.
- B-A translocations are maintained in the heterozygous state by crossing as female by normal.
- Approximately one third of the progeny are again heterozygous for translocation.

Explain in brief about Robertsonian translocations.

- Robertsonian translocations are those translocations, where two acrocentric chromosomes may fuse at centromere to give rise to a metacentric or sub metacentric chromosome. This phenomenon of fusion is also called *centric fusion*.
- The role of Robertsonian translocations in animals, particularly in *Drosophila* and in *Bovoidea* is well known.
- In plants - Robertsonian translocations - less frequently known, although they might have played an equally significant role.

- In *Nothoscodum inodorum*, a karyotype 2n= 13M+6A is believed to have been derived from 2n=16M.
- Dr. Keith Jones emphasised on the role of Robertsonian translocations in the evolution of some members of family *Commelinaceae.*

  E.g. *G. schiedeana* was a tetraploid with 2n=16 (x=4)

  G. karwinskyana was a tetraploid with 2n=20 (x=5)

  G. schiedeana – diploid race with 2n=10 (x=5)
- Karyotypic studies revealed that basic complement with x=4 consisted of 3 metacentric and one acrocentric (3M+1A) and x=5 consisted of 2M+3A.
- Robertsonian translocation took place through centric fusion in cytotypes with x=5 with 2M+3A, it will give rise to x=4 with 3M+1A.

Explain the uses of translocations in crop improvement?

Translocations can be used for solving various problems of cytogenetics and plant breeding.

1. Study of chromosome behaviour during meiosis :
2. Assignment of centromere position: Reduction in recombination values in the interstitial region is used to determine the relative position of breakpoints, marker genes and the centromere.
3. Interchanges as genetic markers: Interchange breakpoints can be plotted on linkage maps in the form of partial sterility loci.
4. Determination of unknown locus of a gene and chromosome mapping: Interchanges with known breakpoints may be used to determine the unknown locus of a gene.
5. Association of linkage groups or genes to specific chromosomes: In several crops, E. g. barley, maize and peas.
6. Testing of the independence of linkage groups.
7. Association of different linkage groups: Thus new linkage groups may be created using translocations.
8. Determination of the initiation of chromosome pairing.
9. Production of duplications: Duplications are obtained in the F_2 generation when two interchanges involving the same chromosomes but with different break position are intercrossed.
10. Evolution of allopolyploids: If two lines belonging to a single species, each containing a series of different translocations, are crossed and the chromosomes of the F_1 are doubled.
11. Gamete selection: In 1946, Burnham proposed the "*Oenothera* method" of gamete selection for the isolation of inbred lines with the help of multiple interchanges.

12. Enlarging the genome: Interchanges may be used to increase the chromosome number of a species. A barley (2n = 14) line with 8 pairs of chromosomes (2n = 16) was reported by Wiebe in 1975.
13. Evolution of karyotype: Changes in chromosomes structure and change in chromosome number.
    - Change in chromosome structure: Interchanges may alter the centromere position.
    - Reduction in the genome size: Centric fusion (Robertsonian translocation) involving two acrocentric chromosomes produce a large metacentric chromosome and a very small chromosome containing a centromere with or without minute arms.

The small chromosome is generally devoid of genes and may be eliminated from the cell resulting in a reduction in the number of chromosomes in the genome.

14. Interchanges as source of trisomics: Trisomics are produced when 3:1 disjunction occurs in a ring or chain of 4 chromosomes resulting in the formation of gametes containing n+1 chromosome.
15. Interchanges may be used for pest control. Males heterozygous for multiple interchanges will produce nonfunctional sperms. Therefore, such males may be released in large numbers for insect control.
16. To study the somatic pairing in Diptera. They may also be used to transfer desired character into an inbred.

What is meant by inversions? And explain its types?

Inversions: An inversion is a condition where the order of gene changes on a chromosome. It is produced when two breaks occur in a chromosome and the intercalary segment reunites in reverse order. Inversion can be of two types.

1. Paracentric inversion: The inverted segment does not involve the cenromere of chromosome
2. Pericentric inversion: The inverted segment does involve the centromere.

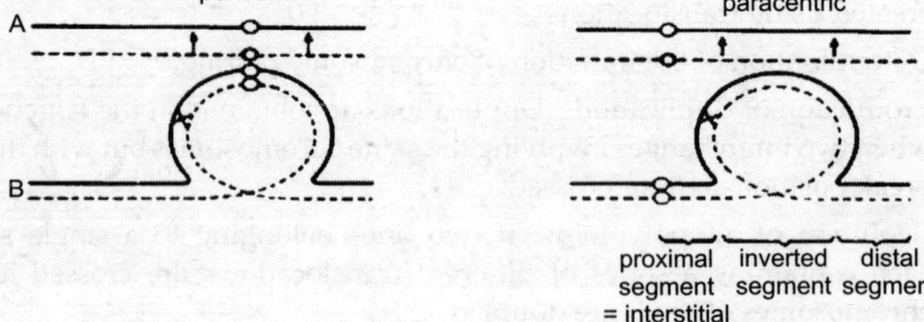

In case of paracentric inversions, no changes in karyotypes are brought about, but in pericentric inversions karyotype changes are introduced.

Explain in brief about the genetic consequences of inversions.

Genetic consequences of inversions: Due to peculiar structure formation at meiosis after crossing over and seperation of chromatids, the two chromatids resulting from crossing over would have deficiency and duplication, thus gametes are nonfunctional and results into genetic or zygotic lethality.

Type of inversion	Meiosis and constitution of chromatids	Effect on fertility
Pericentric inversion	No bridge, chromatids have deletions and duplications	In plants: embryo sac and pollen abortion In drosophila: zygote abortion and no sperm abortion
Paracentric inversion	Bridge and acentric fragments, chromatids have deletions and duplications	In plants: pollen abortion but embryo abortion low or absent In drosophila: zygote abortion low or absent and no sperm abortion

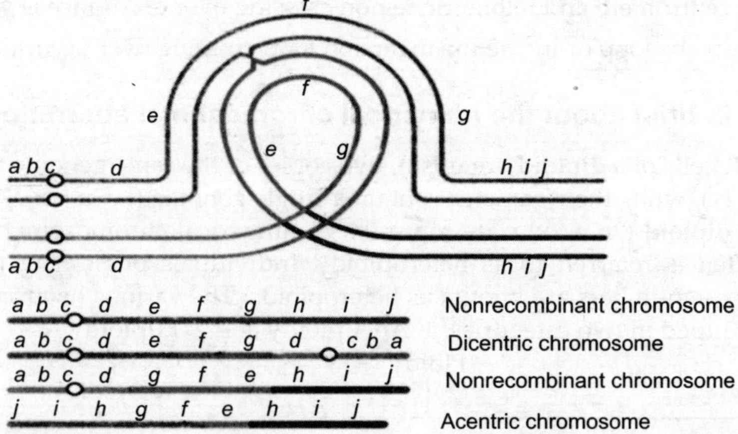

Inversions are also called as cross over suppressors. Substantiate your answer.

Inversions are called crossover suppressor because the products of single cross over will not function and the only cross over recovered are double cross over, thus observed frequency of recombination between any two genes in question are reduced. The reduction in crossing over is not the actual reduction in cytological crossing over but is due to lack of recovery of the products of single cross overs. This property of inversion was utilized by H.J.Muller for detection of sex linked lethal mutation by producing CLB stock in drosophila.

- Cytologically crossing over occurs in inversion which leads to production of non functional gametes because of chromatids with deletions and duplications.

- Near the break point, crossing over occurance is suppressed because of non homology between chromosomes.
- Recombination values are very less between genes.
- E.g. In drosophila, recombination frequency values found out.

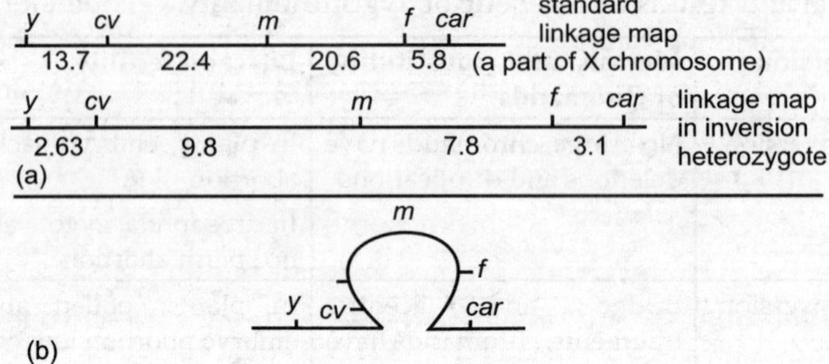

- Recombination frequency reduced in inversion heterozygote linkage map. So we go for modified linkage map.
- In centromere and telomeric region crossing over occurance is very less.
- Near the loop or inside the inversion loop crossing over occurance is low.

Explain in brief about the numerical chromosomal aberrations?

In somatic cells of a diploid organism, two copies of the same genome are present (i.e. 2n = 2x), while their gametes contain a single genome (i.e. n = x). A deviation from the diploid (2n = 2x) state represents a numerical chromosomal aberration which often is referred to as heteroploidy. Individuals possessing the variant chromosome numbers are known as heteroploids. The various heteroploid states can be grouped in two categories. 1. Aneuploidy and 2. Euploidy.

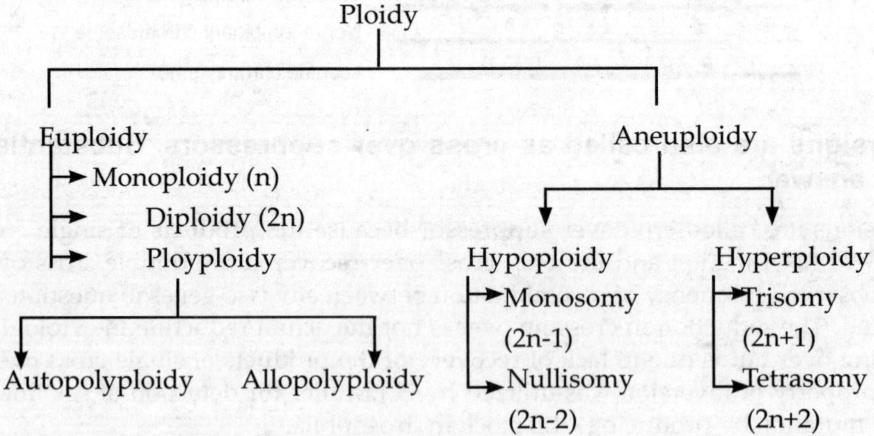

Heteroploidy – Change from the 2x state.

1. Euploidy – Number of genomes different from two.

   (*a*) Monoploidy (x) – Only one genome present

(*b*) Haploid (n) – Gametic chromosome number of the concerned species present.

(*c*) Diploid (2n) – Somatic chromosome number of the concerned species present.

(*d*) Polyploidy - More than two copies of the same genome or more than one genome present.

2. Aneuploid – One or few chromosomes extra or missing from 2n.

(*a*) Hypoploidy – Lacking one or two chromosomes from the diploid complements.

- Monosomy – One chromosome missing
- Nullisomy – One chromosome pair missing

(*b*) Hyperploidy – Addition of one or two chromosomes to one pair or two different pairs.

- Trisomy – One chromosome extra
- Tetrasomy – One chromosome pair extra

What is meant by haploidy? Explain the methods of production of haploidy?

The haploid chromosome number of a diploid species is known as the basic number and haploid individual belonging to a diploid species is therefore, also called a 'monoploid'. The haploid set of chromosome in such a diploid species is called a 'genome'.

The individuals with gametophytic chromosome numbers in polyploids are better called 'polyploids' to distinguish them from haploids in the diploid species which are better called 'monoploids'.

Classification of haploids:

1. Euhaploids (Derived from euploids)

(*i*) Monohaploids (monoploid – derived from diploids)

(*ii*) Polyhaploids (derived from polyploids)

(*a*) Allopolyhaploids (derived from allopolyploids)

(*b*) Autopoluhaploids (derived from autopolyploids)

2. Aneuhaploids (derived from aneuploids)

(*a*) Disomic haploids (derived from tetrasomic line)

(*b*) Addition haploids (derived from alien addition line)

(*c*) Nullisomic haploids (derived from nullisomic line or deficient in chromosome)

(*d*) Substitution haploid (one chromosome replaced by that of another chromosome)

(*e*) Misdivision haploids (presence of telocentric or isochromosomes)

Haploidy production system:

Generally these are two methods haploidy production, *in vivo* and *in vitro*

In vivo: Parthenogenesis and apogamy chromosome elimination.

In vitro: Cultural method.

Examples:-1.Spontaneous:- Maize and Linum

2. Artificial introduction by external stimulus
   (*a*) Irradiation- Crepis and Wheat
   (*b*) Wounding/injury- Oenothera, Tobacco and Maize
   (*c*) Temperature- Datura and Secale
   (*d*) Chemical- Populus and Capsicum
3. Delayed pollination- Maize
4. Wide hybridization- S. *tuberosum*
5. Alien cytoplasm- Wheat
6. Haploid inducing genes
   (*a*) Independent gametophyte - Maize
   (*b*) Haploid initiator gene (hap) - Barley
7. Semigamy – Cotton and Arabidopsis
8. Pseudogamy – Potato and Strawberry
9. Chromosome elimination:
   1. Somatic reduction E. g. Sorghum

      Chlorophenicol and paraflurophenyl alanine induction of haploidy
   2. Barley haploidy by crossing with H.bulbosum.

      Also other Hordeum species wheat (Barley)
   3. Wheat haploid from wheat X maize cross.

Cultural method:

1. Anther and microspore clture.

   Anther culture in Datura inoxia by Guha and Maheswari
2. Ovule culture:In barley by San Noeum.

Explain the role of haploidy in crop improvement?

Applications:

1. Production of homozygous lines: Used directly as cultivar (or) in breeding programme eg. marglobe tomato (doubled haploid) Chase used-DH lines as inbred line in hybrid maize programme.
2. Hybrid sorting: Refers to selection of recombinant superior gametes.

   eg. F-211 (tobacco), Haayul, 2(Rice) Yuhna 1, 2 (wheat), *Mais haptona* (Brassica napus).

3. Short breeding cycle: Only 4-5 years as compare to 10 years in conventional method.
4. Analytical breeding: Selection at haploid level and improved dihaploids used for chromosome doubling eg. Potato.
5. Generation of exclusive male Asparagus (commercially male are preferred).
6. Mutation research: eg. Tobacco-cell line resistance to methionine sulphoxamide and wild fire (*Pseudomonas tabaci*).
7. Cytogenetic research: Determining basic chromosome number and polyploidy and production of alien addition and substitution line (wheat).
8. Evolutionary studies.
9. Genetic studies: Hybrid development.
10. Indirect resistance- Rice- Hwacheongbye.

How haploids are produced in *invitro*? Compare the double haploids with purelines produced through conventional breeding methods? Why double haploids are not much helpful for production of inbred line in cross pollinated species?

The four important methods for invitro production of haploids are:

(*i*) Anther culture

(*ii*) Isolated microspore culture

(*iii*) Unpollinated ovary culture

(*iv*) Embryo rescue from wide crosses

  (*a*) Bulbosum method

  (*b*) Wheat haploids from wheat × maize crosses

Comparison between DHs and purelines produced through conventional breeding:

- In self pollinated crops, long period is required to recombine desirable gene combinations from different sources in homozygous form. Generally, it takes 8-10 years to develop stable, homozygous and ready to use material from a fresh cross of two or more lines.
- Due to inbreeding depression in cross pollinated crops, it is difficult to obtain vigorous inbreds for hybrid seed production programmes.
- Haploids possessing gametic chromosome number are very useful for producing instant homozygous true breeding lines i.e. by colchicines treatment.
- Haploids constitute an important material for induction and selection of mutants particularly for recessive genes.
- Period required to develop homozygous true breeding line is less through haploid breeding as compared to conventional breeding.

- In conventional breeding, the early segregating generation population involves variation due to both additive as well as nonadditive effects whereas DH lines exhibit variation only due to additive gene effects along with additive × additive type of epistasis which are easily fixable through one cycle of selection.
- Elimination of dominant effects and availability of sufficient seed of each doubled haploid line for replicated testing leads to high heritability in such populations.
- In contrast to relatively large sized segregating populations in conventional methods, less number of DH lines is required for the purpose of selection of desired recombinants through the production of haploids.
- The DH populations derived through anther culture from highly inbred parents show variability for quantitative as well as reduced vigour.
- DH approach is increasingly being used for rapid development of populations for QTL mapping and construction of genetic linkage map for traits of interest.

What is meant by polyploidy? Write the classification of polyploidy?

Polyploidy is nothing but organism having more than two basic sets of chromosomes. And it can be classified mainly into

1. Autopolyploidy – Individual having more than two copies of single genome
   (*a*) Autotriploidy – Three copies of the same genome (3x)
   (*b*) Autotetraploidy – Four copies of the same genome (4x)
   (*c*) Autopentaploidy – Five copies of the same genome (5x)
   (*d*) Autohaxaploidy – Six copies of the same genome (6x)
2. Allopolyploidy – Polyploids having two or more distinct genomes
   (*a*) Allotetraploidy – Two copies of each two distinct genomes ($2x_1+2x_2$)
   (*b*) Allohexaploidy - Two copies of each three distinct genomes ($2x_1+2x_2+2x_3$)
   (*c*) Allooctaploidy - Two copies of each four distinct genomes ($2x_1+2x_2+2x_3+2x_4$)
3. Segmental allopolyploidy – Partial homology of chromosome sets that are combined in an allopolyploid
4. Autoallopolyploidy – Duplicated set from one species plus at least one set from different set present.

   Amphidiploid: It is an allopolyploid, derived from an interspecific or intergeneric F_1 hybrid between two parents.

   - If two parents are diploids, the amphiploids derived from them are described as amphidiploids. Which should, therefore be tetraploids.
   - Alternatively if one or both parents are polyploids, the derived polyploids will be called amphipoyploids.

What is meant by autopolyploidy? Explain the autopolyploids with origin, morphology and cytology and role in crop improvement?

Autopolyploids: Origin

1. Spontaneous: Certain genes *ps*-parallel spindle (Potato), *os*- omission of II meiotic. division.
2. Production of adventitious buds E.g. *Solanaceae*- most common 6-36%. 1% IAA-increases callus development.
3. Physical agents- Heat or cold; centrifugation and X-rays, γ-rays.
4. Regeneration *in vitro* - callus and suspension culture.
5. Colchicine treatment – $C_{22}H_{25}O_6N$. Other chemicals- N_20 (laughing gas), Fungicides and herbicide (Oryzoline), Acenaphthene, 8-hyrdoxyquinoline.
6. Somatic hybridisation.

Morphology and Cytology:

1. Large cell size, number of stomata per unit area - lower; guard cell of stomata are larger.
2. Pollen grain - larger.
3. Slower growth and later in flowering.
4. Large and thick leaves, large flower and fruits, less in number.
5. Increase in general vigour and vegetative growth.
6. Reduced fertility.
7. Lower dry matter.

Autotriploids: Mostly sterile,

- Eg. Watermelon, Banana, Apple and Sugarbeet.
- Highly fertile-spinach
- Some more vigrous than diploid like sugarbeet.

Watermelons: 4x (female) , 2x (male) are used to produce triploids. Pollination is essential for fruit setting. Commercial planting involve 1 diploid with 5 triploids for proper fruit development.

Problems with triploids

- Genetic instability at 4x level
- Irregular fruit shape
- Tendency toward hollowness of fruits
- Production of empty seeds
- Laborious in seed production.

Sugarbeet: Large roots

Seed production (4 x) × (2 x) -low seed yield but high triploids,

(2 x) × (4 x) -high seed yield but low triploids

Triploid (3x) clone of tea - Variety TV-29

Sesquidiploid are hybrids that are diploid for the chromosome of one species (usually the cultivated) and haploid for the other, they are moderately fertile.

Autotetraploid: Fertility can be improved by selection eg. Petkus rye.

Applications

1. Breeding - Banana
2. Improving quality - Maize 43% more carotenoid and Vit A.
3. Overcoming Self Incompatibility - Tobacco, white clover. *Petunia*
4. Making distant crosses.
5. Used as variety - Pusa Giant Barseem.

Limitations of autopolyploidy:

1. Higher water content
2. High sterility and poor seed set.
3. Slow progress under selection
4. Maintenance of 3x clonally
5. Undesirable feature of autoploidy
6. Effect cannot be predicted.

Recent studies using GISH analysis revealed Coffee and groundnut to be allopolyploid.

* *Arachis hypogea* (4 x = 40) = *A. villosa* x *A. ioaensis*

* *Coffea arabica* (4x = 40) = C. *congensis x C. engenioides*

What is meant by double reduction? And write the requirements for double reduction to occur?

Double reduction: This phenomenon is called as equational non disjunction by Bridges. Double reduction is also called as post reduction or equational segregation at a particular locus.

- Double reduction is one of the evidence that crossing over takes place at chromatid stage.

Requirement for double reduction:

- Multivalent formation
- Crossing over between gene and centromere
- Homologous chromosomes carrying products of cross over should move to same pole at anaphase I
- At anaphase II, the separation of chromatids should be such that the segments of sister chromatids carrying the locus under consideration enter the same gamete.
- Frequency of double reduction will also be influenced by the orientation of multivalent because an alternate orientation will not allow the same pole.

So relative frequencies of alternate and adjacent orientation will influence the frequency of double reduction.

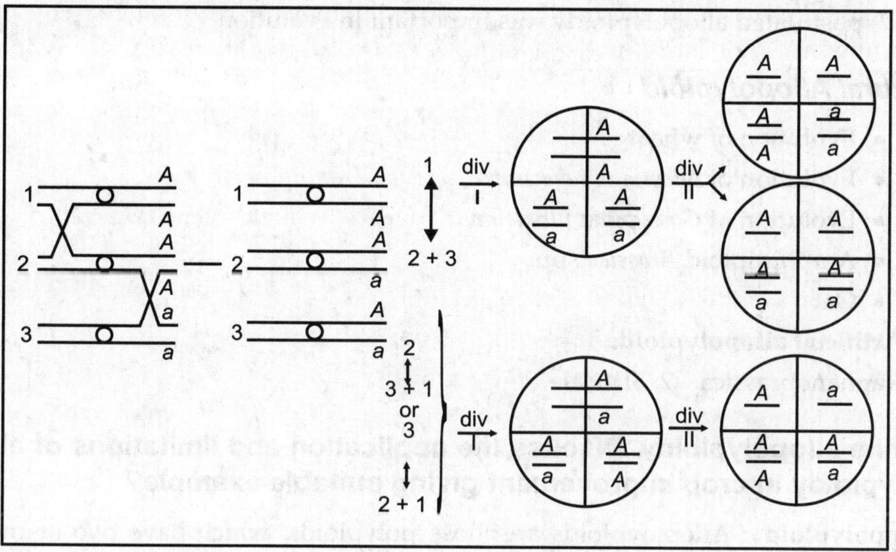

What is meant by allopolyploidy? Explain allopolyploids with origin, morphology, cytology and role in evolution?

Allopolyploidy:

Allopolyploidy is a type of polyploidy which contains genome from two or more species, the species is called allopolyploid and it is related with creation of new species

Origin and Production of Allopolyploid

1. Irregular mitotic cell division

   E.g. F_1 from cross *(Primula verticillata* × *P. floribanda)* produced fertile branch with twice the chromosome number *(P. kewensis).*

2. Irregular meiosis: unreduced gametes, E.g. Raphanobrassica
3. Experimental production by doubling the chromosome number of distant hybrids by colchicine treatment.

Morphological and Cytological feature

- Difficult to predict precise combination of parental characters.
- Natural distribution different
- Hardier than parent
- Most of them Apomictic e.g.- *Taraxacum, Rubus.*
- Concept of Homeologus chromosome is involved e.g. Wheat 5B, Oat, Tobacco
- Diploidization

Role in evolution

Allopolyploid are more successful as crop species than autopolyploids-Winge (1917) postulated allopolyploidy was important in evolution

Natural Allopolyploid

- Evolution of wheat
- Evolution of *Nicotiana tabacum*
- Evolution of *Gossypium hirsutum*
- Amphidiploid *Brassica* spp.
- Oat

Artificial allopolyploid:

Raphanobrassica 2. Triticale

Define allopolyploidy. Discuss the application and limitations of allopolyploidy in crop improvement giving suitable example?

Allopolyploidy: Alloployploids are those polyploids, which have two or more than two genomes, each genome from a different and distinct species.

Application in crop improvement:

1. Creation of new species.

   Eg: Triticale. In India TL419 has been release and several promising lines are being elevated.
2. Other promoting synthetic allopolyploids that may succeed as crops are Raphanobrassica, Festuca - lolium hybrids and triploids obtained by crossing *Brassica napus* with *Brassica campestris*.
3. Allopolyploids may be synthesized to widen the genetic base of allopolyploid crops, which have rather narrow genetic base.

   Eg: Genetic base of *Brassica napus* is sought to be widened by synthesizing it from the parental diploid species (*Brassica oleraceae* and *Brassica campestris*)
4. Synthetic allopolyploids serve as a bridge species between two diploid species, which do not cross with each other, or those which do cross but their F_1 hybrid is sterile.

   Eg: The F_1 hybrid from cross between *Nicotiana sylvestris* and *Nicotiana tobacum* is sterile, but the amphidiploid *Nicotiana digluta* produced by chromosome doubling of the F_1 is reasonably fertile. The hybrid from cross between *Nicotiana digluta* and *Nicotiana tobacum* is partially fertile and can be easily backcrossed to *Nicotiana tobacum*

   In this manner resistance to tobacco mosaic virus was transferred from *Nicotiana sylvestris* to Nicotiana *tobacum*, so in this synthetic allopolyploid acts as bridging species.

5. Polyploidy is one of the sources for variation and thus helps in evolution and allopolyploid played a key role in natural evolution of number of plants like wheat, cotton, galeopsis and primula etc.
6. Allopolyploidy is one of the effective mechanism for maintaining hybrid vigour in successive generations, because as a result of allopolyploidy, different genomes are brought together and are passed from one generation to other generation
7. Interspecific gene transfer.

Limitations:

- Effect can't be predicted
- Many defect associated with allopolyploid like low fertility, cytogenetic and genetic instability.
- Synthetic allopolyploids have to be improved through extensive breeding at polyploidy level.
- Only small proportions are promising.

How do you distinguish autoployploids from allopolyploids? Explain.

Autopolyploids distinguish from allopolyploids by using

1. Morphological characters
2. Biochemical studies
3. Cytological studies

Morphological characters:

- In autopolyploid, morphological characters are same, because these are derived from the same species, while in allopolyploids from different species.
- In allopolyploids sometimes new phenotypes is expressed irrespective of parental genotypes.
- In some species like, *Bromus carinata* (2n=56) and B. *oryzinicus* (2n=84), both they share 21 pairs of chromosomes. So they are keeping in one morphological line. But always morphological lines are not reliable.
- In case of wheat, (Sankar and Stebbins) use morphological lines to distinguish the 'B' choromosome.

Biochemical studies:

- By using phenolic compound content, seed content and also isozymes in specific enzymes, distinguish autopolyploids from allopolyploids.
- Disadvantages here are enzymes are stage dependent and it requires technical skills and knowledge.

Cytogenetic studies: It is by mitosis and meiosis.

- Mitosis: Pairing of different chromosomes in same type
- Meiosis: Pairing of different type of chromosomes

Eg. In cotton (2n=56); (n=26 are long chromosomes and other 26 are short chromosomes). These long chromosomes belong to 'A' genome. Short chromosomes belong to 'D' genome.

In wheat, observe the four pairs of 'sat' chromosomes, of these four pairs, one pair belongs to 'A' genome, two pairs belongs to 'B' genome and another one pair belongs to 'D' genome. For searching 'B' genome, search for two pairs of 'sat' chromosomes in wheat.

- And also in meiosis,
  - Autoployploids forms the multivalent, but in allopoplyploids forms the bivalents.
  - So polyhaploids from autopolyploids form the bivalents and ployhaploids from alloployploids forms the univalents.
  - But segmental allohaploids forms the both univalents and bivalents.
- If pairing is observe in interspecific or intergeneric hybrids,
  - From same species, is called autosynthetic pairing
  - From two species is called allosynthetic pairing
  - But in segmental allosynthetic pairing, observed some amount of autosynthetic pairing.

Explain inheritance pattern in autoployploids using different types of segregation in duplex individuals.

Polysomic segregation ratios are expected to differ depending upon whether the gene is located close to the centromere or away from the centromere, thus determining the low or high probability of crossing over between the centromere and the gene.

Depending upon the intensity of equational separation of alleles carried on sister chromatids in the first meiotic division, the segregation in polyploids can be classified as

1. Chromosome segregation – No crossing over occurs
2. Random chromatid segregation – 50% crossing over occurs
3. Maximum equational segregation – 100% crossing over occurs

Segregation ratios in duplex (AAa) individuals: The gametic ratios in duplex (AAa) genotypes, due to chromosome segregation, chromatid segregation and maximum equational segregation based on 50% and 25% transmission rates of 'x+1' type of gametes. The frequencies of 'A' and 'a' types of gametes with x or n chromosomes are based on the genotype. For in stance, in genotype AAa the ratio will be 2A:1a, which by multiplying with the common factor will be brought at par with the total of the three genotypes (AA, Aa, aa) of 'x+1' type of gametes, in case of 50% transmission. In case of 25% transmission, the relative frequencies of A:a calculated as above will be multiplied so as to make it three times (75% 'x' as compared to 25% 'x+1') the total of three genotypes in the class 'x+1'.

Chromosome segregation: In chromosomal segregation, frequencies of 'x+1' type of gametes will be calculated on the basis of all possible paired combinations. For instance AAa will give 1AA:2Aa.

For 50% transmission frequency of 'x+1' type of gametes:

	Functional gametes of male	
Functional gametes of female	2A	A
1AA	2AAA	AAa
2Aa	4AAa	2Aaa
2A	4AA	2Aa
1a	2Aa	Aa

In this, gametic (dominant to recessive) ratio is 17:1.

For 25% transmission frequency of 'x+1' type of gametes:

	Functional gametes of male	
Functional gametes of female	2A	A
1AA	2AAA	AAa
2Aa	4AAa	2Aaa
6A	12AA	6Aa
3a	6Aa	3aa

In this gametic (dominant to recessive) ratio is 11:1.

Random chromatid segregation: In chromatid segregation, relative frequencies of A and a types in gametes with 'x' chromosomes remains the same, but the relative frequencies of AA, Aa and aa would change due to random chromatid segregation.

For instance, AAa will be treated as AAAAaa with six chromatids which have to form pairs at random, so that 4A's will give n(n-1)/2 = (4×3)/2 = 6AA; 4A's and 2a's will give 4×2=8Aa and two a's will give only 1aa. This gives a ratio of 6AA:8Aa:1aa. In order to bring equality of 'x+1' and 'x' types of gametes, 2A:1a will become 10A:5a.

For 50% transmission frequency of 'x+1' type of gametes:

	Functional gametes of male	
Functional gametes of female	4A	2a
6AA	24AAA	12AAa
8Aa	32AAa	16Aaa
1aa	4Aaa	2aaa
10A	40AA	20Aa
5a	20Aa	10aa

In this, gametic (dominant to recessive) ratio is 14:1.

For 25% transmission frequency of 'x+1' type of gametes:

Functional gametes of female	**Functional gametes of male**	
	4A	2a
6AA	24AAA	12AAa
8Aa	32AAa	16Aaa
1aa	4Aaa	2aaa
30A	120AA	60Aa
15a	60Aa	30aa

In this, gametic (dominant to recessive) ratio is 41:4.

Maximum equational segregation: Here, clear distinction is made between two A's by calling them A_1 and A_2. Pairing and crossing over events among three homologous chromosomes, in a triploid with genotype AAa showing the crossover products in three different orientations of a trivalent.

		Segregation type		
		Alternate disjunction	Adjacent I disjunction	Adjacent 2 disjunction
1	Centromeres dijunction	(1+3)/2	(1+2)/3	(2+3)/1
2	Chromatid pairs at poles (AI)	$(A_1A_2 + A_1a)/A_2a$	$(A_1A_2 + A_2a)/A_1a$	$(A_2a + A_1a)/A_1A_2$
3	(x+1) gametes (AII)	A_1A_1, A_1a, A_1A_2, A_2a	A_1A_2, A_1a, A_2A_2, A_2a	A_1A_2, A_1a, A_2a, aa
	Total gametes (x+1)	5AA:6Aa:1aa		
4	(x) gametes (AII) Multiply the total values in '4' by 2 in order to equalize (x) and (x+1) for 50% transmission	A_2, a 2 A_2, 2a	A_1, a 2 A_1, 2a	A_1, A_2 2 A_1, 2 A_2
	Total gametes	8A : 4a		
	Total gametic ratio (x, x+1)	A_1A_1: A_2A_2: aa: 3 A_1A_2: 3A_1a : 3A_2a : 4A_1 : 4A_2 : 4a = 5AA:6Aa:1aa:8A:4a		

Here, gametic (dominant to recessive) ratios are 67:5 (50%) and (25%).

In case of simplex genotypes (Aaa), expected gametic segregation ratios are

	50%	25%
Chromosomal segregation	2:1	11:7
Random chromatid segregation	29:16	3:2
Maximum equational segregation	23:13	

Breeding at lower ploidy level is easy compared to breeding polyploids. Substantiate your answer?

Breeding at lower ploidy level is easy compared to breeding polyploids because

- In some autoploypoids, especially odd number polyploids like 3n, 5n and 7n, where univalents are formed during meiosis.
  - Eg. In polyploids like AAa, at meiosis will lead to formation of univalents like A and a.
  - These univalents arrange in random manner at equatorial plate.
  - In some cases i.e. there is no random manner, they are lagging behind, so it forms laggards in that case.
  - And also attaining of homozygosity is also difficult in that case and also identification of heterozygous genotypes is also difficult.
  - In later generations, they show complex ratios.
- But lower ploidy level like diploids, no univalent formation, no complex ratio are observed.

Hyperploids are more common in diploids while hypoploids are common in polyploids.

Hyperploids are more common in diploids while hypoploids are common in polyploids:

- In case of diploids, contain two sets of chromosomes, so if any addition of chromosomes, they can tolerate that genetic load. And also diploids are not ready to lose their chromosomes, so hyperploids are morer common in diploids.
- But in case of polyploids, they have already high genetic load, so if any addition, it cannot tolerate. And also polyploids are always ready to lose their chromosomes. So hypoploids are more common in diploids.

Write the evolutionary pattern of the following crops?

1. Wheat 2. Cotton 3. Tobacco 4. Brassica spp. And 5. Potato.

Wheat: The genes in Triticum, to which common bread wheat (2n = 6x = 42) belongs, and the closely related genes Aegilops, each consists of species having three different ploidy levels (2x, 4x, 6x) with x = 7.

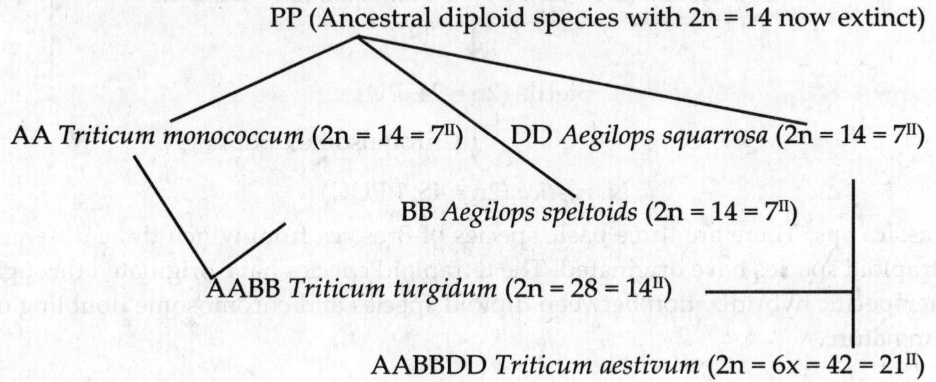

The tetraploid species developed as amphidiploids between two diploid species, and hexaploid species originated from a cross between tetraploid and diploid species. It is believed that tetraploid species *Triticum turgidum* evolved as amphidiploid between *Triticum monococcum* (AA) and *Aegilops speltoids* (BB). The hexaploid bread wheat originated as an amhidiploid between *Triticum turgidum* (AABB) and *Aegilops squarrosa* (DD).

Cotton: Probable origin of the new world cotton (*Gossypium hirsutum*) from two diploid species of Gossypium.

Gossypium herbaceum (Old world AA) × *Gossypium thurberi* (New world DD)

2n = 26 = 13^{II} (Longer chromosomes) ↓ 2n = 26 = 13^{II} (Shorter chromosomes)

F_1 hybrid (AD)

2n = 26 = 26^{I} (13 longer + 13 shorter)

↓ Chromosome doubling

2n = 52 = 26^{II} AADD (13 longer + 13 shorter)

According to Bacsely (1940), a possible origin of *Gossypium hirsutum* is from cross between Asiatic cultivated cotton (*Gossypium herbaceum*) and American wild species (*Gossypium thurberi*) followed by chromosome doubling of F_1 in nature.

Tobacco: It is believed that *Nicotiana tabacum* has originated as amphidiploids between two diploid species viz., N. *sylvestris* and N. *tomentosa*. Similarly, N. *rustica* is believed to be amphidiploids between wild species N. *paniculata* and N. *undulata*.

N. *sylvestris* (2n = 24, SS) × N. *tomentosa* (2n = 24, TT)

↓

Sterile (2n = 24, ST)

↓ Chromosome doubling

N. *tabacum* (2n = 48, SSTT)

N. *paniculata* (2n = 24, PP) × N. *undulata* (2n = 24, UU)

↓

Sterile (2n = 24, PU)

↓ Chromosome doubling

N. *rustica* (2n = 48, PPUU)

Brassica sps: There are three basic species of Brassica from which three different tetraploid species have originated. The tetraploid species have originated through interspecific hybridization between diploid species and chromosome doubling of F_1 in nature.

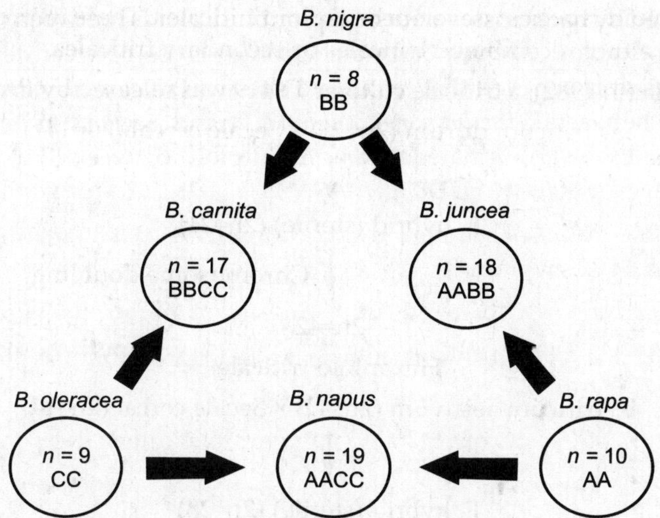

Potato: *Solanum tuberosum* is a amphidiploid from crosses between cultivated diploid species S. *stenotomum* and S. *pureja*.

S. *stenotomum* (2n = 24) × S. *phureja* (2n = 24)

↓

Sterile (2n = 24)

↓ Chromosome doubling

S. *tuberosum* (2n = 48)

Write short notes on the following?

1. Triticale
2. Raphanobrassica

Triticale:

- Triticale is the first man made crop, developed by A. Muntzing in 1951and it is resulted as an allopolyploid derived by crossing Wheat (*Triticum*) and Rye (*Secale*).
- Depending upon whether tetraploid (2n = 4x = 28) or hexaploid (2n = 6x = 42) wheat is utilized for the synthesis, one would get hexaploid triticale (2n = 6x = 42) or octaploid triticale (2n = 8x = 56) respectively. But in each case, only diploid rye (2n = 2x = 14) was used.
- The hexaploid triticales, which are initially produced as amphiploids are called primary triticales.
- These primary triticales have many drawbacks like shrivelled grains, meiotic instability, poor yield and preharvest sprouting.
- For improvement, these primary triticales are regularly intercrossed among themselves and also with either hexaploid wheat or octaploid triticales

followed by backcrosses with hexaploid triticales. These improved triticales can be either recombined triticales or secondary triticales.

- In India (in 1982), a triticale cultivar TL419 was released by PAU, Ludhiana.

a. Triticum durum (2n=28) × Secale cereal (2n=14)

↓

F_1 hybrid (sterile) (2n=21)

↓ Chromosome doubling

2n=42

Hexaploid triticale

b. Triticum aestivum (2n=42) × Secale cereal (2n=14)

↓

F_1 hybrid (sterile) (2n=28)

↓ Chromosome doubling

2n=56

Octaploid triticale

Raphanobrassica: The amphididiploid *Raphanobrassica* ($2n = 36 = 18^{II}$ = RRCC) was obtained by Karpechenko (1928), due to spontaneous chromosome doubling of hybrids between *Raphanus sativus* ($2n = 18 = 9^{II}$ = RR) and *Brassica oleracea* ($2n = 18 = 9^{II}$ = CC).

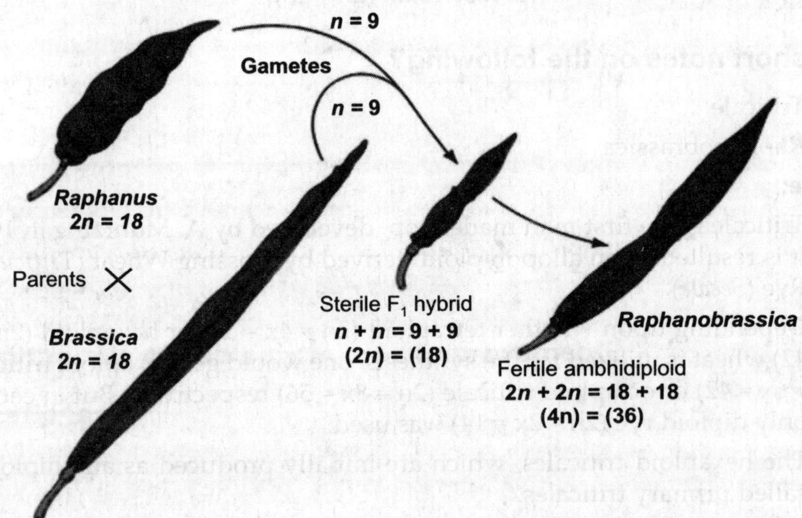

What is meant by aneuploidy? Explain the use of aneuploidy in crop improvement?

Aneuploidy (Numerical changes of chromosomes):

Aneuploidy involves a gain or a loss of one or a few chromosomes. Aneuploidy commonly arises as a result of irregular meiotic mechanisms such as non-disjunction leading to an unequal distribution of chromosomes to opposite poles. Consequently, gametes resulting from such aberrant meiosis may have loss or gain of chromosome. Also chromosome additions often cause chromosome imbalance and reduced plant vigour.

Application of aneuploidy:

1. **Chromosome additions:** Chromosome addition lines are developed by backcrossing the synthetic alloploid (F_1) as seed parent to a cultivated species as pollen parent. For example, E.R. Sears transferred the resisance to leaf rust *of Aegilops umbelhdata* to *Triticum aestivum* (bread wheat) via bridge cossing with *T. dicocoides.*

   The addition of chromosomes from other species (called alien addition lines) has been explored in interspecies crosses such as wheat x rye. Chromosome addition lines may be unable enough to be developed as cultivars.
2. **Chromosome deletions:** Chromosome deletion leads to a loss of function. The consequence of a deletion depends on the functional role of the genes of the chromosome that is lost. Monosomics (2n-l) may be used just like trisomics to assign genes to chromosomes in a polyploid species. This requires the development of monosomics for the entire existing chromosome pairs in the species, E.g. E.R. Sears developed monosomics for the Chinese spring cultivar of wheat.
3. **Chromosome substitution:** A chromosome substitution entails recipient species with an alien chromosome. Intervarietal and interspecific chromosome substitutions are more important in plant breeding than the addition of chromosomes, E.g. Substitutions of chromosome 1B of wheat and chromosome IR of rye. Other application of Aneuploidy are as follows:
   - Studies on effect of loss or gain of entire chromosome or chromosome arm
   - Locating genes on chromosome, linkage group
   - Study of Homology
   - Study of chromosome involved in translocation.

What is meant by trisomics? Explain the classification, meiotic behavior and origin of different types of trisomics?

Trisomics:

Blakeslee (1921) coined the term 'trisomic' to signify that a particular chromosome was present in three doses. So trisomics are those organisms which have one extra chromosome (2n+1). Since the extra chromosome may belong to any one of the different chromosome pairs, the number of possible trisomics in an organism will be equal to the haploid chromosome number. For instance, in barley (2n = 14) the haploid chromosome number is n = 7.

An individual having two extra chromosomes each belonging to a different chromosome pair is called double trisomic (2n + 1 + 1).

Trisomics classification:

Depending on the nature of extra chromosome, simple trisomics are of three types.

1. **Primary trisomics:** The additional chromosome is normal one in primary trisomics. So 'n' (haploid) numbers of different primary trisomics were possible.
2. **Secondary trisomics:** Trisomics having isochromosome as additional chromosome. So '2n' (diploid) numbers of different secondary trisomics were possible.
3. **Tertiary trisomics:** When additional chromosome in a trisomic is translocated one (non homologous chromosome), it is known as tertiary trisomic. So many numbers of tertiary trisomics were possible.

Meiotic behaviour in trisomics:

- In case of primary trisomics, most common configuration is chain trivalent followed by fry pan and y shaped configurations.
- In case of secondary trisomics, most common configuration is ring trivalent followed by chain and fry pan configuration.
- In case of tertiary trisomics, most common configuration is dumb bell shape (pentavalent)

somatic chromosomes	metaphase I configurations
(a) primary trisomics	(i) (ii) (iii) (iv)
(b) secondary trisomics	(i) (ii) (iii) (iv)
(c) tertiary trisomics	(a pentavalent)

Origin and sources of trosomics:

Primary trisomics:

(*a*) Unequal disjunction – Due to either non-congression or non disjunction of chromosomes of a bivalent to separate at AI

(*b*) Asynapsis or desynapsis

(*c*) Triploidy – Most important and dependable source for the production of trisomics.

(*d*) Autotetraploidy

(*e*) Haploidy

(*f*) Translocation heterozygotes

(*g*) Tetrasomics

Secondary trisomics:

(*a*) Monosomics

(*b*) Primary trisomics

(*c*) Asynapsis, desynapsis, triploids and haploids

(*d*) Monotelodisomics, telotrisomics, and tertiary monosomics

(*e*) Monoisodisomics

Tertiary trisomics:

(*a*) Translocation heterozygotes

(*b*) Secondary and primary trisomics

(*c*) Tertiary monosomics

Explain the characterization and identification of trisomics using interchange testers?

If interchange testers (i.e. interchange homozygotes with known chromosomes involved in interchange) are available in a particular case, these can be used for crosses with the trisomics. Only the trisomic plants from the F_1 progeny from these crosses are analyzed at meiosis for identification of chromosomes involved in the trisomic condition.

- If a pentavalent is formed at meiosis in F_1 trisomic plants, it suggests that one of the two chromosomes involved in the interchange tester stock used in the cross, should be involved in the corresponding trisomy.
- Contrary to this if instead of a pentavalent, a quadrivalent and a trivalent are observed, this suggests that none of the two chromosomes involved in the interchange is involved in the trisomy.

Along with this, many traits like morphology of individual, karyotypic changes of chromosomes and genetic markers/linkage groups were used for characterization and identification of trisomics.

Explain the use of trisomics in chromosome mapping?

- The trisomic ratios utilized for locating genes on specific chromosomes and finding out distances of these genes from centromere utilizing a technique called chromosome mapping.
- If linkage groups are already established in an organism, trisomics can be effectively used for assigning these linkage groups to specific chromosomes.

- Since the segregation ratio for genes located on the chromosome involved in trisomic condition differs from that genes located on other chromosomes, through these ratios, one can find out on which chromosome a particular gene is located.
- If instead of primary trisomics, secondary trisomics or telotrisomics are used, one can also find out on which of the two arms of a specific chromosome, the gene is located. This technique has been successfully utilized in plant species such as maize, tomato, barely, *Datura, Sorghum* and *Arabidopsis*. More recently it has been successfully used for rice.
- For monogenic traits, 3:1 ratio in F_2 and 1:1 ratio in test cross are obtained, these ratios are significantly modified if the gene is located on a chromosome found in trisomic condition.
- The trisomic ratio, however, cannot be predicted with certainty, since depending upon the gene position one of the following three kinds of segregation can be observed.

  1. Chromosome segregation 2. Chromatid segregation 3. Maximum equational segregation.
- Chromosome segregation is observed for genes located close to centromere. For other genes, the ratios observed may lie in between those expected in the two extremes. Despite the lack of definite expected ratio, the observed ratio still deviates so much that conclusions about location of genes on specific chromosome is not difficult to make.

Trisomics are more useful for cytogenetic studies in diploid crops than in the polyploids, why?

1. Aneuploids used for cytogenetic studies in diploids are usually different from these used in polyploids.
2. Although, hypoploids i.e. deficient aneuploids (e.g. monosomics) are more important than hyperploids (e.g. trisomics), they cannot be conveniently obtained in diploids, since such defeciencies, ordinarily cannot be tolerated by them.
3. Diploids cannot tolerate the loss of chromosomes while polyploids can tolerate the loss of chromosomes (or) polyploids cannot tolerate the addition of chromosome because already load is there in plant.
4. Therefore, in diploids mainly trisomics and in polyploids mainly the monosomics are used for cytogenetic study.
5. But in recent years, maize has proved to be an exception in tolerating monosomy despite being a diploid.

   A complete set of monosomics has been produced in maize and is being utilized for a variety of purposes including chromosome mapping.

Explain in brief about balanced tertiary trisomics (BTT)?

Balanced tertiary trisomic described by Ramage (1965) provide an alternative method for the continuous availability of male sterile line to be used as female parent for commercial hybrid in barley.

BTT conditions:

- Extra chromosome should be translocated one.
- Nuclear gene should be very close to break point to have chromosome segregation.
- One or two marker gene should be associated with extra chromosome.
- Transmission should be from egg only and not from male side.
- BTT should be vigorous to produce more seed for hybrid seed production.
- Male sterile gene should be stable and express in all environmental conditions.
- Here maintenance of male sterile plant is not separately done.
- For maintenance of trisomics only, go for selfing and select the male fertile and discard the sterile.
- In seed production block, only select sterile plants and discard fertile plants.

Progeny of selfed balanced tertiary trisomics consists of 70% homozygous recessive diploids, 1% homozygous recessive primary trisomics and 30% heterozygous balanced tertiary trisomics.

Write short notes on tetrasomics?

- Addition of two chromosomes to one pair or two different pairs is known as tetrasomy and such individuals are known as tetrasomics (2n+2).
- Among polyploids, tetrasomics have been obtained in wheat, oats and cotton, but a complete set of tetrasomics has been obtained only in wheat.
- Tetrasomics have never been used for chromosome mapping.
- Main utility of tetrasomics is to produce 'compensating nullisomic-tetrasomic lines' suggesting homeologous relationships.
- Eg. In wheat, all 42 compensating nullisomic-tetrasomic lines have been obtained making it possible to place 21 wheat chromosomes in seven homeologous groups.

What is meant by monosomy? And write the classification of monosomics with metaphase I configuration?

Monosomy: An individual lacking one chromosome from a diploid set (2n-1) is called monosomic and such condition is known as monosomy.

Monosomics classification (chromosome formulae of different types of aneuploids produced in bread wheat)

Aneuploid	Chromosome no. (2n)	Metaphase I configuration
Monosomic	6x-1 = 41	$20^{II}+1^{I}$

Nullisomic	6x-2 = 40	20^{II}
Monotelodisomic	6x = 42	20^{II}+1hetero^{I}
Monoisodisomic	6x = 42	20^{II}+1hetero^{I}
Monotelosomic	6x-1 = 41	$20^{II}+1t^{I}$
Ditelosomic	6x = 42	$20^{II}+1t^{II}$
Tritelosomic	6x+1 = 43	$20^{II}+1t^{III}$
Monoisosomic	6x-1 = 41	$20^{II}+1i^{I}$
Diisosomic	6x = 42	$20^{II}+1i^{II}$
Triisosomic	6x+1 = 43	$20^{II}+1i^{III}$
Teloisodisomic	6x = 42	$20^{II}+1ti^{II}$
Monotelomonoisosomic	6x = 42	$20^{II}+1t^{I}+1i^{I}$
Double monotelosomic	6x = 42	$20^{II}+2t^{I}$
Ditelo monotelosomic	6x+1 = 43	$20^{II}+1t^{II}+1i^{I}$
Double monoisosomic	6x = 42	$20^{II}+2i^{I}$

Enlist the methods of production of monosomics? And explain how monosomics are produced through backcrossing of interspecific hybrids to its parents?

1. From haploids: The production of monosomics in wheat was followed by E. A. Sears (1939) with an accident production of two haploids in the progeny of crosses between rye (*Secale cereale*) and avariety of spring hexaploid wheat.

Chinese spring wheat (♂) × rye (♀)

Triticum aestivum (n=21) ↓ *Secale cereale* (n=7)

Haploid (2n=21), × T. *aestivum* (n=21)

(Restitution nucleus with n=20) ⟶

$2n=20^{II}+1^{III}$ (trisomics) + $2n=20^{II}$ (nullisomic) ⟵ $2n=41$ ($20^{II}+1^{I}$)

Monosomics

Selfed

↓

$2n=20^{II}+1^{IV}$ (tetrasomics)

2. From backcross of interspecific hybrids: Monosomics have been conveniently produced in some crops by crossing the polyploidy species with one of its progenators followed by a backcross with the polyploid crop for which monosomics are intended to be isolated.

T. *aestivum* (AABBDD) (n=21) × T. *dicoccum* (AABB) (n=14)

↓

14^{II} (AABB) + 7^{I}(D) (n=14 to 21) × T. *aestivum* (AABBDD) (n=21)

↓

20^{II} + 1^{I} (D) (monosomic for D)

3. From partially asynaptic plants
4. Irradiation treatment
5. Spontaneous production etc. is used for the production of monosomics.

Write short note on meiotic behaviour and breeding behaviour of monosomics?

Meiotic behaviour of monosomics: Monosomics usually form bivalents in addition to a solitary univalent at diakinesis and metaphase I.

- Rarely, trivalent may be formed due to association of a univalent with a non-homologue.
- Similarly, more than one univalent may be found due to failure of association of one or more pairs of homologous chromosomes into bivalents.
- These aberrations may give rise to other monosomics in the progeny of a particular monosomic. This phenomenon is called as 'univalent shift'.

Breeding behaviour of monosomics:

- The progeny of monosomics consist of disomics (normal), monosomics, nullisomics in ratio of 24:73:3; their frequencies vary according to the frequency of functioning of $n - 1$ gametes male and female side.
- The transmission of n -1 gametes through male is very low because the n -1 pollen is unable to compete with normal pollen.
- In wheat, the functioning of $n - 1$ (20 chromosomes) gamete was estimated to be around 4% through male and 75% through female.

	Functional male gametes	
Functional female gametes	n (96%)	n-1 (4%)
n (25%)	2n (24)	2n-1 (1%)
n-1 (75%)	2n-1 (72%)	2n-2 (3%)

Explain how monosomics are useful in locating genes chromosomes?

Locating genes chromosomes:

Locating dominant genes: Locating dominant genes by absence of expression in the nullisomics. i.e. appearance of recessive phenotype in F_1 indicates that gene present on missing chromosome.

Eg. In wheat chromosome 3D was found to carry a gene for red kernel.

Critical line	No-critical line
Monosomic with dominant (A-) × Variety with recessive (aa) Gametes A (25%) and - (75%) → (a) 100% Aa:a- (1:3)	Monosomic with dominant (AA) × Variety with recessive (aa) ↓ Chromosome in monosomic condition does not carry the gene Aa (All)

Locating recessive genes: When monosomics are crossed with a variety having contrasting dominant character no segregation in F_1 will be observed. Because all the F_1 progeny will inherit dominant allele from normal disomic parent i.e. F_1 will not differ in the critical and non-critical lines.

Therefore, an analysis of F_2 resulting from a selfing of monosomics of F_1 (A-) will give a phenotypic ration of 97:3 (dominant: recessivee) in F_2.

Critical line	**No-critical line**
Monosomic (a-) × Normal variety (AA) Gametes (a) and (-) → (A) F_1: A-(75%) Self → F_2 97:3 (75%) 72.75:2.25 Aa (25%) Self → 3:1 (25%) 18.75:6.25 Pooled ratio 91.50:8.50	Monosomic (aa) × Normal variety (AA) Chromosome in monosomic condition does not carry the gene F_1 Aa (All) F_2 3:1

What is meant by alien addition and alien substitution lines? And explain their method of production with specific example?

Alien addition line: It has one pair of chromosomes from a related wild species in addition to the normal chromosome complement (2n) of the species.

Eg. Rye chromosome additions to wheat: An amphiploid (2n = 56) between wheat and rye is first produced using normal method of crossing the two species followed by doubling of the chromosome number in F_1 hybrid.

- The amphiploid is crossed back to wheat giving a heptaploid with $21^{II}+7^{I}$, where bivalents belong to wheat and univalents belong to rye.
- On selfing these heptaploids (2n = 7x), monosomic ($21^{II}+1^{I}$ rye) and disomic addition lines ($21^{II}+1^{II}$ rye) are obtained.
- When monosomic additions are available, these may be selfed to get the disomic addition lines.

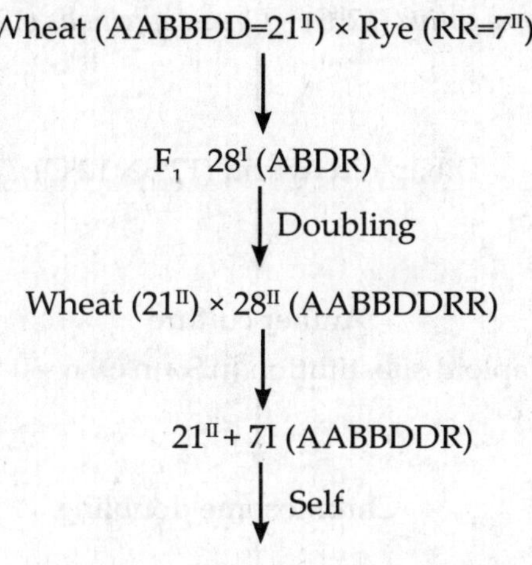

Alien substitution lines: It has one chromosome pair from a different species in the place of the chromosome pair of the recipient species.

- The first report of alien substitution in wheat involved rye chromosome 1, which was spontaneously, substituted for a wheat chromosome during wheat and rye crosses (O' Maro, 1940).
- For systematic production of alien substitution lines, the addition lines each with 22^{II} were backcrossed to different monosomics used as a female parent.
- The F_1 plants with $21^{II}+2^{I}$ (1W + 1R) were selected, which either on selfing or on backcrossing to the same addition line gave monosomic or disomic substitution lines.

Monosomic (20^{II} + 1^{I}) × alien addition line [21^{II}(W)+1^{II}(R)]

↓

20^{II} (W) + 1^{I} (1W+1R) × 21^{II} (W) + 1^{II} (R)

↓ Self ↓

Select 21^{II} — 21^{II} + 1^{I}

(21^{II}W or 20^{II}+1^{II}R) — (21^{II}W+1^{I}R) or (20^{II}W+1^{II}R+1^{I}W)

↓

20^{II}W+1^{II}R

Write the schematic representation for production of alien substitution and alien addition lines through anther culture in rice?

Chu (1982) outlined a scheme involving anther culture for the development of substitution and addition lines.

Production of alien substitution lines:

Oryza sativa (12^{II}S) × *Oryza officinalis* (12^{II}O)

↓

Dihaploid hybrid (12^{I}S + 12^{I}O)

↓

Anther culture

Haploid substitution (n^{I}S + n^{I}O, n = 0-12)

↓

Chromosome doubling

↓

Diploid substitution (n^{II}S + n^{II}O, n = 0-12)

Production of alien addition lines:

Oryza sativa (12^{II}S) × *Oryza officinalis* (12^{II}O)

↓

Dihaploid hybrid (12^{I}S + 12^{I}O)

↓

Chromosome doubling

↓

Amphididiploid (12^{II}S + 12^{II}O) × *O. Sativa* (12^{II}S)

↓

Backcross hybrid (12^{II}S + 12^{I}O)

↓

Anther culture

↓

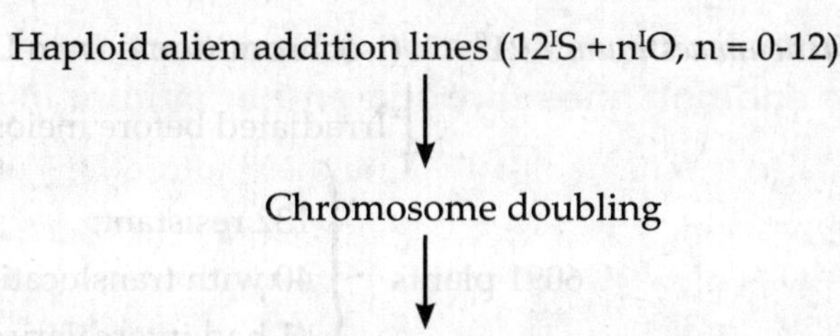

Explain the steps followed by E. R. Sears in 1956 for development of leaf rust resistance in wheat by translocation method?

E. R. Sears in 1956, successfully first transferred a segment from *Aegilops umbellulata* chromosome 6U, carrying resistance to wheat leaf rust (Lr9), to chromosome arm 6BL of wheat.

1. For this, a cross of hexaploid wheat with *Aegilops umbellulata* was not successful, an amphiploid (T. *dicoccoides* × A. *umbellulata*) with $2n = 21^{II}$was used as a bridge species.
2. This amphiploid was crossed with Chinese spring followed by two backcrosses with Chinese spring and selection for leaf rust resistance.
3. This gave a plant with 2n = 43, in which the extra chromosome was from A. *umbellulata* and carried the gene for rust resistance.
4. This plant was given a heavy dose of X-rays before meiosis. Pollen from this irradiated plant was used for pollinating Chinese spring wheat.
5. Among 6091 offspring, 132 were resistant, of which 40 had translocations and only one of them being intercalary.
6. Plants for this intercalary translocation resembled Chinese spring except for rust resistance.

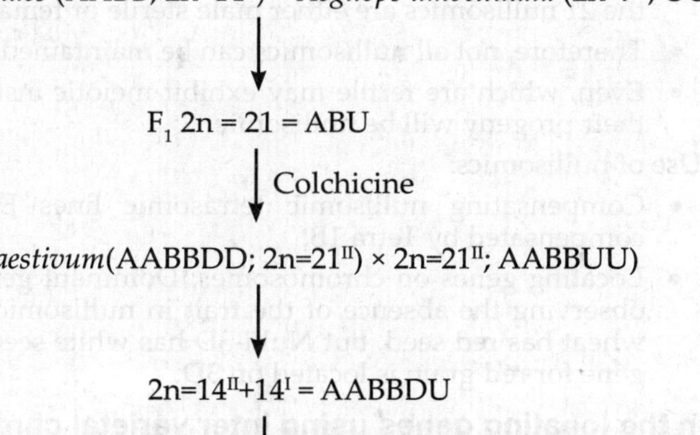

Contd...

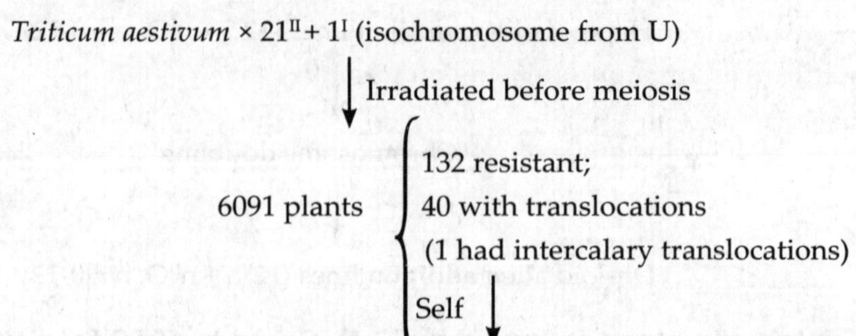

Explain in brief about nullisomic analysis.

1. Nullisomics are those organisms which lack one pair of homologous chromosomes (2n-2).
2. Complete set of nullisomics available in wheat and oat only.
3. Source of production:
   - Monosomics on selfing will give nullisomics in a small frequency. Eg. The frequencies of nullisomics in the selfed progenies of monosomics range from 0.2% to 7.6% with a mean of 3.8% in wheat, while in oqts it ranges from zero to 82.9%.
   - Monotelosomics or monoisosomics also produce nullisomics in their progeny.
4. Identification of nullisomics: In wheat, most nullisomics identified using morphological traits Eg. Shape of spike. And in oats, identified by the vigour at the seedling stage.
5. Meiotic behaviour: In nullisomics of wheat, one should expect the presence of 20 bivalents and no univalents at metaphase I of meiosis.
6. Breeding behaviour:
   - The nullisomics are not always fully fertile and in wheat, about half of the 21 nullisomics are either male sterile or female sterile.
   - Therefore, not all nullisomics can be maintained by selfing.
   - Even, which are fertile may exhibit meiotic instability, so that not all their progeny will be nullisomic.
7. Use of nullisomics:
   - Compensating nullisomic tetrasomic lines Eg. Nulli 1A can be compensated by Tetra 1B.
   - Locating genes on chromosomes: Dominant genes can be located by observing the absence of the trait in nullisomics. Eg. Chinese spring wheat has red seed, but Nulli-3D has white seeds suggesting that the gene for red grain is located on 3D.

Explain the locating genes using inter varietal chromosome substitutions?

Locating genes on specific chromosomes has been achieved by substituting individual specific chromosome for the corresponding chromosome of another variety having the contrasting character.

Inter varietal substitution using monosomics:

Eg. Wheat

Recipient variety (Monosomic) × Donor variety (disomic)

↓

F_1 select monosomics ($21^{II} + 1^{I}$)

↓ Self

Recurrent parent × $20^{II} + 1^{II}$

↓

Select monosomics

↓

Self and backcross repeatedly

- If monosomic series should be available in the recipient variety and each of monosomic is crossed with the donor variety for making 21 individual chromosome substitutions.
- In the F_1, the monosomics are cytologically selected and selfed to get the disomics for the univalent chromosome derived from donor variety.
- These disomics are backcrossed with original monosomic and the cycle is repeated to reconstitute the genotype of recipient variety in which one pair of chromosome is derived from the donor variety.
- If the substitution leads to a major change in the morphology for the character under investigation, this will suggest that one or more genes for this character are located on the specific chromosome involved in the substitution.

Inter varietal substitution using nullisomics:

Recipient variety (Nullisomic) × Donor variety (disomic)

↓

F_1 ($20^{II} + 1^{I}$) × nullisomic (20^{II})

↓

$20^{II} + 1^{I}$

↓

After 4 or 5 generations, with one selfing results in
Disomic, monosomic and nullisomic are in ration of 24:73:3

- Here, nullisomic recipient variety is crossed with disomic donor parent
- In F_1, we can observe the $20^{II} + 1^{I}$ (univalent belongs to donor variety). And it produces the gametes 21^{I} and 20^{I}.
- So, backcross this F_1 to nullisomic (recipient) variety. If 21^{I} as female will pollinated by nullisomics resulting $20^{II} + 1^{I}$.
- After 4 or 5 generations of repetition, one self result in disomic, monosomic and nullisomic are in ration of 24:73:3.
  - In monosomics, selfing is done because to increase the transmission rate by the formation of bivalents due to selfing.
  - Nullisomic methos is easy compared to monosomic, but monosomics are more used, because in nullisomics, some are fertile and some are not fertile.
  - And also in monosomics method, monosomic is used as female in backcross and in nullisomics, it is used as male because it has less transmission frequency.

Explain in brief about chromosomal engineering.

Chromosomal engineering is carried out for creation of alien addition, alien substitution, alien recombination and induced translocation line. It is also used to promote chromosome pairing.

1. Manipulation of 5B- chromosome: Ph gene present on long arm of 5B, prevents homologous pairing. Manipulation can be done by two way:

(*a*) Suppressing the chromosome 5B- activity

- Yellow rust resistance transferred to *Triticum aestivum* variety Chinese spring (Riley)
- *Aegiolops speltoides* used as bridge species
- New cultivar Compair developed.
- E. A. Sears transferred brown rust resistance from *Agropyron elongatum* to *T. aestivum.*

(*b*) By eliminating chromosome- 5B.

- Riley reported hybrid derivative of nullisomic 5B of *T. aestivum* and diploid *(Aegilops bicornis)* leading to allosynthetic recombinant chromosome.
- Hold fast cultivar *(T. aestivum)* developed by transfer of mildew resistance from *Aegilops bicornis.*

(*c*) Induced translocation: Development of addition line through colchicine induced amphidiploidy.

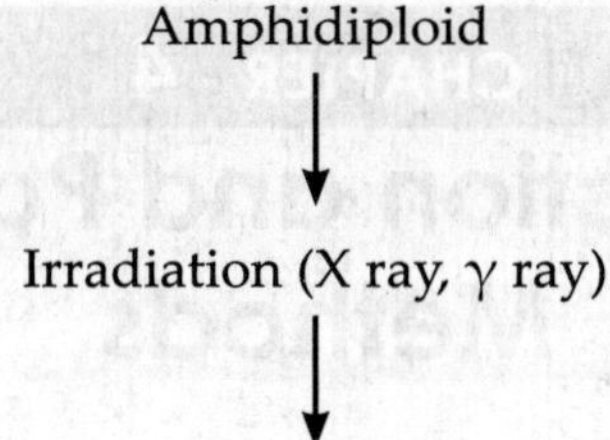

Desirable Translocation detected in M_2 generation.

(*d*) Use of Monosomic

- Rust resistance from rye to wheat (Joshi and Singh *et al.*)

Monosomic- 5B x Russian rye (rust resistant)

↓

Rust resistant cultivar

(*e*) Use of modified method: Alien recombination and Alien substitution. Bhowal suggested

- Method of alien genetic recombination
- Method of alien chromosome substitution.

Explain the significance of ph locus on 5B chromosome of wheat.

- Despite the close relationship between three genomes of wheat (AABBDD), meiosis in hexaploid wheat is characterized by the formation of bivalents only.
- It is due to the 'diploidizing system' i.e. genetic control of bivalents formation at meiosis.
- During the production of monosomics and nullisomics, it is found that, when aperticular chromosome was absent in specific nullisomic wheat plant, there was complex meiotic pairing leading to multivalent formation.
- Among these multivalent a particular multivalent involves not only homologous chromosomes, but also homeologous chromosomes.
- The most potent gene for this behavior is Ph1 (pairing homeologous) which is located on the long arm of chromosome 5B.
- In wheat the diploidizing system (Ph1) functions even in hemizygous (monosomic) condition. There are diploidizing systems that are hemizygous ineffective (Festuca) or hemizygous insufficient (Hordium spp).
- So in presence of Ph1 locus, homeologous chromosomes remain separated and therefore usually do not pair, but in its absence they are randomly distributed and therefore are more closely associated thus giving them an opportunity to pair.

CHAPTER - 4

Reproduction and Pollination Methods

Explain in detail about the flower development in higher plants (*Arabidopsis tahliana*).

A flower is a modified shoot bud having the parts: calyx, corolla, androecium and gynoecium or their modifications.

- Flower having both androecium and gynoecium is called bisexual or hermaphrodite flowers.
- Unisexual flowers have either androecium or gynoecium. Flowers having only androecium is staminate, only gynoecium are pistillate flowers.

Flower development:

- Flowers of *Arabidopsis* are composed of four types of organs arranged in concentric rings or whorls.
- In the developemnt of flower, each whorl gives rise to different floral organs
- Whorl 1 yields the sepals (the green, outermost floral leaves), whorl 2 the petals (the white, inner floral leaves), whorl 3 the stamens (the male organs, which form pollen) and whorl 4 the carpels (which fuse to form ovary).

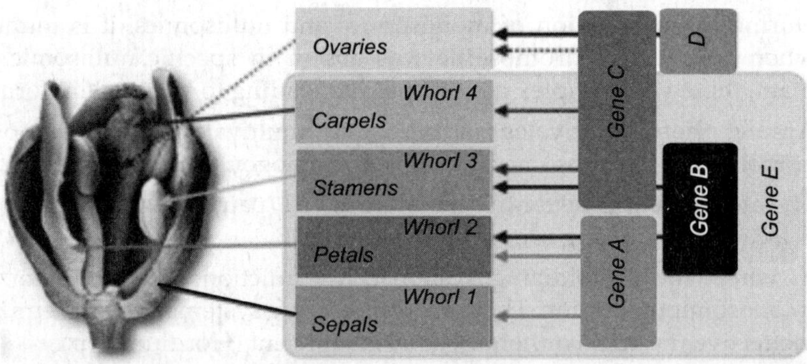

Genetic control of flower developemnt: Mutations that affect floral development fall into three major classes, compared with wild type flower, one class lacks sepals and petals, another class lacks petals and stamens and the third class lacks stamens and carpels.

On the basis of crosses between homozygous mutant organisms, these classes of mutants can be assigned to different cpmplementation groups, each of which defines a different gene.

- The phenotype lacking sepals and petals is caused by mutations in the gene *ap1* (*apetala-1*).
- The phenotype lacking stamens and petals is caused by a mutation in either of two genes, *ap3* (*apetala-3*) or *pi* (*pistillata*).
- The phenotype lacking stamens and carpels is caused by mutations in the gene ag (*agamous*).
- So these genes encode a transcription factors, transcription factors encoded by *ap3, pi* and *ag* are members of *MADS box* family of transcription factors.
- Each member of this family contains a sequense motif of 58 amino acids
- *MADS box* transcription factors are very common in plants and are also found less frequently in animals.

Floral development in mutants of Arabidopsis				
	Whorl			
Genotype	**1**	**2**	**3**	**4**
Wild type	Sepals	Petals	Stamens	Carpels
ap1/ap1	Carpels	Stamens	Stamens	Carpels
ap3/ap3	Sepals	Sepals	Carpels	Carpels
pi/pi	Sepals	Sepals	Carpels	Carpels
ag/ag	Sepals	Petals	Petals	Sepals

Combinatorial determination of floral organs: The role of the *ap1, ap3, pi* and *ag* transcription factors in the determination of floral organs can be inferred from the phenotypes of mutations. The logic of the inference is based on the observation that mutation in any of the genes eliminates two floral organs that arise from adjacent whorls.

- *ap1* is necessary for sepals and petals, *ap3* and pi are both necessary for petals and stamens, and ag is necessary for stamens and carpels.
- It may be inferred that *ap1* is expressed in whorls 1 and 2, *ap3* and *pi* are expressed in whorls 2 and 3, and that *ag* is expressed in whorls 3 and 4.
- The model of gene expression suggests that floral development is controlled in combinatorial fashion by the four genes.
  - Sepals develop from the tissue in which only *ap1* is active
  - Petals are evoked by the combination of *ap1, ap3* and *pi*.
  - Stamens are determined by the combination of *ap3, pi* and *ag*.
  - Carpels derive from tissue in which only *ag* is expressed.

Domains of expression of genes determining floral development		
Whorl	**Genes expressed**	**Determination**
1	*ap1*	Sepal
2	*ap1* + *ap3* and *pi*	Petal
3	*ap3* and *pi* + *ag*	Stamen
4	*ag*	Carpel

This model of floral determination is often called the flower ABC model because the wild type activity of *ap1* was originally designated A, that of *ap3* and *pi* acting together as B, and that of *ag* as C. Therefore, the combination of activities present in each whorl would be represented as A in whorl 1, AB in whorl 2, BC in whorl 3, and C in whorl 4.

What is meant by sporogenesis and gametogenesis and explain their types?

Sporogenesis: Production of spores is known as sporogenesis.

- Spores are haploid (n) cells produced by meiosis in androecium or gynoecium; they ultimately produce male and female gametes.
- Sporogenesis is of two types. 1. Microsporogenesis and 2. Megasporogeneisis

Microsporogenesis:

1. Microspores are produced in anthers.
2. Each anther has four pollen sacs that contain numerous pollen mother cells (PMC's).
3. Each PMC undergoes meiosis to produce four haploid microspores.
4. Each microspore matures into pollen grains.

Megasporogenesis:

1. Megaspores are produced in ovules present inside the ovary of a pistil.
2. A single cell in each ovule differentiates into a megaspore mother cell, which undergoes meiosis to produce four haploid megaspores.
3. Three of the megaspores degenerate leaving only one functional megaspore per ovule.

Gametogenesis: Production of male and female gametes is known as gametogenesis.

- It occurs within micro and mega spores, and involves only mitotic division of their haploid (n) nuclei.
- It is of two types 1. Microgametogenesis and 2. Megagametogenesis.

Microgametogenesis: Production of male gametes through mitotic division of haploid nuclei of microspores is called as microgametogenesis.

1. Microspore nucleus divides mitotically to produce a generative and a tube or vegetative nucleus.

2. Generative nucleus divides once to produce two sperms.
3. Pollen together with the pollen tube constitutes the microgametophyte.

Megagametogenesis: Production of female gamete or egg cell through mitotic division of haploid megaspore nucleus is known as megagametogenesis.

- Megaspore nucleus divides three times to yield eight haploid nuclei.
- Three nuclei move to one pole, produce one egg cell and two synergids.
- Three nuclei go to other pole, produce three antipodals.
- Two nuclei remain in the middle, and fuse to form a secondary nucleus (2n).
- Megaspore consists of egg cell, two synergids, three antipodals and a secondary nucleus is known as embryo sac or megagametophyte.

What is meant by reproduction? And write the types of reproduction?

Reproduction refers to the process by which living organisms give rise to the offspring of similar kind (species). or Reproduction is the process of multiplication of living beings.

The modes of reproduction are of two types, 1. Asexual reproduction and 2. Sexual reproduction.

I. Asexual reproduction: Multiplication of plants without fusion of male and female gametes.

1. Vegetative reproduction: Multiplication by vegetative plant parts.
   (*a*) Natural: Multiplication by Rhizome, tuber, corm, bulb, runner, sucker, stolon and bulbils.
   (*b*) Artificial: Multiplication by stem and root cuttings, grafting, layering and budding.
2. Apomixis: Development of seed without sexual fusion
   (*a*) Parthenogenesis: Development of embryo from egg cell without fertilization
   (*b*) Apogamy: Development of embryo either from synergids or antipodal cells
   (*c*) Apospory: Origin of embryo from diploid egg cell of another embryo sac developed from diploid tissue.
   (*d*) Adventive embryony: Origin of embryo sac directly from diploid cells belonging to either nucellus or integument.

II. Sexual reproduction: Multiplication of plants by fertilized embryos

Autogamy: Development of seed by self pollination

Allogamy: Development of seed by cross pollination

What is meant by asexual reproduction? And write the significance of asexual reproduction in crop improvement?

Asexual reproduction: Multiplication of plants without the fusion of male and female gametes is known as asexual reproduction.

Significance in crop improvement:

1. Asexual reproduction produces progeny plants identical in genotype to the parent plant.
2. Any genotype, whether homozygous or heterozygous, is maintained indefinitely. Hence any desirable plant can be used as a variety.
3. Promising individuals occurring at any stage in a breeding programme can be easily picked up and maintained.
4. It makes use of desirable bud mutations.
5. Diploid apomixis can be utilized to maintain/ multiply the hybrid varieties.
6. Haploid apomixis, especially through *in vitro* ovary culture and through anther culture, produces haploid plants. Chromosome doubling of such plants allows rapid isolation of homozygous lines from heterozygous plants.
7. Major disadvantage of asexual reproduction is that it does not enable the genes present in two different individuals to be combined into a single clone.

What is meant by apomixis? And write the genetics of apomixis?

Apomixis refers to the development of seed without sexual fusion (fertilization).

Obligate apomixis: Reproduction in some species is only by means of apomixis; this is termed as obligate apomixis.

Facultative apomixis: Sexual reproduction occurs in addition to apomixis, is termed as facultative apomixis.

- Term apomixis was coined by Winkler.

Genetics of Apomixis: Apomixis typically involves two developmental changes from the normal reproduction i.e. a suitable modification/ failure of meiosis during megasporogenesis, and initiation of mitosis in the unfertilized diploid egg cell or some other diploid cell of embryo sac.

- Gametophytic apomixis is controlled by dominant gene or a few closely linked genes on the same chromosome
- Apospory is due to dominant gene in *Ranunculus spp.*
- In *Panicum maximum*, dominant, mono factorial control of apomixis reported by Savidan.
- Nuclear adventive embryony in citrus is controlled by single dominant gene.

- At least four recessive genes in *Parthenium* (two being needed for apomeiosis and the other two are concerned with parthenogenesis).

Explain the techniques used to screen apomixis.

Technidues to screen apomixis

- Cytological analysis of embryo sac development is commonly used. Eg:- Aposporus embryo sac is generally 4-celled and lacking antipodals.
- Callose fluoscence in combination with clearing of ovules can be used to detect diplospory.
- Uniformity of progenies from cross pollinated parents is the best indication. Also:
  1. Maternal phenotype among the progeny among the progeny of F_1 crosses.
  2. Limited or no genetic variances in F_2 progeny of two distinct parents.
  3. High seed frequently in aneuploid, triploid and wide crosses.
  4. Multiple seedlings per seed.
- Marker – two molecular markers (UGT 197 and OPC-04) that co-segregate with the apomictic mode reproduction in *pennisetum*.
- Progeny test to determine the degree of apomixis.

Explain the role of apomixis in crop improvement.

Apomixis in crop imrprovement:

1. Rapid multiplication of genetically uniform individuals can be achieved without risk of segregation .
   - Maintaining heterozygote advantage generation after generation.
   - Genetic uniformity maintained over generation for both homozygosity and heterozygosity is the choice of goal.
2. Efficient exploitation of maternal effect from generation to generation. Eg: Horticulture crops, especially fruit trees and ornamental plants
3. Used for fixation of hybrid vigour to produce true breeding hybrids
   - Crops where hybrids are more productive but male sterility and fertility restoration system are not known/ not developed.
   - Production of hybrids seed is so expensive.
   - In crops where commercial hybrids are presently produced as it eliminate the need for isolate planting for

     (*a*) Production of F_1 hybrid and

     (*b*) To maintain and increase the parents of F_1

An apomictic hybrid could be grown and multiplied as a single line. So obligate apomixis would eliminate out crossing to be handled.

4. Superior apomictic plants would be ready for immediate performance testing.
5. Genetic variability continuous to increase in apomictics through mutation and hybridization in sexual plants.
   The Agamospermous reproduction through the seed has allowed mutations to accumulate in various genotypes.
6. Production of phenotypically stable population in the form of 'vybrids'
7. Adventive embryny for uniform root stock and virus free scion in citrus, Merion variety in *Poa pratensis*.

What is meant by self pollination? Write the mechanisms promoting self pollination and genetic consequences of self pollination?

Pollination of a flower by its own pollen is called as self pollination or autogamy.

Mechanisms promoting self pollination:

1. **Bisexuality:** Presence of male and female organs in the same flower.
2. **Homogamy:** Maturation of anthers and stigma of the flower at the same time.
3. **Cleistogamy:** Pollination and fertilization occurs in unopened flower bud eg. Wheat, barley and oat.
4. **Chasmogamy:** Fertilization after opening of flower. eg. Wheat, barley, rice and oats.
5. Position of anthers:
   - Stigmas are surrounded by anthers eg. Tomato and brinjal
   - Stamens and stigmas are enclosed by petals eg. Green gram, black gram, soya bean and chickpea.

Genetic consequences of self pollination: Continuous inbreeding results in progressive loss of genetic variability until each individual has become homozygous for almost all genes. It leads to fixation of alleles and genotypes.

1. Each variety is a highly inbred line distinctly differentiated for gene frequency i.e. proportion of alleles.
2. Intra-varietal genetic variation is absent
3. Inter-varietal genetic variation is large due to random fixation of different alleles in different lines.
4. Genetic correlations between relatives are gradually enhanced with the progress of inbreeding till it approaches unity.
5. In absence of directional dominance, prolonged inbreeding promotes the discrete groupings of non-interbreeding individuals within population.

6. Inbreeding followed by selection favouring only one type of homozygotes tends to freeze the genetic variance because of rapid fixation of homozygotes.
7. Self pollinated species do not show inbreeding depression, but may exhibit considerable heterosis.

What is meant by cross pollination? Write the mechanisms promoting cross pollination and genetic consequences of cross pollination?

When a flower is pollinated by pollen produced on another flower is called as cross pollination.

Mechanisms promoting cross pollination:

1. Unisexuality: Unisexual flowers
   - **Monoecy:** Both staminate and pistillate flowers occur in the same plant eg. Maize, mango and castor.
   - **Dioecy:** Male and female flowers present in different plants eg. Papaya, spinach and asparagus.
2. **Dichogamy:** Maturation of anthers and stigmas of same flower at different times.
   - **Protogyny:** Pistil mature before anthers eg. Pearl millet
   - **Protandry:** Anthers mature before pistil eg. Maize and Sugarbeet.
3. **Heterostyly:** Styles and filaments in a flower are of different length. eg. Linseed.
4. **Herkogamy:** Hindrance to self pollination due to some physical barriers like hyline membrane around the anthers eg. Alfalfa.
5. **Self incompatibility:** Inability of fertile pollens to fertilize the same flower. E.g. Brassica, Nicotiana and Radish.
6. **Male sterility:** Pollen grains are either absent or non functional. eg. Sunflower, Bajra and Maize.

Genetic consequences of cross pollination: Frequent cross pollination creates and maintainance of tremendous amount of genetic variability, chiefly because of high amount of heterozygosity in population.

1. Each variety is a highly random pollinated population usually maintained at genic equilibrium in absence of selection.
2. Immense amount of genetic variability floats in crops at both intra and inter population level due to frequent random gene flow among genotypes.
3. Role of dominance is potential, release of recessive alleles on selfing.
4. To high degree of panmixia (random pollination) leads to rapid non-discrimination of population characteristics.
5. Reduction of genetic correlations among progenies

6. Cross pollinated species show severe to high inbreeding depression and high heterosis. Eg. Hybrid varieties.
7. Hybridization between two inbreeds usually leads to recovery of vigour lost by inbreeding.

Describe the procedure to determine the mode of pollination of a species and the amount of cross pollination?

Determination of mode of pollination:

1. First step in determining the mode of pollination is to examine the flowers critically for the mechanism like dioecy, monoecy, protogyny, protandry and cleistogamy.
2. Second step consists of isolation of single plants and recording seed set under isolation,
   - Space isolation i.e. individual plants grown at sufficient distances to prevent cross pollination, is preferable to isolation by bags or cages since later may create an environment unfavourable for pollination and seed set.
   - Failure to set seed in isolation proves the species to be cross pollinated; and setting of seeds is only indicative of self pollination.
3. Finally, the effects of selfing on the vigour of plants should be studied.
   - Loss in vigour due to inbreeding is common in cross pollinators, but self pollinators show no inbreeding depression.

Determination of amount of cross pollination:

The amount of cross pollination is determined by planting two strains of concerned species in a mixed strand. One strain is homozygous for a dominant character, preferably an easily recognizable seed or seedling character. While other strain has recessive form of the character.

The two strains are planted in such a manner that each plant of the recessive strain is surrounded by plants of the dominant strain to provide abundant pollen. Seeds from the plants of only recessive strain are harvested.

The percentage of these seeds carrying the dominant allele represents the percentage of cross pollination in the species. The frequency of cross pollination varies greatly with the variety, weather conditions and location.

What is meant by self incompatibility? Explain the classification of self incompatibility suggested by Lewis?

Self incompatibility refers to the inability of a plant with functional pollen to set seeds when self pollinated. Lewis in 1954 has suggested a simple classification of self incompatibility as follows:

1. Heteromorphic system and 2. Homomorphic system.

Heteromorphic system: Self incompatibility is associated with differences in floral morphology. So it is due to differences in the lengths of styles and stamens.

It is of two types, distyly and trystyly.

Distyly: Refers to two types of styles (short and long) and stamens (low and high). E.g. Family *Primulaceaa.* In *Primula,* there are two types of flowers,

(*i*) thrum type (Ss) – short style and long anthers

(*ii*) pin type (ss) – long style and short anthers

Crosses are compatible only between style and stamens of matching length. i.e. crosses are compatible between pin × thrum or thrum × pin but not between pin × pin and thrum × thrum flowers.

Results of four crosses in Primula		
Crosses	**Genotypes**	**Results**
pin × pin	ss × ss	Incompatible
thrum × thrum	Ss × Ss	Incompatible
thrum × pin	Ss × ss	1 thrum : 1 pin
pin × thrum	ss × Ss	1 thrum : 1 pin

Tristyly: Anthers and styles have three different positions i.e. short, medium and long.

Eg. *Lythrum sativum*

Homomorphic system: Self incompatibility results due to physiological causes rather than differences in flower morphology. It is of two types, gametophytic and sporophytic system of self incompatibility.

Gametophytic system: SI is controlled by the genetic constitution of gametes.

- East and Mangeldorf (1925) discovered in *Nicotiana sanderae.*
- It is governed by single gene S with large number of multiple alleles.
- Pollen tube growth is usually inhibited in the style or ovary.
- This system gives rise to three types of pollinations, fully incompatible (S_1S_2 × S_1S_2), half incompatible (S_1S_2 × S_1S_3) and fully compatible (S_1S_2 × S_3S_4)
- Permits recovery of male parent only in partial fertile crosses. E.g. S_1S_2 × S_1S_3
- Plant species of this system have binucleate pollen.
- This system operates with wet stigmatic surface.
- This system is suppressed by polyploidy, X- ray or γ- irradiation.
- Eg. Red clover, white clover, tobacco and bajra etc.

Sporophytic system: Self incompatibility is governed by the genotype of pollen producing plant.

- East and Babcock (1950) discovered in *Crepis foetida* and Gerstel (1950) in *Parthenium argentatum* (Guayule).

- It is governed by single gene S with large number of multiple alleles.
- Exhibits the inhibition of pollen tube growth on the stigma of the same flower.
- This system contains a form of dominance in which S_1 is dominant over all others, S_2 is dominant over others except S_1. i.e. $S_1 > S_2 > S_3 > S_4$.
- In this system crosses are either fully sterile or fully fertile.
- Exhibits the reciprocal differences. Eg. $S_1S_2 \times S_2S_3$.
- This system permits the production of homozygotes. Eg. $S_2S_3 \times S_1S_2$
- Permits the recovery of parental genotypes.
- Plant species of this system have trinucleate pollen.
- This system operates with dry stigmatic surface.
- Eg. radish, cabbage and cauliflower etc.

Pollen tube-style interaction:

Pollen stigma interaction: These interactions occur just after the pollen grains reach the stigma and prevent the pollen germination.

- In gametophytic system, the stigma surface is plumose having elongated receptive cells and is commonly known as wet stigma. Incompatible pollen grains generally germinate on reaching the stigma. The incompatibility reaction occurs at later stage within the stigma, due to clear cut serological differences among the pollen grains with different S genotypes.
- In sporophytic system, stigma is papillate and dry, and is covered with a hydrated layer of proteins known as pellicle; and is the site of incompatibility reaction. Once the pollen crosses the stigmatic barrier, there is no further inhibition of pollen tube growth.
- The pollen and stigma morphology, and pollen cytology appear to be correlated with the site of inhibition rather than with the mode of genetic control of self incompatibility reaction.
- In gametophytic system, pollen grains germinate and pollen tube penetrates the stigmatic surface. But inincompatible combinations, growth of pollen tube is retarded within the stigma eg. in *Oenother* a or a little later in the style eg. *Lilium*.
- In the later case there is cessation of protein and polysaccharide synthesis in the pollen tubes, which leads to the degeneration of tube wall and the bursting of pollen tube.

Pollen tube ovule interaction:

- The pollen tubes reach the ovule and effect fertilisation. eg. Theobroma cocoa.
- In incompatible combination, embryos degenerate at an early stage of development.

Methods for temporary suppression of SI: A temporary suppression of the SI reaction is referred to as pseudofertility.

1. Bud pollination
2. Surgical techniques
3. Irradiation
4. High temperature
5. High Co_2 concentration
6. Sodium chloride spray
7. End of season pollination
8. Double pollination
9. *Invitro* pollination

Write the significance of self incompatibility in crop improvement? And also mention the disadvantages of self incompatibility?

Significance of SI in crop improvement:

1. Where male sterility is non-existent, SI can alternatively facilitate the production of F_1 hybrids in seed crops, forage and grasses. A good deal of success has been achieved in breeding for hybrids employing SI in crops, particularly crucifers and forage crops.
2. Double cross hybrids and synthetics can also be produced with help of SI. Eg. S_1S_2 and S_3S_4 hybrids can be crossed ($S_1S_2 \times S_3S_4$) to produce double cross.
3. Self fertility be induced temporarily or permanently by mutation of S-allele to S_f through artificial irradiation in clonally propagated crops like cherries and apples.
4. Seedless varieties where desirable, such as in pineapple and grapes etc., can be evolved if SI is present.

Disadvantages of SI:

1. Variation in seed set due to poor fertility
2. Poor preservation of genetic purity of improved varieties since cross pollination is non restricted.
3. Difficulties in development and maintenance of homozygous lines (inbreeds) which can be utilized for hybridization
4. Uneven quality of fruits in horticultural crops because of mixed planting of different varieties based on their cross-compatibility.

Write the mechanisms of self incompatibility? And enlist the methods used for temporary suppression of self incompatibility?

Mechanisms of self incompatibility: Three categories of self incompatibility mechanisms are pollen-stigma interaction, pollen tube style interaction and pollen tube ovule interaction.

What is meant by male sterility? Explain the different types of male sterility with suitable diagrams?

Male sterility: It refers to a condition in which pollen is either absent or non functional in flowering plants.

Types of male sterility: There are five types of male sterility

I. Genetic male sterility (GMS): It is governed by nuclear genes.

- Discovered by Lewis and Crow in 1956
- Ordinarily governed by single recessive gene (ms), but dominant gene in safflower, cotton and wheat.
- It consists of A line (male sterile line) and B line (maintainer line)
- A male sterile line is maintained by crossing it with heterozygous male fertile plant (Ms ms).

Crosses between		Resulting genotype	Remarks
Male sterile (A line)	Male fertile		
[msms]	[MsMs]	[Msms]	Male fertile
[msms]	[Msms] B- line	[Msms] & [msms]	Male fertile (50%) & Male sterile (50%)

Merits:

- Fertility restoration in hybrid and crossing plants are relatively easy.
- Used for production of hybrid seed in both seed and vegetatively propagated crops.
- Does not have undesirable agronomic characters.

Demerits:

- It is less stable – TGMS and PGMS
- During maintainance of male sterility, 50% plants are fertile which have to be removed every year. It increases the cost of hybrid seed.

Thermo sensitive genetic male sterility (TGMS): Complete male sterility is produced by male sterility gene at higher temperature (Eg. 23.3°C or higher for rice TGMS line Pei-Ai645) but at temperature below critical point it exhibit normal fertility, Eg. Indian lines- SM-5,38, JP-2, JP-385.

Photosensitive genetic male sterility (PGMS): Expression of male sterility gene drastically affected by the prevailing photoperiod provided the temperature is within a critical range (23-28°C for rice PGMS).

II. Cytoplasmic male sterility (CMS): It is controlled by cytoplasmic genes or plasma genes.

- Discovered by Welch and Grimball.
- Male sterility can be maintained by crossing it (A line) with fertile maintainer line (B line).
- Progeny of such male sterile plants would always be male sterile because cytoplasm of zygote comes primarily from egg cell.
- Main advantage is that it is highly stable and not influenced by environment.
- Disadvantage is that it cannot be used for development of hybrids where seed is economic product.
- And it can be used only in asexually propagated species such as sugarcane, potato and forage crops.

Crosses between		Resulting genotype	Remarks
Male sterile (A line)	Male fertile (B- line)		
[ ] S	[ ] F	[ ] S	Male sterile

III. Cytoplasmic genetic male sterility (CGMS): It is governed by both cytoplasmic and nuclear genes.

- Discovered by Jones and Devis (1944) in onion.
- Male sterility can be maintained by crossing it (A line) with fertile maintainer line (B line).
- For hybrid seed production A line can be crossed with R (Restorer line).

Merits:

- Widely used for hybrid seed production in both seed propagated and vegetatively propagated species.
- It is highly stable and reliable
- And it is not affected by environment.

Crosses between		Resulting genotype	Remarks
Male sterile (A line)	Male fertile		
[msms] S	[msms] F (B- line)	[msms] S	Male sterile
[msms] S	[MsMs] S	[Msms] S	Male fertile
[msms] S	[MsMs] F (R-line)	[Msms] S	Male fertile
[msms] S	[Msms] S	[Msms] S & [msms] S	Male sterile & Male fertile
[msms] S	[Msms] F	[Msms] S & [msms] S	Male sterile & Male fertile

Demerits:

- Undesirable effects of the cytoplasm.

  Eg. T- cytoplasm of maize-Helminthosporium leaf blight toxin produced by fungus.
- Unsatisfactory fertility restoration.
- Unsatisfactory Pollination eg. Capsicum, Rice.

  In rice they can be overcomed by regular tripping of the ears in the morning using rope.
- Modifying genes.
- Contribution of cytoplasm by sperm nuclei.
- Environmental effects-(in Maize, Bajra, Jowar).
- Non availability of a suitable restorer.
- Spontaneous reversion.

IV. Transgenic male sterility: Male sterility which is induced by the technique of genetic engineering is called transgenic male sterility.

It comes under GMS and it is heritable.

E.g. It is induced in tobacco and rapeseed by transferring a gene from *Bacillus amyloliquifaciens*. In this case, one gene barnase causes male sterility, hence this gene integrated in A line. Another gene barstar suppresses the male sterility gene *barnase* and leads to restoration of fertility. Hence it is transferred to restorer line.

V. Chemically induced male sterility: It is induced by treatment with chemicals, and it therefore confined to the generation of chemical treatment (i.e. nonheritable). These chamicals are called *male gametocides*.

Eg. NAA, MH, GA3 and Ethrel etc.

What is meant by autoplasmic and alloplasmic CMS line? Why it is difficult to restore fertility in alloplasmic line and how to correct the same?

Euplasmic/ Autoplasmic CMS line: CMS where mutation takes place in native cultivated species, such that there is incompatibility between organelle and nuclear genome and leads to male sterility.

- Restoration of this system is easy and high.

Alloplasmic CMS line: Cytoplasm of wild species is transferred to genome of cultivated species resulting into CGMS called alloplasmic male sterility.

Eg. Transfer of sterile cytoplasm from *Brassica tournifortii* to Varuna variety by repeated backcrosses.

Wild (*Brassica tournifortii*) A line rr/S × B line rr/F (Varuna)

↓

F_1 rr/S × rr/F (Varuna)

BC_1F_1 rr/S × rr/F (Varuna)

BC_2F_1 rr/S × rr/F (Varuna)

After 5-6 backcrosses, resulting in male sterile line in Varuna variety.

But problem with alloplasmic line is availability of restorer line is very difficult i.e. restorer gene are in very low in frequency. So create problem in hybrid seed production.

But in majority of the cases, the wild species which is donor of sterile cytoplasm also contains some genes which can restore fertility in its nuclear genome. So they can be used for getting 'R' genes.

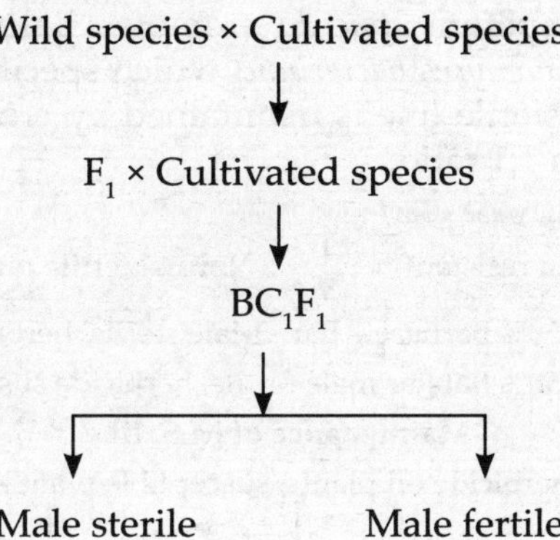

Male fertile is because of some 'R' genes in nuclear genome, but cytoplasm with 'S' genes. So used for getting 'R' genes.

Male fertile × other species (Pusa jagannath but not with varuna)

Iso cytoplasmic restorer line (New pusa jagannath)

(Cytoplasm is similar with male sterile line)

Here, do not transfer both nuclear (R) and cytoplasmic (ms) genes into single/ same variety (say into varuna) because there will be no heterosis, so both are transferred in to different varieties i.e. in to different genetic backgrounds.

What is meant by transgenic male sterility? Explain in detail its types available for use in hybrid seed production?

The male sterility which is induced by the technique of genetic engineering is called transgenic male sterility.

Different types are:

1. Barnase-barstar system
2. Modified glucanase gene
3. Inhibition of flavonoid biosynthesis
4. Through hormone engineering

Among all these system, Barnase-Barstar is most important and was developed by Mar et al. in 1990.

Barnase- Barstar system:

Gene utilised (i) barnase (it is bacterial RNAase from *Bacillus amyloliquefaciens*) barstar (it is RNAase inhibitor, from same source) (iii) promoter used was TA 29 from tobacco which is tapetum specific promoter. Barnase gene is tagged with another gene (from *Streptomyces bacterium*) which specifies resistant to herbicide phosphinothricin. Male sterile line is maintained by crossing with a male fertile line. Progeny obtained in 1:1 ratio.

Barnase -Bar/- × -/-

(Male sterile phosphinothricin resistant) ↓ (Normal fertile phosphinothricine susceptible)

50% Barnase – Bar - Male sterile, herbicide resistance

50% Barstar male fertile, herbicide susceptible

Maintenance of M.S. line

Spray of herbicide on plants, susceptible plant killed only

↓

Barnase-Bar/- × Barstar /Barstar

↓

50% Barnase-Bar/Barstar: Malefertile

50 % -/Barstar: Male fertile

Modified glucanase genes: Worall *et al.* (1992) used a tapetum specific modified glucanase gene from tobacco. The tapetum wall gets dissolved by enzyme callase which is a β-l, 3 glucanase. Plants showed premature dissolution of the callose wall, causing moderate to complete male sterility.

Inhibition of flavonoid biosynthesis: Flavonoids are important in plant reproduction and senescence related mechanism. Vander Meer *et al.* (1992) induced MS by antisense inhibition of flavonoid biosynthesis. This gene shows tapetum specificity. In maize and Petunia, the pollen lacking flavonoids were unable to produce functional pollen tubes.

Hannone engineering: Changes in endogenous level of auxins result in male sterile phenotype. Transgenic plants having the *rol C* gene of *Agrobacterium rhizogenes* under the control of CaMV35S promoter shows male sterility. Male fertility was restored by using antisense construct of *rol C*.

Explain the physical detection methods for different cytoplasmic male sterility systems?

Whether a particular sterile genotype belongs to cytoplasmic, genetic and cyplasmic-genetic system can be detected only by its progeny performance on crossing with a few normal (fertile) genotypes. In subsequent generations, any one of three patterns in crossed progenies of each genotype (grown in separate rows) may be distinguished.

Trend I: All the progenies in all the rows may be sterile.

Trend II: Some rows may consist of all fertile plants. And in some rows, sterile and fertile plants may occur in 1:1 ratio.

Trend III: Some rows may have all fertile plants, some all sterile plants, and yet others have fertile and sterile plants in 1:1 ratio.

Based on all the possible kind of genotypes for fertility/sterility that can exist in a population, we can readily arrive conclusion that;

Trend I is possible only in case of CMS situation (Cytoplasmic sterility).

Trend II can appear only when sterility is controlled by genetic factors alone i.e. genetic sterility (GMS).

Trend III will be encountered only when both nuclear genes and specific cytoplasm involved i.e. cytoplasmic-genetic sterility (CGMS).

Explain the phenotypic classification of male sterility?

Irrespective of the genetic makeup and nuclear and/or cytoplasmic gene constitution, male sterile exhibit certain specific phenotypic characteristic especially in anther form and/or function, which readily differentiate them from the male fertile.

Phenotypic classification of male sterility

Male terility type	Androecium	Phenotypic manifestation	Breakdown time and type	Influence on male sex organs	Effect on growth regulators	G × E interaction	Natural frequency
Structural male sterility	Absent, reduced or malformed	Stamens either absent, misformed or transformed into nonandroecial structures. Lack of microsporogenous tissue.	Staminal differentiation and development before micro-sporogenous onset (pre meiotic gene action)	Both form and function ability impaired completely	Exceedingly high	Extensive	High
Sporogenous male sterility	Form normal to abnormal but function fully impaired	Apparently normal, faulty or abnormal microsporogenous tissue or anomalous male meiosis or micro spore formation.	Either immediately prior or just after microsporogenesis (Meiotic gene action)	High to moderate	Low, meiotic breakdown time delayed or evaded	Low to moderate	Exceedingly high, most prevalent
Functional male sterility	Form normal, development normal	Pollen function ability impaired	Just prior to or during pollen liberation stage (post meiotic gene action)	Pollen function ability completely impaired.	Breakdown time considerably delayed	Exceedingly high	Low

Explain the steps involved in development of male sterile and restorer lines.

Transfer of male sterility is achieved by repeated backcrosses of existing male sterile line (non recurrent parent) possessing desirable characteristics.

Eg. Transfer of cytoplasmic male sterility in onion through six generations of back crossing (Allard, 1966).

Male sterile [msms]S × Male fertile [msms]F

↓ (Cross the male fertile strain to male sterile strain)

F_1 [msms]S (male sterile) × Male fertile [msms]F

↓ Backcross the F1 to fertile strain

BC_1 [msms]S (male sterile) × Male fertile [msms]F

↓ Backcross to fertile strain

Repeated backcrosses

↓

Transfer of cytoplasmic male sterility to a new strain after 6 to 7 backcrosses

Development of restorer line is somewhat indirect. First, a restorer line (R) is crossed with a cytoplasmic male sterile line A. It improves the percentage recovering of restorer genome rather than direct crossing of R line with C line. Resulting male fertile lines are used as a female parent in repeated backcrosses with the strain C used as the recurrent parent, into which the transfer of restorer gene is desired.

In each generation, male sterile plants discarded, and the male fertile plants are used as females for backcrossing to strain C. At the end of backcross programme, a restorer line isogenic to the strain C would be recovered.

Male sterile [msms] S × Restorer line [MsMs]F

↓ (Cross the restorer line to male sterile line)

F_1 [Msms]S (male sterile) × Male fertile [msms]F

↓ Cross the fertle F_1 to male fertile non restore line

[msms]S, [Msms]S × Male fertile [msms]F (discard the [msms]S line)

↓ Backcross to new fertile non restore line

Repeated backcrosses to non restorer line for 5 to 6 generations

↓

[msms]S, [Msms]S

Male sterile lines discarded and fertile lines to be self pollinated

↓

Progeny test: Nonsegregating progenies selected and segregating progenies discarded

Write the reasons and ways for diversification of male sterile lines? And also mention the causes for breakdown of male sterility?

Reasons for diversification of male sterile line:

- One or two male sterile lines severely limit the choice of pollen parents (restorers) due to a narrow range of combining ability.
- A single male sterile line may cause genetic uniformity of the hybrid crop, and thus vulnerability to an outbreak of epidemic is seriously increased.
- A narrow range of adaptation of one or two male sterile lines may restrict their effective utilization.

Methods for diversification of male sterile lines:

1. Identification of new cytoplasmic sources: In hybrid programme, many agronomically promosing genetic stocks are often rejected because of the lack of good fertility restoration. So the availability of more than one cytoplasm might restrict this discard.

   Eg. A_1 (Tift 23A), A_2 (L66A) and A_3 (L67A), three types of cytoplasms are identified in pearl millet (Burton and Athwal, 1967).
2. Rectification of existing male sterile lines: Back crossing and mutational breeding are two most successful approaches to rectification of existing male sterile lines for one or other defects like disease vulnerability.

   Eg. Tift 23A, of pearl millet was found to be highly susceptible to downy mildew, but mutational rectification of Tift 23B (maintainer line of Tift 23A), produced two male sterile lines 5069A, 5071A with disease resistance.
3. Use of cytogenetic configurations: If cytoplasmic male sterility system is nonexistent, it may be possible to devise cytogenetic configurations- a genetic male sterility (Mounder, 1972).

   Eg. Selection of plants possessing deficient chromosome carrying R (restorer) gene. So it gives sterility.

Causes for breakdown of male sterility:

1. Very often, seeds of B line which are grown together in a maintenance block are mechanically mixed up with those of A line. They will multiply, if not roughed out before pollen shedding.
2. Due to an interaction between the cytoplasm and the genome influence the precision of male sterility system, thus adversely affecting the stability of the male sterility system.
3. Male sterile line may have differential behaviour under diverse environmental conditions. Thus, pollen shedders may appear in one environment but not in other.
4. Particulate transference of maintainers cytoplasm to the male sterile line through pollen grains also results in a weakening of the sterile mechanism.
5. Sometimes, B line may carry some impurity of genotypes which causes an imperfect maintenance of sterility.

Explain the role of PGMS and TGMS in hybrid seed production?

In GMS, maintenance of male sterility suffers from a serious defect that 50% of the plants in the male sterile line are male fertile (Msms); these plants must be identified and eliminated before they shed pollen.

This defect is corrected entirely by the use of a photoperiod-sensitive (PGMS) or temperature-sensitive genetic male sterility (TGMS). For example, a TGMS rice line is fully male fertile at temperatures below 28°C, while it is completely male sterile at temperatures above 30°C. The TGMS line is multiplied in a zone where temperatures during the critical period of growth are below 28°C. In this environment, the TGMS plants will be male fertile, will produce selfed seed and yield all TGMS progeny. For *hybrid seed production*, the TGMS line is grown in an area where the temperature during the critical growth period is above 30°C; as a result, all the plants in this line will be male sterile. The TGMS line is pollinated by any male fertile line, which combines well with this line to produce on excellent F_1 hybrid.

This (TGMS) system is the most efficient system of hybrid seed production

1. The TGMS line does not contain male fertile plants (which is the case with normal GMS) and the plants are male sterile.
2. TGMS line is maintained by selfing since TGMS plants are fully fertile in specific areas.
3. Any line combining well with the TGMS line can be used as the male parent.

In contrast, CGMS system requires a maintainer line (B line) for maintaining the CMS line, and a suitable restorer line has to be developed and used as the male parent of the hybrid.

Thus use of CGMS requires 3 lines (A, B and R lines) and conversion of the male parent of the hybrid into an R line (*three-line hybrid seed production*). In contrast, TGMS or PGMS requires only two lines (TGMS or PGMS line and a male fertile line), and any male fertile line can be used as the male parent of the hybrid (*two-line hybrid seed production*).

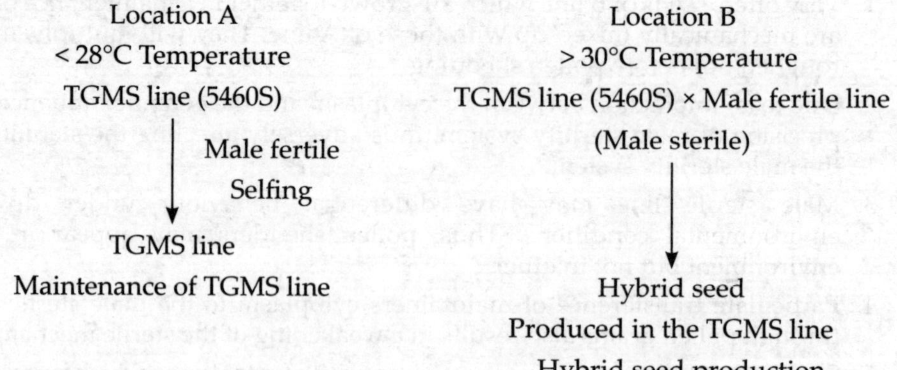

This is the scheme for hybrid seed production using TGMS. TGMS line (5460S) is maintained/multiplied at location A, which has <28°C temperature during the critical growth period. As a result, TGMS line is fully fertile and is maintained by selfing. Hybrid seed production is done at a separate location B where temperature during the critical growth period is above 30°C, as a result, TGMS is completely male sterile.

Write short notes on XYZ system of male sterility for production of hybrid wheat?

XYZ system of male sterility for production of hybrids in wheat was proposed by Driscoll in 1972.

X = Disomic addition line 2n + 1IIR wheat genome carry gene for male sterility Rye chromosome has gene for fertility restorer. This is closely linked with marker hairy peduncle. They are male fertile.

Y = This component has only one dose of alien rye chromosome so that it produces two types of male gamete (with alien - non functional; without alien-functional)

Similarly two kinds of female gamete are produced: [(i) with alien chromosome which on fertilization give rise to a 'Y' plant and (ii) Without alien, which is non kairy and is sterile].

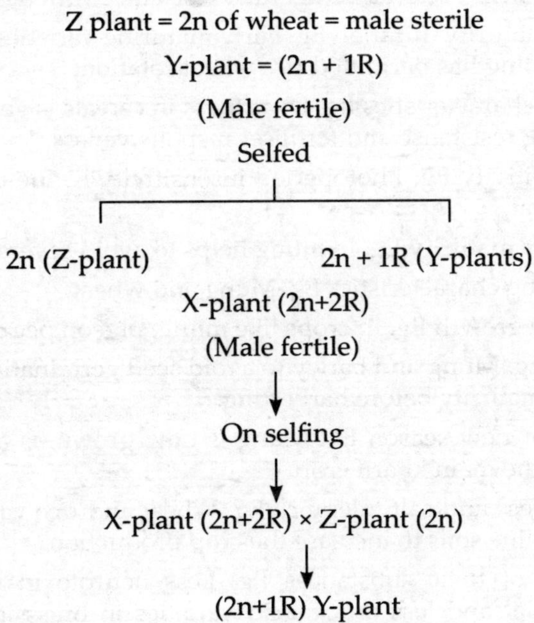

- Z-plant are male sterile, it bear the hybrid seed.
- X and Y needed for the production of female parent of the hybrid.

CHAPTER - 5

Plant Breeding Methods

Define plant breeding and write the various objectives and activities of plant breeding?

Plant breeding is concerned with changing the genotype of plants so that they become more useful to man; and consists of the principles and methods required for favorably changing the genetic constitution of crop plants in terms of agronomically as well as economically.

Objectives of plant breeding:

The chief objectives of plant breeding are:

- Higher yields Eg. Hybrid varieties of maize.
- Improved quality Eg. Grain size/ colour, milling and baking qualities in wheat and cooking quality in rice.
- Disease and insect resistance Eg. Rust resistance varieties in wheat.
- Change in maturity duration Eg. Early maturing varieties in wheat suitable for late planting has permitted rice wheat rotation.
- Agronomic characteristics Eg. Dwarfness in cereals is generally associated with lodging resistanse and fertiliser responsiveness.
- Photoinsensitivity Eg. Photoperiod insensitivity is due to genes *Ppd1* and *Ppd2* in wheat.
- Synchronous maturity Eg. In mung helps to avoid several pickings.
- Nonshatlering characteristics Eg. Mung and wheat.
- Determinate growth Eg. In crops like mung, pigeon pea and cotton etc.,
- Dormancy Eg. Mung and barley to avoid seed germination during rains at the time of maturity before harvesting.
- A variety for new season Eg. Mung is now grown as a summer crop in addition to the main kharif crop.
- Moisture stress and salt tolerance Eg. Wheat and rice varieties for rainfed areas and saline soils to increase the crop production.
- Elimination of toxic substances Eg. Less neurotoxin (BOAA) varieties of Khesari dal and less erucic acid varieties in brasssica, which help in increasing the nutritional values of these crops.

Activities in Plant breeding: The various activities involved bringing the desirable changes in genotypes of crop species and the consequent benefits to farmers are creation of variation, selection, evaluation, multiplication and distribution.

- Creation of variation: Genetic variation is a prerequisite for any crop improvement and created by domestication, germplasm collection, mutation, polyploidy, somaclonal variation and genetic engineering.
- Selection: It consists of identification and isolation of plants based on phenotype, which are having the desirable combination of characters, and growing their progeny.
- Evaluation: The newly selected lines or strains are tested for yield and other traits and their performance is compared with the existing best varieties called checks and conducted at several locations for two or more years under the concerned coordinated crop improvement programme. If the new lines or strains are superior to the checks, it is released and notified as a new variety and its seed can now be multiplied and, more importantly, certified by a seed certification agency for quality.
- Multiplication: It consists of large-scale production of certified seed of the released and notified variety. Seed production is usually done by seed production agencies in a step-wise manner, and the seed is certified by a seed certification agency.
- Distribution: Certified seed is ultimately sold to the farmers who use it for commercial crop cultivation and to reap the benefits in the form of an enhanced, stable production of superior produce and at a lower cost.

What is meant by Pre breeding? And explain its role in crop improvement?

The process of the initial introgression of traits from an undomesticated source (wild) or agronomically inferior source, to a domesticated or adapted genotype is called **pre-breeding** or **germplasm enhancement**. The process varies in complexity and duration, depending on the source, the type of trait, and presence of reproductive barriers.

- In recent time, a modern cultivation practice for development of uniform high yielding varieties (HYVs) has reduced the crop genetic diversity.
- So it's time to diversify the gene pools of their crop to reduce genetic vulnerability. Also it is compulsory for breeder to look beyond the advanced germplasm pool to find desirable genes.
- The desired genes may reside in unadapted gene pools. But the transfer of genes from unadapted genetic resource often associated with linkage drags and unwanted characters.
- Hence these exotic germplasm often cannot be used directly in cultivar development.
- Instead the materials are gradually introduced into the cultivar development programme through crossing and selection for intermediates with new traits, while maintaining a great amount of the adapted traits.

- The traditional techniques used are hybridization followed by backcrossing to elite parent, or the use of cyclical population improvement technique.
- The modern tools of molecular genetics and other biotechnological procedure are enabling gene transfer to be made into elite lines without linkage drag.

The major uses of germplasm enhancement or pre breeding are as follows:

1. Prevention of genetic uniformity and consequences of genetic vulnerability.
2. Potential crop augmentation
3. Introduction of new quality traits
4. Introduction of disease and insect resistance genes
5. Introduction of environment resistance gene.

Examples:

B-carotene in tomato: *Lycopersico hirsutum*

Vitamin C in tomato: *L. pimpenellifolium* & *L. peruvianum*

Fibre strength in cotton: *Gossypium thurberi*

Explain how the method of reproduction and mating system influence the constitution of population released as variety and the method of breeding followed?

Method of reproduction:

Asexual reproduction: (Development of zygote without the union of male and female gametes).

- In asexually reproducing crops, entire genotype will be passed on to its progeny.
- So in this, release either homozygous or heterozygous population as a variety.

Sexual reproduction: (Development of zygote with the union of male and female gametes).

1. **Self pollinated crops:**
    - Because of selfing, it leads to increase of homozygosity and reduction of heterozygosity
    - So in this, develop homozygous population and release as a variety.
2. **Cross pollinated crops/ Open pollinated crops:**
    - Because of crossing, it leads to increase of heterozygosity and reduction of homozygosity.
    - So in this, develop high potential equilibrium population and release as a variety i.e. gene and genotypic frequencies are constant throughout generations.

Variety	Genotypic frequency		
	A_1A_1	A_1A_2	A_2A_2
Self pollinated crops	1	0	0
	0	0	1
Cross pollinated crops	0.36	0.48	0.24
	0.25	0.5	0.25

Mating system:

Self pollinated crops: Develop homozygous population and release as a variety.

Cross pollinated crops: Develop high potential equilibrium population and release a variety.

Methods of breeding		
Self pollinated crops	**Cross pollinated crops**	**Asexually propagated crops**
1. Plant introduction	1. Plant introduction	1. Plant introduction
2. Pureline selection	2. Mass and progeny selection	2. Clonal selection
3. Mass selection	3. Back cross method.	3. Mass selection (rarely used)
4. Pedigree method	4. Heterosis breeding	4. Heterosis breeding
5. Bulk method	5. Synthetic breeding	5. Mutation breeding
6. Single seed descent method	6. Composite breeding	6. Polyploidy breeding
7. Backcross method	7. Polyploidy breeding	7. Distant hybridization
8. Heterosis breeding	8. Distant hybridization	8. Transgenic breeding
9. Mutation breeding	9. Transgenic breeding	
10. Polyploidy breeding	10. Mutation breeding (rarely)	
11. Distant hybridization		
12. Transgenic breeding		

Explain in brief about different types of genetic populations expected in crop improvement?

Different types of genetic populations in plant breeding		
Population	**Nature**	**Examples**
Homogeneous	Genetically similar population	Purelines, inbred line, F_1 hybrids progeny of a clone
Heterogeneous	Genetically dissimilar population	Land races, composites synthetics and multilines.
Homozygous	Non-segregating populations	Purelines, inbred lines mass selected autogamous varieties and multilines

Heterozygous	Populations segregate on selfing	F_1 hybrids, composites, synthetics and a clone
Combinations		
Homogeneous and Homozygous	Genetically similar and non segregating populations	Purelines and inbred lines
Homogeneous and heterozygous	Genetically similar but segregating on selfing	F1 hybrids between inbred lines and progeny of a clone
Heterogeneous and homozygous	Genetically dissimilar but non segregating populations	Multilines and mass selected varieties in autogamous species
Heterogeneous and Heterozyous	Genetically dissimilar and segregating populations	Composites and synthetics

What is meant by selection? And explain the genetic basis of selection?

Selection and its genetic basis:-

Falconer:- Genetic properties of a population can be changed by

(1) Selection and (2) Controlling the mating system

Selection:- Lerner coined the term selection and it is non-random differential reproduction of genotype or gradual heritable changes in the genetic makeup of genotypes without mutation or hybridization.

- Selection operates on phenotype, so correspondence between genotype and phenotype is good for advance under selection.
- Linkage and epistasis may adversely influence the selection advance.

Basic principles of selection:-

- Operates on existing variability (not create variability)
- Acts only through heritable differences
- Works because some are favoured at the expense of other.

Purpose of selection:-

- To modify the mean: Directional selection
- To reduce variability: Stabilizing selection
- To extend the range in one direction: Directional selection
- To secure uniformity: Population improvement programme
- To diversify the population (disruptive selection)

Genetic basis of selection:-

Gene frequency-selection acts not to adapt the gene frequency at a given locus to the background or residual genotype but to co-adapt all loci on all chromosome, so that a coherent gene pool is created and perpetuated.

The rate of change of gene frequency depends on-

- Existing gene frequency
- Genotype
- Selection intensity
- Degree of dominance.
- Unit of selection for intra-population eg: - Alleles individual family
- Unit of selection for inter-population eg: - Lines, Strains, Breeds
- Additive genetic variance responds to selection which is equivalent to breeding value that is deviation of progeny mean from population mean

$$R= Sh^2$$

$$R=i\sigma_p h^2$$

- Selection is effective when p = q = 0.5 and it gradually becomes in effective as gene frequency drift towards 0 or 1.

Explain in detail about the elements of selection?

Elements of selection:

- Selection intensity
- Selection differential
- Selection response
- Heritability

Selection intensity: It refers to amount of selection applied expressed as proportion of population favored and inversely proportional to percentage proportion selected.

- Large 'i' means stringent selection pressure.
- In real selection experiment, the desired alleles becomes preponderant after each cycle of selection so 'i' also change
- Small 'i', large population to be handled
- Large 'i', allelic erosion due to genetic drift
- i = 2.06 -1.76 is most appropriate

The limit of selection is set by two factors

1. Population size
2. Inbreeding

Under natural selection 'i' is expressed as relative number of offspring produced by different genotype and known as selection co-efficient $s = i/\sigma_p$. Where, i = intensity of selection, s = selection differential and σ_p = phenotypic standard deviation.

Selection differential (s): It is average superiority of selected individuals over the mean of population of their origin in same parental generation before selection is made.

For selection differential, an arbitrary culling level 'k' is fixed for a trait and individual beyond that level are selected.

$$S = x - \mu, \text{ Here } x = \text{Average of selected individuals}$$

The wider the phenotypic variability, greater the's'. The standardised selection differential i.e., s/σ_p is also known as intensity of selection. So, $S=i\sigma_p$. It means's' is directly proportional to 'i' and 'σ'_p'.

Selection response or genetic advance under selection:

It is deviation of the progeny mean ($\overline{x}$) from the original (unselected) population mean (μ). It is also known as genetic advance or genetic gain under selection.

$$R = \overline{X} - \mu.$$

The basic idea of selection response stems from the fact that when a population with considerable genetic variability is subjected to selection, change in gene frequency is obvious and resulted in shift in population mean. The shift in population mean reflected in next generation in form of selection advance.

$$R = \overline{X} - \mu = \text{(realized gain)}$$

$$\text{or } R = 2\sigma_p h^2 \text{ (predicted gain)}$$

$$\text{or } R = b\sigma_p s = h^2 S = h^2 i\sigma_p$$

Where 'b' is breeding value.

Response to selection is governed by three factors:-

1. High heritability (h^2) – can be done by minimising V_E. Individual having high $b_{\overline{op}}$, have high 'R' since $b_{\overline{op}} = h^2$
2. High selection intensity (i):- As 'i' increases would keep 'x' high, therefore response to selection is high.
3. Large selection differential:- High 's' from high 'i', but both i and s are dependent upon standard deviation (σ_p) $s = i\sigma_p$

$$R = b_{\overline{op}} s = h^2 s = h^2 \sigma_p$$

The generalized selection differential $R/\sigma_p = (s/\sigma_p)\, h^2$

Variability is important here because as high value of 's' generally result from high selection intensity (i), but both i and s dependent on phenotypic standard deviation (σ_p). 'σ_p' represents the range of variability since $i = s/\sigma_p$ and $s = i\sigma_p$

- So large the 's' greater the possibility of 'R' being large
- 'R' consists of three components h^2, i and σ_p.

$$R = b_{\overline{op}}s = h^2s = i\sigma_p h^2$$

R is, so to compare different traits in different population is possible by expressing it in terms of 'σ_p'. Thus a generalized measure of 'R'= R/σ_p. This can also be expressed in terms of generalized selection differential

$$R/\sigma_p = (s/\sigma_p)\, h^2$$

I. Impact of selection response:
   1. Change in the proportion of existing genotypes: Directional selection – changing in population mean.
   2. Appearance of new genotypes: Heterozygotes polymorphism. Natural consequence of disruptive selection.
   3. Changes in the variance of the population.

II. Selection response for monogenic and polygenic additive traits:

   Selection is effective when p = q = 0.5, this happens when genes are acting additively. Selection for an additively governed metric trait would be slow but steady progress.

III. Selection response for non-additive trait:

   Trait governed by 'non-additive gene action' is essentially an 'inter-population' selection, since intra –population selection is most effective in the utilization of non- additive genetic variance.
   - Selection for heterozygote – utilization of heterosis – overdominance
   - Selection for epistasis warrants the inbreeding followed by selection or selection between inbred families is most effective approach.

IV. Response to selection is inversely related to past history of selection (because it is already undergone selection, so additive gene action depleted).

Patterns: It depends upon the availability of free variability arised from potential variability.

The five main types of response to selection for various traits were elegantly given by Allard (1960) they are.

(*a*) Rapid initial gain under selection
(*b*) Slow, steady and long continued response
(*c*) Slow response culminating a plateau
(*d*) Little or no response to selection
(*e*) Selection response after plateauing

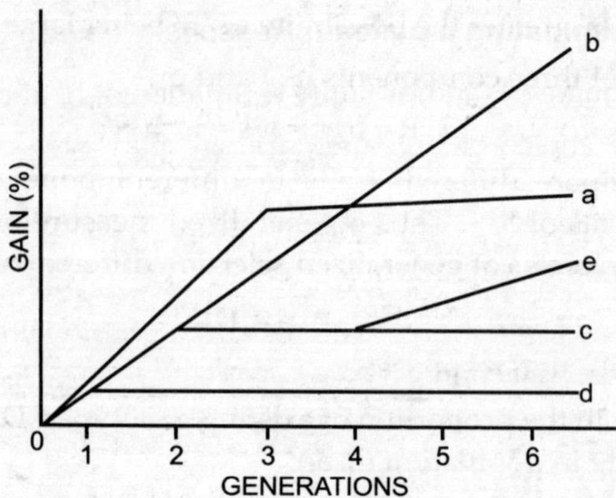

Patterns of selection response (a–e) for metric traits

Rapid initial gain under selection:

- Selection for adjustment of quasi-quantitative traits (simply inherited but not necessarily monogenic).

  Eg. Resistance, pigmentation and leaf angle etc.
- The cause for rapid gain underlies the increase in the frequency of major gene, which is expected to be large until these genes are fixed by selection.
- Subsequently, slow progress due to slow increase in minor gene frequency coupled with further small changes in the frequency of major genes.
- And also, here the genetic variability is reduced due to the fixation of major genes.

Slow, study-long continued response:

- Selection for quantitative traits is not rewarded as well as for the quasi-quantitative trait.
- Gradual shift in the population mean, followed by little shrinkage of genetic variability.
- Extreme phenotypes (transgressive segregants not found in the original population).
- Selection response mainly due to prolonged availability of variability over generations.
- Selection does not seem to increase or decrease variability very drastically.

  Eg. 50 generation selection experiment for oil and protein content in maize (Wood worth *et al* 1952).

Slow response culminating a plateau: *Reasons:*

- Lack of adequate σ^2_A and/or faulty estimation of h^2 and genetic advance.
- Unfortunate choice of breeding strategy may also contribute to such a pattern.

Little or no response to selection:

Directional selections for complex characters like grain yield leads to practically little or no response to selection.

Reasons:-

- Lack of genetic variability
- Limited exploitation of σ^2_A by ear to row method or low h^2 for yield

Selection response after plateuing:

This situation represents a typical behaviour of mendalian population with respect to storage and release of variability.

The linked loci carry variability in a population in two forms.

1. **Free variability:** If free variability is available, response to selection will be reflected in the actual initial gain. After it is exhausted, response will assume plateuing. This period continues till the
2. **Potential variability** is converted in to free variability by segregation. This follows a second flush of response to selection.

Write short notes on the following.

1. Asymmetrical repose to selection
2. Correlated response to selection
3. Selection for multiple traits

Asymmetrical response to selection:

It refers to unequal response to selection in the opposite direction. And it is characteristic feature of two way selection and disruptive selection.

Causes: The loci contributing negatively to the co-variance and having $p \neq q$. Also said as sensitivity of genetic variance to changed gene frequency. i.e., loci showing gene frequency other than 0.5 i.e., $p \neq q$.

Eg. 1. Two way selection for high and low protein in maize

2. Disruptive selection: Flowering time in *Brassica*, lateness gives more response than earliness.

Falconer put forth following causes:

(*a*) Different selection differentials (s) – may differ between positive and negative lines (unidirectional natural selection, differential fertility, i, changed variance σ^2)

(*b*) Genetic asymmetry: Base population.

- Directional dominance: p=q=0.5 – Response greater in direction of recessiveness
- Directional gene frequency.

(*c*) Selection for heterozygote in one direction and homozygote in other direction causes unequal response.

(*d*) Inbreeding depression

(*e*) Maternal influence.

Correlated response to selection:

It refers to natural consequence of selection for a character selected, accompanied by change in other related characters, not subjected to selection pressure.

- The phenomena observed much earlier by Darwin
- This causes for homeostatic regulation and co adaptation of gene pool

  Eg. Selection for chaetae number in drosophila and disruptive selection for flowering time in brassica.
- Correlated response to selection is the basic property of quantitative traits. Whereas no correlated response in qualitative traits.

Genetic cause of such response explained by following theory:

1. Genetic correlation theory on quantitative basis: Characters genetically related to each other i.e. with some value of 'r_g' tend to move in same direction under selection. There are two mechanisms that cause genetic correlation, hence correlated response.
   1. Pleiotrophy – developmental correlation
   2. Linkage
2. Polygenetic inheritance theory on purely qualitative basis. Selection favouring alleles in certain block affect the frequencies of alleles that are linked to them. Association of loci and blocks located on different chromosome will also have similar effect.

Measurement of Correlated Response:

$$\text{Correlated response of 'y'} = \text{Selection response of 'x'} \times b_{yx}$$

$$\text{CR of 'y'} = i_x\sigma_{Ax}.h_x \times [\text{Cov}_{Axy}/\sigma_{Ax}.\sigma_{Ay}]$$

$$= [i_x h_x r_{Axy}.\ \sigma_{Ax}.\sigma_{Ay}]/\sigma_{Ax}$$

$$= i_x h_x r_{Axy} \sigma_{py} h_y$$

$$CRy = i_x \sigma_{py} h_x h_y . r_{Axy}$$

CRy depends on genetic correlation (r_A) and heritabilities of two characters.

Indirect selection/Direct selection = $CRy/Ry = [i_x \sigma_{py} h_x h_y . r_{Axy}]/[i_y \sigma_{py} h^2_y]$

$= [i_x h_x / i_y h_y] r_{Axy}$

This ratio tells us to how far the indirect selection comparable to direct selection.

Selection for multiple traits:

These are three basic strategies to simultaneously select multiple traits: tandem selection, independent culling and selection index.

Tandem selection: In this, the breeder focuses on one trait at a time (serial improvement). One trait is selected for several generations, thus another trait is focused on for the next period. It is effective when genetic correlation does not exist between the traits of interest, or when the relative importance of each trait changes throughout the years.

Independent culling: Also called truncation selection, independent culling entails selecting for multiple traits in one generation.

Eg. Traits A, B and C, and breeder may select 50% plants per family for A on phenotypic basis and from the group select 40% plants per family based on trait B; Finally, from that subset, 50% plants per family are selected for trait C, giving a total of 10% selection intensity (0.5 × 0.4 × 0.5).

Selection index: The problem with this approach is that as more traits are selected for, the less the selection pressure that can be exerted on any one trait.

In selecting on a multivariate phenotype, the breeder explicitly or implicitly assigns a weighting scheme to each trait, resulting in the creation of a univariate trait (an index) that is then selected. The index is the best linear prediction of an individual's breeding value.

End product of mass selection is a mixture of pure lines. Justify your answer?

- Mass selection (maximum inbreeding and minimum selection pressure)
- Mass selection is most effective in cross pollinated crops, but it can be applied in self pollinated crops with reasonable degree of success.
- In self pollinated crops, mass selection employed for purification of existing improved varieties through selection by truncation (removing of extreme type such as rogues etc). It assumes the role of stabilizing selection
- Usually, large numbers of single plants selected from such a local variety are sown, and progenies are examined critically for off-type plants or rogues, disease susceptible plants.
- After eliminating these undesirable lines, seeds from the remaining lines are harvested in bulk to recapitulate the purity of local variety.

- So the final end product gives the mixture of pure lines. So by this we can produce the nucleus seed of pure line varieties.

What is meant by segregating generation? And write a brief note on the choice of handling generation (choice of breeding method) after hybridization depends upon the target genotype?

Segregating generation: The F_2 and the subsequent generations obtained through selfing a hybrid between two or more strains.

Target type of choice based breeding:

1. If choice is 50:50 type i.e. 50% of characters from one parent and 50% from another parent.

   Eg. AABBccdd × aabbCCDD

   ↓

   AaBbCcDd

   - In segregating generations, 50:50 types are most common.
   - In this Bulk, Pedigree and Single seed descent methods of breeding will be most effective.

2. If choice is 75:25

   Eg. AABBCCdd × aabbccDD

   AaBbCcDd

   - After one backcross (BC_1), 75:25 are more in population than 50:50 types.
   - In this limited backcross followed Pedigree, Bulk and Single seed descent type of breeding in BC_1F_2 – BC_1F_5 generation will be most effective.

3. If choice is 99:1

   - After 5 – 6 backcrosses, the population will be in 99:1 type.
   - In this, backcross method will be most effective.

4. If choice is 100:0

   Eg. Male sterility

   - In this, transmit cytoplasm from donor parent to recurrent parent by backcrossing and eliminate the nuclear alleles of donor parent by repeated backcrosses.

   Note: Before backcrossing or hybridization, first observe the plants i.e., complementarity for desirable alleles after that only starts the hybridization.

What is the logic and advantage behind the single seed descent method?

- The single seed descent method is more efficient than bulk method in so far as the retention of the large size of variability (representative) in concerned in many crops. Eg. soyabean and oat (Brim, 1966).
- Because according to Gouldens proposal taking one seed from each plant of F_2 to advance the generation under green house conditions to reach the F_5/F_6 generation that amounts to representation of F_2 spectrum of genetic variation in homozygous form.
- So by F_5/F_6 generation, in SSD method leads to the production of large number of such homozygous progenies without sorting to selection within progenies. So selection depends on between line varieties.
- However like bulk method, this method also suffers from a lack of the provision of critical selection in F_2 and F_3 generations.

Explain how loci showing different types of gene action respond to selection practiced in segregating generations derived after hybridization:

Loci showing additive gene action:

- Additive gene action is important for breeder because it favours the selection of desirable homozygotes and elimination of both heterozygotes and undesirable homozygotes.
- So if loci showing additive gene action than selection starts in early generations.

  Eg. Selection in F_2 generations in pedigree method of breeding.

Loci showing complete dominance gene action:

- If loci showing complete dominance gene action, it favours the selection of heterozygotes and dominant homozygotes compared to recessive homozygotes.
- So in this, breeder has to practice selection in advanced generations, by that time entire heterozygosity will converted in to homozygosity.

  Eg. Selection in F_6/ F_7 generations in bulk and single seed descent methods of breeding.

Loci showing over dominace gene action:

- If loci showing over dominance gene action, superior plants selected by virtue of over dominance i.e., heterozygosity and it is disruptive in (early generation) F_2 generation.

- Here also, wait upto advanced generations (F_6/ F_7) generation and practice the selection.
- But it is most useful in hybrid development.

What is meant by backcross method of breeding? And write the similarities and differences between genetic consequences of selfing and backcrossing?

In the backcross method, the hybrid and the progenies in the subsequent generations are repeatedly backcrossed to one of the parents of the F_1. The objective is to improve or correct one or two specific defects of a high yielding variety, which is well adapted to the area and has other desirable characteristics.

Similarities:

1. Proportion of homozygosity and proportion of homozygous individuals will be same in the population subjected to either selfing or backcrossing.
   - Proportion of homozygosity after 'm' number of generations of selfing/ backcrossing = (2^m-1/ 2^m)
   - Proportion of homozygous individuals after 'm' number of generations of slefing/ backcrossing for 'n' number of alleles = $(2^m\text{-}1/ 2^m)^n$
2. Proportion of heterozygosity and proportion of heterozygous individuals will be same in the population subjected to either selfing or backcrossing.
   - Proportion of heterozygosity after 'm' number of generations of selfing/ backcrossing = (1/ 2^m)
   - Proportion of heterozygous individuals after 'm' number of generations of slefing/ backcrossing for 'n' number of alleles = $(1/ 2^m)^n$

Dissimilarities:

1. Proportion of alleles contributing from parents is different.
   - In selfing, same allelic frequency comes from both the parents i.e. 50:50
   - In backcrossing, proportion of alleles comes from recurrent parent goes on increasing and from donor parent goes on decresing.
2. In backcrossing, alleles comes from recurrent parent in current generation equal to proportion of homozygosity in the next generation i.e. alleles comes from recurrent parent 'm' generation is equal to proportion of homozygosity in (m+1) generation.
   - Proportion of alleles from recurrent parent = (2^{m+1} –1/ 2^{m+1}). This is also called as rate of elimination of alleles from donor parent.
   - Proportion of alleles from donor parent = 1/ 2^{m+1}

3. Minimum population size required will be different in selfing and backcrossing.

Number of genotypes (n = number of segregating loci)			
Selfing		Backcrossing	
F_2	3^n	BC_1	2^n
F_6/F_7	2^n	BC_5	1

- In backcrossing, less population size is required as the proportion of segregation (1:1) is less compared to selfing (1:2:1).
- In selfing, 2^n is the proportion of homozygotes in F_6/F_7 population (i.e. selection in advanced generation) but in backcrossing only one type of homozygote is observed i.e. resemble to recurrent parent.

Enlist the requirements and applications of a backcross method of breeding? Describe the procedure of backcross method for the transfer of a recessive gene. Discuss the merits and demerits of a backcross method of breeding?

Requirements of a back cross programme:

- A suitable recurrent parent must be available/which lacks in one or two characteristics.
- A suitable donor or nonrecurrent parent must be available, which should have the character or the characters to be transferred in an intense form.
- The character to be transferred must have high heritability and, preferably, it should be determined by one or few genes.
- A sufficient number of backcrosses (6-7) should be made so that the genotype of the recurrent parent is recovered in full.

Applications of back cross method:

- Intervarietal transfer of simply inherited characters as well as quantitative characters.
- Interspecific transfer of simply inherited characters
- Transfer of cytoplasm
- Transgressive segregation
- Production of near iso-genic lines
- Germplasm conversion

Procedures for transfer of a recessive gene by back cross method of breeding:

Eg. Transfer of rust resistance gene (rr) from donor or non recurrent parent to recurrent or recipient parent with genotype RR.

Generation	Crossing pattern	Remarks
Parents	rr × RR ↓	Crossing between recurrent and nonrecurrent parents
F_1	Rr × RR ↓	F1 is backcross to recurrent parent
BC_1	RR, Rr ↓	Self pollinate the plants for expression of recessive character
BC_1F_1	RR, Rr, rr × RR ↓	Discard the susceptible plants (RR, Rr) and backcross resistant plants to recurrent parent
BC_2	Rr × RR ↓	Progeny is backcross to recurrent parent
BC_3	RR, Rr ↓	Self pollinate the plants for expression of recessive character
BC_3F_2	RR, Rr, rr × RR ↓	Discard the susceptible plants (RR, Rr) and backcross resistant plants to recurrent parent
BC_4	Rr × RR ↓	Progeny is backcross to recurrent parent
BC_4	RR, Rr ↓	Self pollinate the plants for expression of recessive character
BC_5F_2	RR, Rr, rr	Discard the susceptible plants (RR, Rr), self the resistant plants and harvest the seeds separately.
BC_5F_3	Individual plant progeny rows grown ↓ Preliminary and multilocation yield tirals ↓ Variety proposal	

Merits of backcross method:

- The genotype of new variety is nearly identical with that of the recurrent parent, except for the genes transferred.
- It is not necessary to test the variety developed by the backcross method in extensive yield tests, because the performance of the recurrent parent is already known.
- The backcross programme is not dependent upon the environment, except for that needed for the selection of the character under transfer. Therefore,

off-season nurseries and green-houses can be used to grow 2-3 generations each year. This would drastically reduce the time required for developing the new variety.

- Defects, such as susceptibility to disease, of well adapted variety can be removed without affecting its performance and adaptability and much smaller populations are needed in the backcross method than in the case of pedigree method.
- This is 'the only method for .interspecific gene transfers, and for the transfer of cytoplasm.
- It may be modified so that transgressive segregation may occur for quantitative characters.

Demerits:

- The new variety generally cannot be superior to the recurrent parent, except for the character that is transferred.
- Hybridization has to be done for each backcross, which is often difficult, time taking and costly.
- Undesirable genes closely linked with the gene being transferred may also be transmitted to the new variety.
- By the time the backcross programme improves it, the recurrent parent may have been replaced by other varieties superior in yielding ability and other characteristics.

Explain the backcross methodology for transfer of more than one character form different donor parents in to single recipient parent?

Transfer of More than one character: This method can also be employed for transfer in more than one character. In such situation three strategies can be used.

1. Stepwise transfer
2. Simultaneous transfer
3. Stepwise but parallel transfer.

1. Step wise transfer:

- In this approach, the recurrent parent is first improved for one character.
- The improved recurrent parent is then used as the recurrent parent in afresh backcross programme for the transfer of the second character.
- This approach takes much longer time for the transfer of two or more characters.

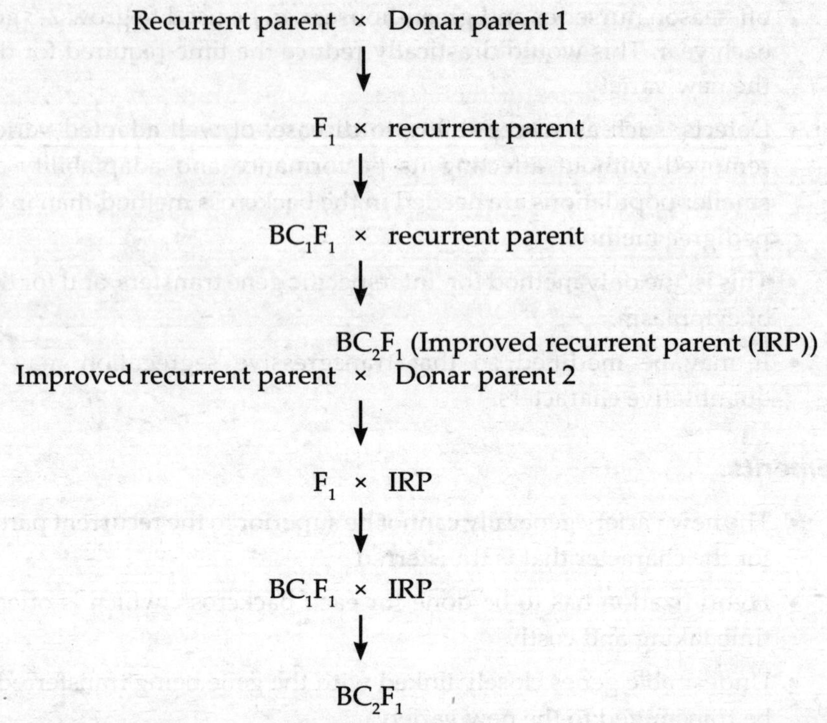

2. Simultaneous method:

- In this approach, genes for the different characteristics may be transferred simultaneously in the same backcross programme.
- The characters to be transferred are brought together in to the hybrid by successively crossing each of the nonrecurrent parents to the recurrent parent or the hybrid thus produced.
- But in this approach, a larger backcross population would be needed than in the case of transfer of a single character.

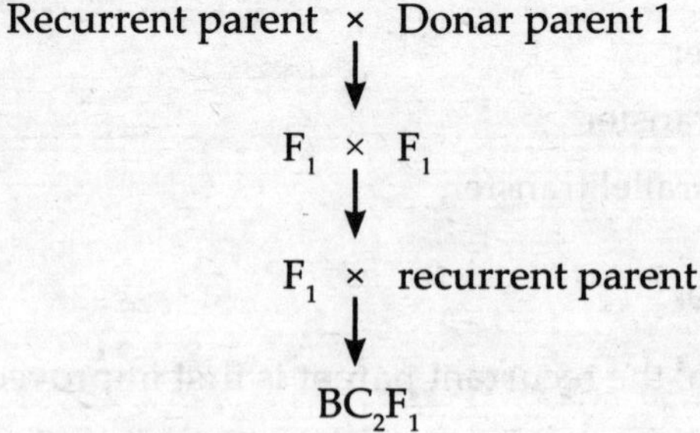

3. Simultaneous but separate transfers:

- Each character is transferred into the same recurrent parent in simultaneous but separate backcross programme.

- The improved versions of the recurrent parent resulting from the different backcross programmes are then crossed together.
- Homozygous lines for the characters being transferred are then selected from the segregating generations using the pedigree method.
- This approach appears to be the most suitable of the three strategies.

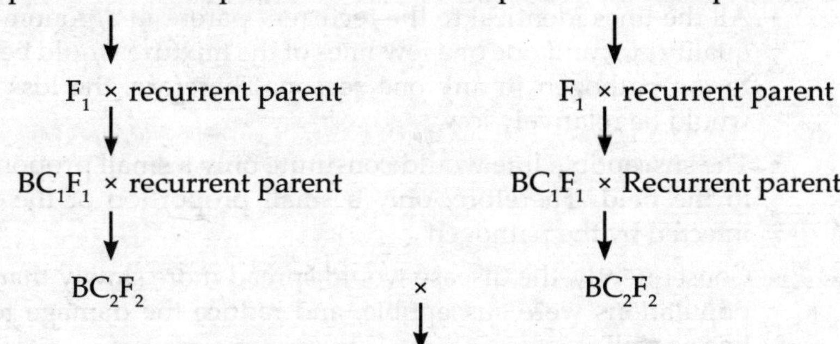

Define multilines and characteristics of a good multilane. Discuss their development, merits- and demerits.

Definition: Jenson (1952) - Term refers to a mixture of near isogenic lines (NILs) for resistance to a specific race of the pathogen. N .E. Borlaug outlined systematic procedure to obtain multilines

Characteristics of a good multiline:

- Multiline has genetic diversity for vertical resistance genes for the concerned disease.
- Each vertical resistance gene should be strong enough to be effective for a reasonable period of time;
- It should have normal resistance to the other important diseases of the concerned region.
- The components of a multiline should be uniform for agronomic and other features, except for the concerned disease resistance.
- The multiline should have yield advantage to the extent that is possible.

Development of multiline varieties:

A multiline variety is usually created by mixing the seeds of several lines that are similar in appearance but have different genes for resistance to a given disease. It involves the steps like

1. Development of component (isogenic lines) lines by backcross breeding programme.

2. Evaluation of grouping of component lines in multilocation trials for yielding ability, morphologic features, agronomic characteristics and produce quality.
3. Releasing of multilines.

Merits of multiline varieties:

- All the lines identical to the recurrent parent in agronomic characteristics, quality etc., and one or a few lines of the mixture would become susceptible to the pathogen in any one season. Therefore, the loss to the cultivator would be relatively low.
- The susceptible line would constitute only a small proportion of the plants in the field. Therefore, only a small proportion of the plants would be infected by the pathogen.
- Consequently, the disease would spread more slowly than when the entire populations were susceptible, and reduce the damage to the susceptible line as well.
- It gives the advantage of horizontal resistance with the ease in breeding for vertical resistance and makes R genes more durable.
- It reduces the risk of homogenizing the pathogen population on a global scale as well as stabilizing and optimizing production on a given farm.
- Multilines may outyield the recurrent parent even in the absence of the concerned disease and also stabilize the race composition of the concerned pathogen population.

Demerits:

- Production and maintenance of multilines is a time-taking job as their component lines have to be maintained separately. It has to be regularly reconstituted, and new lines have to be continuously developed.
- The farmer has to change the seed of multiline varieties every few years depending upon the change in the races of the pathogen. Even seed certification posses difficulties.
- A possibility is there that a new race may attack all the lines of a multiline variety.
- Multilines are based on a popular variety and are produced by backcrossing. Therefore, their potential performance remains the same as that of the recurrent parent. By the time a multiline variety is developed, its recurrent parent may have already been replaced by other superior varieties.

Successful Examples:

1. First commercial multiline: Miramar 62 (ten isogenic line) of wheat for stripe and stem rust.
2. Miramar 65(12) Tumult (wheat)-yellow rust.

Explain in brief about the concept of multilineal hybrid variety?

N. E. Borlaug (1965) believed that the development of a multilineal hybrid variety will offer the best possible means of incorporating a long lasting rust resistance in to a commercial wheat variety. Once an outstanding single-cross hybrid has been developed, it can be converted into a multilineal hybrid.

1. The first step in this process is the development of many lines phenotypically identical with the male sterile female variety A^S (the modified variety of A) through a back-crossing programme.
2. Each of these lines, designated male variety A^n, will be genotypically different from the other in stem rust resistance, and also from that of the male sterile female variety A^S, but none of these lines will possess the genes for restoration of fertility.
3. Therefore, when crossed with the male sterile female variety A^S, the resulting F_1 plants of each cross will be sterile.
4. These male sterile F_1 lines in turn each will be crossed as female with the male variety B^r to provide a heterotic effect, and this second cross will also simultaneously restore fertility to the F_1 commercial hybrid seed formed in the process.
5. The basic commercial variety A and B, both of which are fertile and into which the cytoplasmic sterile and fertility will be incorporated respectively, should also have stem rust resistance gene.
6. Each individual F_1 plant of a hybrid produced by this method would be different in its resistance to stem rust and rust epidemics is not likely to develop in such a population.

	Cross 1 (Primarily for rust resistance)		Cross 2 (Primarily for heterosis)	
Variety A	× ♂ Var. A^n (Stem rust resistance 1)	→	Sterile ♀ F_1s × ♂ Var. B^r	Multilineal
	× ♂ Var. A^n (Stem rust resistance 2)	→	Sterile ♀ F_1s × ♂ Var. B^r	
	× ♂ Var. A^n (Stem rust resistance 3)	→	Sterile ♀ F_1s × ♂ Var. B^r	Hybrid
	× ♂ Var. A^n (Stem rust resistance 4)	→	Sterile ♀ F_1s × ♂ Var. B^r	
	× ♂ Var. A^n (Stem rust resistance 5)	→	Sterile ♀ F_1s × ♂ Var. B^r	
Variety A^S	× ♂ Var. A^n (Stem rust resistance 6)	→	Sterile ♀ F_1s × ♂ Var. B^r	Variety

Scheme for development of multilineal hybrid wheat variety

What is meant by population improvement? And write the principle involved in the following population improvement methods?

1. Ear to row method
2. Modified ear to row method
3. Simple recurrent selection
4. Recurrent selection for GCA
5. Recurrent selection for SCA

6. Reciprocal recurrent selection
7. Population approach in self pollinated crops
8. Diallele selective mating design
9. Convergence – divergence selection

Population: In genetics, population is nothing but a community of individuals which share common gene pool.

Population improvement: Improvement of random mating population through scheme of selection with or without progeny test.

- Population improvement without progeny test. Eg. Mass selection
- Population improvement with progeny test. Eg. Ear to row method and recurrent selections.

Ear to row method:

- This method was proposed by Hopkins in 1896.
- In this, breeding value of selected plants is determined on the basis of open-pollinated progeny test. Since open-pollinated progenies are half-sibs, this method is referred to as "hals-sib (HS) family selection".
- In this, because of progeny test, environment problem can be overcome effectively.

But.....

- It suffers from the defect that plants in the superior progenies are pollinated by those in both the weak and inferior progenies.

Modified ear to row method:

- This method was proposed by Lonniquist in 1964.
- In this, progeny test consists of a replicated yield trials in place of a single row i.e. environmental effects can be separated and the actual value of each progeny can be more accurately estimated.

Simple recurrent selection:

- This method was presented by Sprague and Brimhall in 1950.
- It includes
  (*i*) Selection of plants with desirable phenotype and self pollinate the plants.
  (*ii*) Progeny row evaluation from selfed seeds of selected plants.
  (*iii*) Intercrossing of progenies in all possible combinations.
- This method is effective in increasing the frequency of desirable genes in population i.e. improvement of *perse* performance of resulting population.

Recurrent selection for GCA:

- This method was proposed by Jenkins in 1940.
- It includes

(*i*) Selfing and simultaneous crossing of selected plants with tester with broad genetic base.

(*ii*) Evaluations of test cross progeny in replicated trials.

(*iii*) Based test cross progeny performance, rándom mating among the superior plants obtained from selfed seed.

- This method is effective in increasing the combining ability with open pollinated tester but not the *perse* performance of population i.e. development of reservoirs of germplasm highly suitable for the extraction of superior inbred lines.

Recurrent selection for SCA:

- This method was proposed by Hull in 1945.
- It involves

(*i*) Selfing and simultaneous crossing of selected plants with tester with narrow genetic base.

(*ii*) Evaluations of test cross progeny in replicated trials.

(*iii*) Based test cross progeny performance, random mating among the superior plants obtained from selfed seed.

- This method is effective in improving the combining ability with inbred tester (complementary stock) there by improvement of inbreds and these inbreds directly used to produce commercial hybrids.

Reciprocal recurrent selection:

- This method was proposed by Comstock, Robinson and Harvey in 1949.
- This method involves the improvement of two different heterozygous populations in the ability to combine well with each other i.e. ability of one population to combine with other population.
- In this, each of two populations serve as testers for the plants selected from the other population.
- This method is useful in selecting simultaneously for both GCA and SCA.

Population approach in breeding for self pollinated crops:

- This method was proposed by Palmer in 1953.
- It involves intercrossing of selected F_2 plants; thereby it leads to accumulation of favorable genes in intermated population i.e., chance of recovery of transgressive segregants increase considerably.
- This method is akin to recurrent selection in cross pollinated crops.
- The chief limitation of recurrent selection in self pollinated crops is the difficulty in making the large number of required crosses by hand, which can be overcome by using genetic and cytoplasmic male sterility.
- Thus the use of male sterility effectively ensures intermating among the plants of a population and eliminates the need for tedious and time consuming hand emasculation and pollination.

Diallele selective mating design:

- This method was proposed by Jensen in 1970.
- This method involves selection and intermating in 1970.
- This method provides
  (*i*) Isolation of purelines as commercial varieties through selfing series.
  (*ii*) Improvement of population through intermating among the selected plants i.e. recurrent selection.
  (*iii*) Introduction of new germplasm in the breeding programme by intermating with selected plants at any stage.
- I.e., this provides the retention/ and creation of large amounts of variability for effective selection through several cycles and the introduction of new genes in the breeding material.

Convergence-Divergence Selection:

- This was proposed by Lonnquist in 1972.
- In the method,
  (*i*) Equal amounts of seeds from the base population are sown at five diverse locations or environments and promising genotypes are mass selected naturally or artificially in different environments.
  (*ii*) Random mating among them is arranged by constructing a central 'pollination field' with equal parts of seeds from each location.
  (*iii*) The bulk seeds harvested from the population field represent an improved bulk which is likely to be adapted over all the five environments.
- However, this also suffers from the same drawbacks as mass selection.

Define recurrent selection? And write the procedure involved in reciprocal recurrent selection? And also compare the efficiency of different recurrent selection methods?

Recurrent selection: It is a method of breeding designed to concentrate favourable genes scattered among a number of individuals by selecting in each generation among the progeny produced by mating *interse* of the related individuals (or their selfed progeny) of the previous generation.

- The idea of recurrent selection was first suggested by Hays and Garber in 1919 and independently by East and Jones in 1920.

Reciprocal recurrent selection (RRS):

- Reciprocal recurrent selection was proposed by Comstock, Robinson and Harvey in 1949, with an objective to improve two different populations in the ability to combine well with each other.
- In this scheme, each of the two populations, say A and B, serve as testers for the plants selected from the other population.

Year	Selection features	Remarks
1st year	Population A × Population B ↓ ↓	• Several plants selected in the population A and B • Selected plants self pollinated • Each selected plant from A test crossed with several random plants from B and vice versa • Test cross progeny and selfed seed from each selected plant harvested seperately
2nd year	Test cross progeny of A Test cross progeny of B	• Separated yield trials conducted for test cross progenies from populations A and B • Superior progenies identified
3rd year	Separate intercross blocks for populations A and B	• Selfed seed from plants producing superior testcross progenies planted separately for population A and B • All possible intercrosses made • Seeds from all intercorsses of a population mixed.
4th year	Population A × Population B ↓ ↓	• Composite seed from intercrosses planted separately • And follow the methods as like 1st year
5th year	Test cross progeny of A Test cross progeny of B	• Follow as like 2nd year
6th year	Separate intercross blocks for populations A and B	• Follow as like 3rd year
May be repeated		

- This selection method effects selection for both GCA and SCA.
- Selection of GCA is accomplished because the two testers (populations A and B) have broad genetic base since they are genetically heterogeneous, whereas/ selection for SCA is done as the two populations would be crossed with each other to produce the commercial variety, and the plants in each of the two source populations are selected for their ability to combine well with the gene combinations present in the other population, that is, for SCA with each other.

Comparison of different recurrent selection schemes i.e., Reciprocal selection for general combining ability (RSGCA), Reciprocal selection for specific combining ability RSSCA and Reciprocal recurrent selection (RRS)

- When dominance is incomplete, RRS and RS GCA would be comparable in their effectiveness, and both will be superior to RSSCA.
- When dominance is complete, these three methods would be equally effective.
- In case of overdominance, RRS and RSSCA would be equally effective, but both would be more effective than RSGCA.
- The above relationships are true when there is no epistasis, absence of multiple alleles and linkage disequilibrium. In most .situations, epistasis is important and linkage disequilibrium and multiple alleles are present, where RRS would be superior to RSGCA and RSSCA.

Enlist the various methods of improvement of inbred lines and discuss about gametic selection.

Methods used for improvement of inbred lines are

(*i*) Pedigree selection

(*ii*) Back cross method

(*iii*) Convergent selection/ Double back cross method (Richey, 1927)

(*iv*) Gametic selection

Gamettic selection:

- The basis of this scheme (Stadler, 1944) is the consideration that the frequency of superior gametes (p) in an random mating population is appreciably higher than that of the superior zygotes (p^2).
- In gamete selection, a good inbred line is crossed with a random sample of gametes from an open-pollinated variety and the resulting F_1 plants are selfed as well as crossed to a suitable tester.
- The tester may be an inbred or a population with a wide genetic base, depending upon the objective of the programme.
- The differences in the performance of the test cross progenies evaluated in a replicated trial would be entirely due to the gametes from the open-pollinated variety, since the other parent is an inbred, which would be homozygous for all practical purposes.
- The selfed seeds from those of F_1 plants that produced superior testcross progenies are planted in progeny rows.
- Selfing and selection is continued to develop inbred lines, which are expected to be superior to the inbred used as one of the parents of the initial cross.

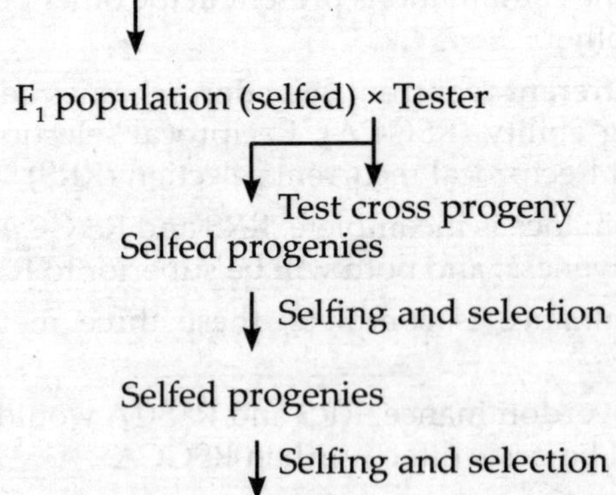

What is meant by inbreeding? And derive the inbreeding coefficient?

Inbreeding: Mating among relatives is called as inbreeding.

Inbreeding coefficient (F): Proportion of autozygous homozygotes (A_1A_1 or A_2A_2) seen in a population or chance that two alleles at a locus are identical.

When inbreeding takes place, the autozygous homozygotes start appearing in population and if goes on increasing generations, their frequency increases at the cost of allozygous genotypes. So proportion of allozygous genotypes (A_1A_2) goes on reducing give rise to autozygous both homozygotes

- Proportion of allozygous genotypes = Panmictic index (1 - F) or P.

In G_0, we can see only allozygous genotypes. After one generation of restricted random mating or inbreeding, in G_1 we can see autozygous homozygotes i.e. inbreeding coefficient.

$$\text{Inbreeding coefficient in } G_1 \text{ generation} = 1/2N$$

It is chance that two alleles at a locus in G_1 individual are replicates of same allele seen in G_0.

$$\text{Inbreeding coefficient of } G_2 \text{ i.e. } (F_2) = 1/2N + (1- 1/2N)\, F_1$$

Here 1/2N is descent from G_1 generation and (1- 1/2N) F_1 from preceding generation (G_0) i.e., genotypes carrying alleles not identical by G_1.

So inbreeding coefficient after't' generations $F_t = 1/2N + (1/2N)\, F_{t-1}$

$$\text{Since, } \Delta F = 1/2N$$

$$F_t = \Delta F + (1-\Delta F)\, F_{t-1}$$

$$F_t = \Delta F + F_{t-1} - \Delta F.F_{t-1}$$

$$F_t - F_{t-1} = \Delta F\,(1- F_{t-1})$$

$$\Delta F = (F_t - F_{t-1})/\ 1- F_{t-1}$$

$$1- \Delta F = 1- ((F_t - F_{t-1})/\ 1- F_{t-1})$$

$$1- \Delta F = 1- F_{t-1} - ((F_t - F_{t-1})/\ 1- F_{t-1})$$

$$1- \Delta F = (1- F_t)/\ 1- F_{t-1}$$

$$1- \Delta F = P_t/\ P_{t-1}$$

$$(1- \Delta F)^2 = P_t/\ P_{t-2}$$

After't' generations

$$(1- \Delta F)^t = P_t/\ P_{t-t}$$

$$(1- \Delta F)^t = P_t/\ P_0 = P_t$$

$$\text{Therefore, } P_t = (1- \Delta F)^t$$

$$1- F_t = (1- \Delta F)^t$$

$$F_t = 1- (1- \Delta F)^t$$

$$F_t = 1-(1-1/2N)^t \text{(Inbreeding coefficient of 't generation')}$$

So it depends on size of population (N) and number of generations of selfing (t). This was given by Sewall Wright.

Write the autozygous and allozygous genotypic frequencies of A_1A_1, A_1A_2 and A_2A_2 genotypes after 't' and 'α' generations based on inbreeding coefficient and panmictic index, if initial frequencies of genotypes are

S. No	Genotypes		
	A_1A_1	A_1A_2	A_2A_2
1	p^2	2pq	q^2
2	D	H	Q
3	0	1	0

Frequency of A_2A_2 genotype after 't' number of generations

$$= q_0^{\ 2} + p_0q_0\,[1-(1-1/2N)^t]$$

$$= q_0^{\ 2} + p_0q_0\,F_t$$

Case 1:

G_0			G_1			G_2		
Total	Autozy-gous	Allozy-gous	Total	Autozy-gous	Allozy-gous	Total	Autozy-gous	Allozy-gous
p^2	0	p^2	$p^2 + p_0q_0F_t$	p_0Ft	$p^2\,P_t$	$p^2 + p_0q_0F_\alpha$	p_0	0
2pq	0	2pq	$2pq - p_0q_0F_t$	0	$2pq\,P_t$	$2pq + p_0q_0F_\alpha$	0	0
q^2	0	q^2	$q^2 + p_0q_0F_t$	q_0Ft	$q^2\,P_t$	$q^2 + p_0q_0F_\alpha$	q_0	0

Case 2:

G_0			G_1			G_2		
Total	Autozy-gous	Allozy-gous	Total	Autozy-gous	Allozy-gous	Total	Autozy-gous	Allozy-gous
D	0	D	$D + (H/2)F_t$	$(D + H/2)F_t$	$D\,P_t$	$D + (H/2)F_\alpha$	$D + (H/2)$	0
H	0	H	$H - H\,F_t$	0	$H\,P_t$	$H - H\,F_\alpha$	0	0
Q	0	Q	$Q + (H/2)F_t$	$(Q + H/2)F_t$	$Q\,P_t$	$Q + (H/2)\,F_\alpha$	$Q + (H/2)$	0

Case 3:

G_0			G_1			G_2		
Total	Autozy-gous	Allozy-gous	Total	Autozy-gous	Allozy-gous	Total	Autozy-gous	Allozy-gous
0	0	0	$2^t-1/\,2^{t+1}$	$2^t-1/\,2^{t+1}$	0	½	½	0
1	0	1	$1/2^t$	0	$1/2^t$	0	0	0
0	0	0	$2^t-1/\,2^{t+1}$	$2^t-1/\,2^{t+1}$	0	½	½	0

What is meant by inbreeding depression or inbreeding change? Derive it. And write the effect of inbreeding on allelic and genotypic frequencies?

Inbreeding depression may be defined as the reduction or loss in vigour and fertility as a result of inbreeding i.e., mating between individuals related by descent or ancestry.

Genotypes	G_0	G_t
A_1A_1	p^2	$p^2 + p_0q_0F_t$
A_1A_2	$2pq$	$2pq - p_0q_0F_t$
A_2A_2	q^2	$q^2 + p_0q_0F_t$

Population mean at equilibrium $(M_0) = a\ (p\text{-}q) + 2pqd$

If population subjected to large random mating, population mean does not change, but if it subjected to inbreeding, population mean will change.

Population mean after 't' generations (M_t)

Genotypes	A_1A_1	A_1A_2	A_2A_2
Frequency after t generations	$p^2 + p_0q_0F_t$	$2pq - p_0q_0F_t$	$q^2 + p_0q_0F_t$
Observed genotypic value	a	D	-a

Therefore population mean

$= [a\ (p^2 + p_0q_0F_t) + d\ (2pq - p_0q_0F_t) - a\ (q^2 + p_0q_0F_t)] / p^2 + 2pq + q^2$

$= a\ (p\text{-}q) + 2pqd - 2pqd\ F_t$

Inbreeding change $= M_t - M_0$

$= [a\ (p\text{-}q) + 2pqd - 2pqd\ F_t] - [a\ (p\text{-}q) + 2pqd]$

$= -2pqd\ F_t$ (at one loci)

$= \Sigma\ -2pqd\ F_t$ (at all loci effect)

This is called inbreeding change or inbreeding depression.

- When population subjected to inbreeding, it affects the population mean either in positive or in negative direction, which depends on direction of dominance (d).
- If loci shows negative dominance, population mean will increases.

Expression of character	Dominant allele	Recessive allele	On inbreeding population mean
Positive dominance	Increases	Decreases	Decreases
Negative dominance	Decreases	Increases	Increases

- On inbreeding, change in population mean depends upon recessive allele.
    1. The recessive allele decreases, the expression of character on inbreeding, the population mean decreases

2. The recessive allele increases, the expression of character on inbreeding, the population mean increases.

- Inbreeding change equal to -2pqd F_t, this product is maximum when q = p = 0.5
- If 50% loci shows negative dominance and 50% loci shows positive dominance, than no change observed because of cancellation effects.
- If loci show high degree of dominance (over dominance), greater change in population mean.
- If loci show additive gene action, no change observed
- If loci show partial dominance, slight effect observed
- If loci show complete dominance, slight higher effect is observed

Explain in brief about the homozygous and heterozygous balance?

Homozygous and heterozygous balance: The self pollinated species are naturally homozygous and have no genetic load, because unfavourable recessive genes become homozygous and are eliminated from the population. These species, therefore develop a genetic constitution, which is adapted to homozygosity, i.e., which does not produce undesirable effects in the homozygous state. This type of genetic organization is known as homozygous balance.

Whereas, in cross-pollinated species, a large number of lethal, subvital and other unfavourable recessive genes, are present with little immediate value to the species. The sum total of these unfavourable genes constitutes genetic load of the species. The harmful effects of such recessive alleles are masked by their dominant alleles as a result of which they are retained in the population. The population, therefore, develops a genetic constitution, which favors heterozygosity; as a result, homozygosity leads to detrimental effects. This type of genetic organization is known as heterozygous balance, because it promotes heterozygosity. The homozygous and heterozygous balances may be visualized as geiie combinations that are adapted to heterozygosity and homozygosity, respectively.

Define heterosis? Prove that $H_{F1}=\sum dy^2$? Briefly explain approach for maximizing heterosis?

The term heterosis was first used by Shull in1914. It may be defined as the superiority of an F_1 hybrid to obtain heterosis. The amount of heterosis is measured as deviation of F_1 over the mid parental value i.e.

$$H=P_1+P_2/2$$

Considering the effect of a single locus A with two alleles A_1 and A_2:

Let the frequency of A_1 in P_1=p

Let the frequency of A_2 in P_1=q

Let the frequency of A_1 in P_2=p′

Let the frequency of A_2 in P_2=q′

Let the difference of gene frequency between the population P_1andP_2 by y so that Y=P-P′=q′-q

Representing P′ and q′ of P_2 in terms of P_1

$$P'=P-y\text{------------ (i)}$$

$$Q'=q+y\text{----------- (ii)}$$

Let the genotype value of

$$A_1A_1=a$$

$$A_1A_2=d$$

$$A_2A_2=-a$$

In both the population, disregarding epistatic interactions, the mean of the two populations may be represented as:

$$MP_1=a(p-q)+2pqd \text{ -----------(iii)}$$

$$MP_2=a(p-q-2y)+2d[pq+y(p-q)-y^2]\text{-----------(iv)}$$

Substituting values of p′ and q′ from (i) and (ii)

The mid parental value M_p is given

$$M_P= 1/2(MP_1+MP_2)$$

$$= 1/2[(iii)+(ii)]$$

$$= a(p-q-y)+d(2pq+y(p-q)-y^2)\text{--------------(v)}$$

Now, when the two population are crossed to produce F_1 individuals taken at random from P_1 are mated to those taken at random from P_2 we may get the following frequency of zygotes in the F_1

Gametes from P_2		Gametes from P_1	
		$A_1(p)$	$A_2(q)$
	$A_1(p-y)$	p(p-y)	q(p-y)
	$A_2(q+y)$	p(q+y)	q(q+y)

The F_1 genotypes formed as follows

Genotypes:	A_1A_1	A_1A_2	A_2A_2
Frequency:	p(p-q)	2pq+y(p-q)	q(q+y)
Genotypic values:	A	D	-a

The mean genotypic value of the F_1 is therefore,

$$MF_1 = ap(p-y)+d(2pq+y(p-q)-a(q(q-y)$$

$$= a(p^2-py-q^2 -qy)+d[2pq+y(p-q)]$$

$$= a(p-q-y)+d[2pq+y(p-q)]\text{---------(vi)}$$

Heterosis expressed difference between between F_1 value and mid parental value is given by subtracting (v) from (vi).

$$HF_1=MF_1-MP=dy^2$$

So heterosis depends on degree of dominance for its occurrence

In the absence of dominance d=0 there is no heterosis

Heterosis also depends on the difference in gene frequency of the two populations (y).

So heterosis may be maximized by increasing level of dominance as well as increasing in the diversity between parents of the crosses.

Heterosis is maximized when q=0.5 and y=1

Experiment of Moll proved that genetic diversity cannot be increased beyond level because of cross incompatibility and breakdown of co-adaptability gene complexes.

Heterosis in F_2:

	A_1	A_2
P_1	P	q
P_2	p-y	q+y

F_2 = a(p-y/2-q-y/2)+2(p-y/2)(q+y/2)d

= a(p-y/2-q-y/2)+2(pq-py/2-qy/2-y^2/4)d

= a(p-q-y)+d(2pq+2py/2-2py/2-2y^2/4)

= a(p-q-y)+d(2pq+py-qy-y^2/2)

= a(p-q-y)+d(2pq+py-qy-y^2/2)

= a(p-q-y)+d(2pq+y(p-q)-y^2/2)

So $HF_2 = F_2 - MP$

$= 1/2dy^2$

$HF_2 = 1/2\ dy^2$

$HF_3 = 1/4dy^2$

Write the different ways of estimation of heterosis? And enlist the methods for fixation of heterosis?

Heterosis is estimated in different ways, 1. Over mid parent, 2. Over better parent and 3. Over commercial cultivar/hybrid.

1. **Average heterosis:** Heterosis is estimated over mid parent i.e. mean value or average of two parents

   Average heterosis = [(F_1 – MP/MP)] × 100

   Where, F_1 is mean value of F_1 and MP is mean value of two parents involved in the cross.

2. **Heterobeltosis:** Heterosis is estimated over superior or better parent.

   Heterobeltosis = [(F_1 – BP/BP)] × 100

   Where, BP is mean value of the better parent of the particular cross

3. Useful heterosis: Superiority of F_1 over the standard commercial check variety.

It is also called as economic heterosis.

$$\text{Useful heterosis} = [(F_1 - CC/CC)] \times 100$$

Where, CC is mean value over replications of the local commercial cultivar

If heterosis is worked out over standard commercial hybrid, than

$$\text{Standard heterosis} = [(F_1 - SH/SH)] \times 100$$

Where, SH is mean value over replications of the local standard hybrid.

Write the methods for fixation of heterosis? Explain how heterosis can be fixed up using apomixes?

Methods for fixation of heterosis:

1. Asexual reproduction eg. Sugar cane, Potato and Banana
2. Apomixis eg. *Hieracium* (Hawck weed)
3. Balanced lethal system eg. *Oenothera Spp*
4. Polyploidy eg. Wheat-rye cross

Fixation of heterosis by apomixes:

There are mainly three methods:

1. Production of completely true breeding F_1 hybrids.
2. Production of partially true breeding F_1 hybrids
3. Production of Vybrid.

Here, heterosis is completely fixed up by use of obligate apomicts.

1. Use of obligate apomicts: Sexual plant should be cross compatible with apomict

   Genetics of apomixis in that species should be known. Crossing of sexual (female) and apomict (male) is done for F_1 production and selfing and backcrossing of sexual hybrids to the male parent is done to recover the recessive trait.
2. If facultative apomict, hetrerosis can be fixed only partially.

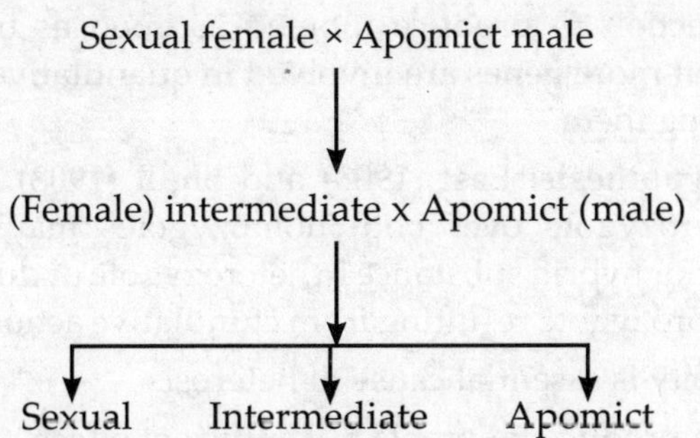

3. Good vybrids are those that have a high frequency of facultative apomicts, a higher degree of heterosis and a minimum negative transgressive segregation.

Theoretically, the performance of vybrids will be intermediate between F_1 and of parental line from which they have been synthesized.

Cross between diverse parents always not result in heterotic combination, Explain?

Extent of heterosis is equal to $\sum dy^2$. Here d is directional dominance and y is the difference in the gene frequency (or) diversity.

- So to increase heterosis we can increase y, but up to an optimal condition.
- Moll *et al* (1962) experiment suggest first linear relationship but beyond optimal distance there is no heterosis , again conformed by moll et al (1965) due to unfavorable interaction of co-adapted gene complex, epistasis or physiological incompatibility cross incompatibility.

Explain in detail about the genetic and molecular basis of heterosis?

Genetic basis: there classical hypothesis

1. Dominance hypothesis
2. Over dominance hypothesis
3. Epistasis hypothesis

Dominance hypothesis: Davenport (1908), Bruce (1910) and Keeble and Pellow (1910) and Jones (1917). According to this hypothesis, heterosis is the result of the superiority of dominant alleles when recessive alleles are deleterious. Here the deleterious recessive genes of one parent are hidden by the dominant genes of another parent and the hybrid exhibits heterosis. And heterosis is directly propotional to the number of dominant genes contributed by each parent.

2 objections:

1. The absence of a skewed F_2 distribution – Jones (1917) and Collins (1921) explained it as more genes are involved and role of environment
2. Failure of section to produce inbreed as good as hybrids- Jones (1917) proposed that more genes are involved in quantitative trait and because of linkage among them.

Overdominance hypothesis: East (1908) and Shull (1908) independently. Here superiority of heterozygote over both homozygotes may arise either due to production of superior hybrid substance in heterozygote or due to greater buffering capacity in the heteorozygote resulting from cumulative action of divergent alleles.

- Heretozygosity is essential cause of heterosis
- A and a both produce favorable but sufficient effects

- Mill borrow – proposed alternative hypothesis for overdominance so many week allele in population and their implication in the genetic adaptaion of the population.

Epitasis: Kings and Jones (1958) significant association between epitasis and heterosis in 80% of the crosses in *Nicotiana rustica* also showed not all heterotic crosses showed non allelic interaction.

Molecular basis of Heterosis:

1. Enzyme activities and protein polymorphism – genes controlling protein, particularly of hybrid vigour.
2. Interaction of gene products: biochemical basis of heterosis is due to supplementary interaction, alternative pathway optional amount and hybrid substances.
3. Intergenomic interaction.
   - Nuclear, mitochondrial, chloroplast complementation (organelle heterosis)
   - Mitochondrial complementation-Mc Daniel and Sarkission.
4. Regulatory genes and dosages dependent effects – Brichler (2003)
   - Up gradation of housekeeping genes.
   - Quantitative traits are expected to be controlled in large part by multiple dosage dependent regulatory loci.
5. Gene expression pattern:
6. DNA methylation – Tsaffaris (1997) hybrids were less methylated than inbreed.
   - In stress condition, F_1 hybrid more resistant to high density induced methylation than inbreds.
   - Epigenetic mechanisms like DNA methylation are heritable and influence diverse development process.

Write a brief note on role of molecular markers in hybrid breeding?

Markers have found significant use in hybrid production by providing a better understanding of mechanism of heterosis and accelerating the overall system of hybrid production. Some of the salient applications are as follows:

1. Molecular profiling and analysis of genetic diversity:
   - Development of core collection
   - Selection of suitable parental line
   - Identification of duplicates Eg. Barley-Lund(2003)
2. Assigning inbred to heterotic groups:
   - Pedigree record supplemented with molecular marker
   - Crossing scheme of maximal efficiency

     Eg. Maize (Dudley *et al.*, 1991); Rice (Xiao *et al.*, 1996).

3. Molecular diversity and prediction of hybrid performance
   - Strong linear correlation between heterozygosity and heterosis
   - More efficient parameter for selecting superior combination, predicting hybrid performance, classifying inbred line, determining heterotic group and pattern.

     Eg. Maize (Hybrid performance Ajmone Marson AFLP r=0.72-Maize-grain yield
4. Understanding genetic basis of heterosis:

   Eg. rice, maize
5. Mapping of genes/QTLs:
   - Fertility restorer gene- Rf_1 to Ri5 - in rice
   - Mapping gene (a) PGMS-pmsl, pms2; (b) TGMS-tmsl, tms2.
6. Introgression of genes/QTL for superior hybrid performance:

   Eg. Maize enhanced inbred

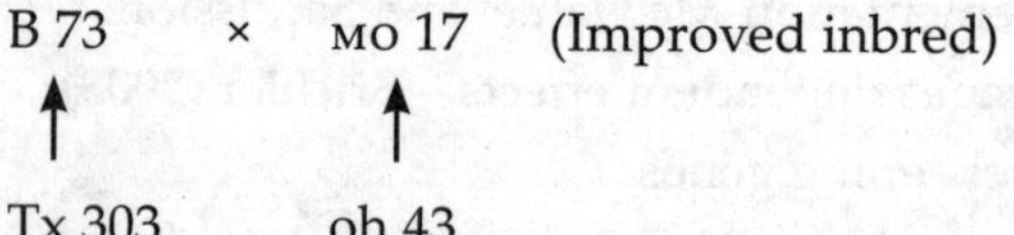

   Rice QTL from *Oryza rufipogon* Yld.1.1, Yid 2.1 from 1st and 2nd chromosome respectively
7. Testing the genetic purity of parental lines and hybrid

   Eg. Yoshita: Used STS marker in rice
8. Protection of parental lines and hybrids
9. Apomixis-Hybrid

Explain the steps involved in development of hybrid varieties? And also mention the merits and demerits of hybrid varieties.

Steps involved in development of hybrid varieties:

Step I: Production of inbreds

- 6-7 generations of continuous selfing of OPV
- Anther culture followed by chromosomal doubling

Step II: Evaluation of inbreds

- Phenotypic evaluation in replicated trials leads to the discarding of inbreds with poor performance
- Top cross test – it provides the reliable estimation of GCA. So that, discard the inbreds with less GCA.
- Single cross test – it provides the reliable estimation of SCA. So, select the inbreds in which hybrid combination with high SCA values.

Step III: Outstanding inbreds used for commercial hybrid seed production

Merits of hybrid varieties:

- Hybrid varieties exploit both GCA and SCA components of heterosis and the produce from hybrid is more uniform.
- Hybrid varieties are the only possible means of exploiting heterosis in self-pollinated species, as open pollinated, synthetic and composite varieties are not possible in them.
- Hybrid varieties are maintained in the form of their parental inbreds, which are grown in isolation and subjected to selfing or sibmating.
- In many self-pollinated crops, hybrid varieties yield 25 - 30 % more than the pureline varieties.

Demerits:

- Farmers have to use new hybrid seed every year, since they cannot produce pureline varieties,
- Hybrid seed production requires considerable technical skill, which makes hybrid seed production a tedious and costly affair.
- The exploitation of full potential of hybrid varieties requires an adequate supply of irrigation water and other inputs, which many farmers are not able to supply.
- In many species, emasculation and pollen dispersal are either unsatisfactory or not practically feasible.
- Hybrid varieties cannot be produced in most of the self pollinated crops on a commercial scale, unless their floral biology is adequately modified through breeding genetic manipulations.
- In most cross-pollinated species, the requirements of isolation are rigid and difficult to fulfill, except on large farms.
- The amount of hybrid seed required to cover the entire area under any crop appears to be impossible to produce without present seed production set-up.
- If the parents of a hybrid variety differ appreciably for grain quality, the produce of the hybrid variety will not be of uniform quality.

Explain why are varietal based hybrids generally less productive than inbred based hybrids?

Varietal based hybrids are generally less productive than inbred based hybrids because

(*i*) Genetic constitution of inbred is a homozygous homogeneous line in cross pollinated crop

- Inbred shows difference in all most all loci i.e. alleles fixed in homozygous condition.

(*ii*) Genetic constitution of hybrid is heterozygous and homogeneous mixture i.e. hybrid shows heterozygosity in all the loci.

(*iii*) Variety means it is equilibrium population with 'q' value varies from 0 to 1.

Genotype	Genotypic frequency		
	A_1A_1 (p^2)	A_1A_2 (2pq)	A_2A_2 (q^2)
Inbred	0 or 1	0	1 or 0
Hybrid	0	1	0
Variety	0.36	0.48	0.24

(*iv*) Pure heterotic hybrid will obtain only when genetic distance between genotypes i.e. y = 1 and allelic frequency (q) must be 0.5.

- When hybrid develops from two varieties, generally cannot expect productive hybrids i.e. getting of y = 1 and q = 0.5 is very rare.
- But hybrids developed from inbreds will be more productive i.e. getting of genetic distance y = 1 and q = 0.5 is most common.

(*iv*) But during selection of inbreds we must select genetically divergence inbreds.

Write short notes on the following.

1. Top cross test
2. Single cross test
3. Triple cross test
4. Double cross test
5. Multiple/ complex cross
6. Poly cross

Top cross test:

- This method was suggested by Devis in 1927.
- It involves a cross between a selection, line, clone etc and a common pollen parent which may be a variety, inbred line, single cross etc. the common pollen parent is called the top cross or tester parent. In corn, a top cross is commonly an inbred-variety cross.
- This method is used for measuring the GCA of parental lines.
  - ❖ A cross between single cross and open pollinated variety is double top cross.

Single cross test:

- It involves a cross between two genotypes, usually two inbred lines, in plant breeding.
- This method is used to measure the SCA of the inbred lines which are selected on the basis of top cross performance.
- With 'n' inbred lines, the possible number of single crosses equal to n(n-1) with reciprocals and n(n-1)/2 without reciprocals.

Triple cross test:

- It involves the cross between single cross and inbred line i.e., it involves three inbred lines.
- With 'n' inbred lines, the possible number of three way crosses are n(n-1)(n-2)/2.
- The main drawback of three way cross is the low production efficiency of male inbred parent.

Double cross test:

- It involves the cross between two single crosses i.e. involves the four inbred lines.
- With 'n' inbred lines, the possible number of double crosses are n(n-1)(n-2)(n-3)/8

Prediction performance of double cross hybrids:

In a double cross hybrid, four inbred parents are involved. Theoretically, the potential of the double cross will be the function of the inbreeding value of these four parental inbreds. Therefore, based on the procedure of testing of the breeding value of inbreds, the performance of a double cross hybrid can be predicted through any of the four methods indicated by Jenkins (1934). Starting with the simplest procedure these methods are:

(*a*) Top-cross testing (one cross per inbred) to know the breeding value to each of the four inbreds (total 4 top-crosses per double cross).

(*b*) Mean of the four non-parental single crosses involved in (AXB) X (CXD) double cross, viz., (AXC), (AXD), (BXC) and (BXD) (total 4 non-parental single crosses per double cross).

(*c*) Average yield performance of all possible six crosses [n(n-1)/2], namely AXB, AXC, AXD, BXC, BXD and CXD (total six crosses per double cross).

(*d*) Average progeny-performance of each inbred can be determined by the mean performance of each inbred in all possible single crosses where it occurs (n-1 crosses per inbred). For instance, the mean performance of AXB, AXC and AXD will determine the average breeding value of the inbred A. Similarly, the mean of AXB, BXC and BXD will indicate the potential of the inbred B and so on (total 12 crosses per double cross).

These procedures of predicting the performance of double cross hybrids have been extensively investigated long ago. The available evidence shows that the method (b), i.e. mean performance of non-parental single crosses, is the most adequate and effective, since there is a close correspondence between predicted and realized yields of double crosses in maize.

Multiple cross/ Complex cross:

- A cross involves more than four inbred lines is referred to as multiple cross.
- Multiple cross is also known as composite cross and is used to combine monogenic characters from different sources in to single genotype.

- Multiple crosses are generally lesser productive than the best double cross combination but they have wider adaptation.

Poly cross test:

- The term poly cross was proposed by Tysdale, Kiesselbach and Westover in 1942 to "designate the progeny from seed of a line that was subject to out crossing with selected line grouping in the same nursery" i.e.
- Open pollination of a group of selected genotypes in isolation from other compatible genotype to promote random mating among selected genotypes is referred to as poly cross.
- In this, every genotype has an equal chance of being pollinated by every other genotype, the genotype isolates should be replicated many times in the isolated natural crossing block in which the out crosses made.
- This method also useful for measuring the GCA of parental lines

What is meant by synthetic variety? And explain how synthetic varieties are advantageous over hybrid varieties?

Synthetic variety: A variety produced by crossing *interse* a number of genotypes selected for good general combing ability in all possible hybrid combinations, with subsequent maintenance of the variety by open pollination.

- Hays and Garber in 1919 were apparently the first to suggest the possibility of the commercial utilization of synthetic varieties.

Merits of synthetic varieties:

Synthetic varieties offer several unique advantages in comparison to hybrid varieties in the exploitation of heterosis.

1. Synthetic varieties offer a feasible means of utilizing heterosis in crop species where pollination control is difficult. In such species, the production of hybrid varieties would not be commercially viable.
2. The farmer can use the grain produced from a synthetic variety as seed to raise the next crop.
3. In variable environments, synthetics are likely to do better than hybrid varieties. This expectation is based on the wider genetic base of synthetic varieties in comparison to that of hybrid varieties.
4. The cost of seed in the case of synthetic varieties is relatively lower than that of hybrid varieties.
5. Seed production of hybrid varieties is a more skilled operation than that of synthetic varieties.
6. Synthetic varieties are good reservoirs of genetic variability. The composites and germplasm complexes also serve as gene reservoirs.

7. There is good evidence that the performance of synthetic varieties can be considerably improved through population improvement without appreciably reducing variability.

Write the factors determining the performance of synthetic varieties? And explain how the performance of synthetic varieties can be improved?

Factors determining the performance of synthetic varieties:

(*i*) Decline in heterozygosity

(*ii*) Decline in F_2 performance

In 1922, Sewall Wright pointed out that "a random-bred stock derived from 'n' inbred families will have 1/nth less superiority over its inbred ancestry than the first cross or a random-bred stock from which the inbred families might have been derived without selection."

This relationship can be expressed as

$$\overline{F}_2 = \overline{F}_1 - (\overline{F}_1 - P/n)$$

Where $\overline{F}_2$ represents the estimated performance of the $\overline{F}_2$ generation. $\overline{F}_1$ is the average performance of all possible single crosses among the parental lines involved, $\overline{P}$ is the average performance of the parental lines, and 'n' is the number of parental lines involved.

- When F_1 hybrids are breed interse, a single cross in F_2 will lose one-half of the excess vigour of the F_1 over the parents.
- When 3, 4, 5 ....n lines comprise the parents of the synthetic variety, 1/3, ¼, 1/5, ...1/n th of the excess vigour will be lost, respectively, in the F_2 generation.

Enhancement of performance of synthetic varieties in advanced generations:

(*i*) **Increasing the number of parental lines:** As per above relationship, with increase in the number of lines would improve the performance of synthetics in advanced generations.

- This improvement would be more, with the smaller values of 'n'. Eg. 1/2, 1/3, 1/4, 1/5 etc. and would become less as the 'n' becomes larger Eg. 1/10, 1/11, 1/12 etc.

(*ii*) **Increasing the performance of F_1 generation:** The performance of F_1 is the average performance of all single crosses among the parental lines and also F_1 performance depends upon GCA of the parental lines.

(*iii*) **Improving the performance of parental lines:** Performance of synthetic varieties in advanced generation can be improved by increasing the performance of parental lines. This can be achieved by

- Second and third cycle inbreds may be developed as their performance would be better than that of first cycle inbreds.

- Inbreds isolated after the population subjected to recurrent selection for GCA.
- Using of short term inbred lines for the production of synthetic varieties.

What is meant by composite variety? And write the differences between synthetic and composite varieties?

Composite variety: Varieties produced by open pollination among a number of outstanding strains usually not tested for combining ability with each other.

Differences between synthetics and composite varieties			
S. No	**Features**	**Synthetic variety**	**Composite variety**
1.	Number of lines	Less number (6-8) of inbred lines	Number of lines are more (even up to 20)
2.	Genetic analysis	GCA of parental lines is tested	No testing for GCA
3.	Prediction of performance	Can be predicted	Cannot be predicted
4.	Genetic base	Broad based	More broad based
5.	Reconstitution of variety	Can be reconstituted	Cannot be reconstituted
6.	Seed replacement of variety	After 4-5 years	After 3-4 years

What is meant by clone? And write the key principle/ features involved in breeding of clonally propagated crops?

Clone: A clone is a group of plants produced from a single plant through asexual production i.e., a group of organisms descended by mitosis from a common ancestor.

Breeding in clonally propagated crops:

The advantage of asexual reproduction is immediately clear; it preserves the genotypeof an individual indefinitely.

- I.e. any genotype is preserved and maintained through asexual reproduction. In contrast, self pollination preserves and maintains only homozygous genotypes giving rise to purelines.
- Since the asexually propagated plants are reproduced by vegetative means genetic segregation and recombination normally do not occur in the successive clonal generations through which they pass while cultivating them.
- This leads to three major consequences.
  (*i*) They gradually accumulate naturally occurring mutations leading to increased heterozygosity.

(*ii*) They progressively accumulate deleterious genes

(*iii*) The constant disuse of sex leads to progressively higher levels of male and female sterility. The occurrence of male sterility is more frequent than female sterility.

- Clonal selection: Improvement of asexually propagated crops by selecting superior clones from local varieties or introduced varieties or intercrossed populations is known as clonal selection.
- And also hybridization followed by selection generally will practice in F^1 generation in clonally propagated crops.
- Breeding in clonally propagated crops suffers from problems like reduced flowering and fertility, difficulties in genetic analysis like GCA and SCA and perennial life cycle etc.

What is meant by ideotype and mention the characteristic features of crop ideotypes in following crops?

1. Rice 2. Wheat 3. Maize 4. Cotton 5. Barley and 6. Pulses

The term ideotype was introduced by Donald (1968). He defined ideotype as "a biological model, which is expected to perform or behave in a predictable manner within a defined environment".

In 1976, Donald and Hamblin proposed the concepts of isolation, competition and crop or commercial ideotypes.

Rice: The concept of plant type was introduced in rice breeding by Jennings in 1964. He suggested that in rice an ideal or model plant type consists of

(*a*) Semi dwarf stature

(*b*) High tillering capacity and

(*c*) Short, erect, thick and highly angled leaves

(*d*) Heavy panicles- number of grains 200-250, test weight – 25g/ 1000 grain

(*e*) Less foliar growth and enhanced translocation of assimilates from leaf to stem during late vegetative and reproductive phase

(*f*) High and descending concentration of tissue nitrogen from top to the base of plant canopy

(*g*) Increase capacity of stem to store assimilate

(*h*) Improved reproductive sink capacity with prolonged period of ripening

(*i*) Increased harvest index > 0.6.

Wheat: The term ideotype was coined by Donald in 1968 in wheat

(*a*) A short strong stem: It imparts lodging resistance and reduces the losses due to lodging.

(*b*) Erect leaves: Such leaves provide better arrangement for proper light distribution resulting in high photosynthesis or CO_2 fixation.

(*c*) Few smaller leaves: Leaves are the important sites of photosynthesis, respiration and transpiration. Few and small leaves reduce water loss due to transpiration.

(*d*) Larger area: It will produce more grains per ear.

(*e*) An erect ear: It will get light from all sides resulting in proper grain development ear: Awns ll get light from all sides resulting in proper grain development

(*f*) Presence of awns, Awns contribute towards photo synthesis.

(*g*) A single culm.

Maize: In 1975, Mock and Pearce proposed ideal plant type of maize.

(*a*) Stiff-vertically oriented leaves above the ear; the leaves below the ear should be horizontally oriented.

(*b*) Maximum photosynthetic efficiency. It seems that selection for high photosynthesis rates in maize is possible.

(*c*) Efficient translocation of photosynthate into grain.

(*d*) Short interval between pollen shed and silk emergence.

(*e*) Ear (cob or spadix) – shoot prolificacy. Prolificacy is the ability of plants to produce a second fertile cob, when spaced widely.

(*f*) Small tassel size: Small tassels would show lower competition for nutrients with the developing ear as well as less shading of the upper leaves.

(*g*) Photoperiod insensitivity: Adequate genetic variation is available in maize for this trait.

(*h*) Cold tolerance of germinating seeds and developing seedlings.

(*i*) Grain filling period should be as long as practically possible.

Cotton: In cotton, genotypes with zero branch, short stature, compact plant, small leaves and fewer sympodia were considered to enhance yield levels. The proposed ideotypes include:

(*a*) Short stature (90-120cm).

(*b*) Compact and sympodial plant habit making pyramidal shape.

(*c*) Determinate in fruiting habit with unimodal distribution of bolling.

(*d*) Short duration (150-165 days).

(*e*) Responsive to high fertilizer dose.

(*f*) High degree of inter plant competitive ability.

(*g*) High degree of resistance to pests and diseases.

(*h*) High physiological efficiency.

Barley: Rasmusson (1987) reviewed the work on ideotype breeding and proposed that in barley, higher yield can be obtained from a combination of

(*a*) Short stature

(*b*) Long awns

(*c*) High harvest index

(*d*) High biomass.

Pulses:

Ideotypes for late sown rainfed condition:

(*a*) Early vigour

(*b*) Plant height of 50 to 60 cm

(*c*) 60-70 pods per plant

(*d*) 2-3 primary branches

(*e*) 9 10 secondary branches

(*f*) 1.7-1.8 seeds per pod

(*g*) 1000 seed weight of 200-250gm and

(*h*) Resistance to diseases like botrytis grey mold ascochyta blight

Ideotypes for drought conditions:

(*a*) Plant with high early vigour

(*b*) Early maturity

(*c*) High HI

(*d*) Relatively few branches

(*e*) Plants with denser-and longer root system and smaller leaf size

Ideotypes for high input conditions:

(*a*) Plants with erect growth habit,

(*b*) Few branches

(*c*) Sown at high density and with longer fruiting zone

CHAPTER - 6

Population and Quantitative Genetics Principles

Define Biometrical genetics? And write the features of Quantitative and Population genetics?

A branch of genetics that utilizes various statistical concepts and procedures to the study of genetical problems is called biometrical genetics. It is integral part of genetics and plant breeding and has wide practical applications in these disciplines. These are two branches of biometrical genetics viz., 1. Quantitative genetics and 2. Population genetics.

Quantitative genetics: A branch of biometrical genetics which deals with the study of polygenic or quantitative characters is known as quantitative genetics. Main features of quantitative genetics are

- It is an extension of Mendelian genetics firmly based on Mendelian principles of heredity
- It provides ways and means for the study of polygenic variation which is not possible by Mendelian genetics.
- It deals with continuous variation
- Its estimates are based on means, variances and covariances.
- It helps in better understanding of genetic principles and better planning of crop improvement programmes.
- Fisher laid the foundation of quantitative genetics in 1918

Population genetics: It deals with the frequency of genes and genotypes in a Mendelian population. Mendelian populations have two important features, viz., 1. Random mating and 2. Equal survival of all genotypes. Main features of population genetics are

- It deals with oligogenic variation or discontinuous variation
- Its cannot be used for analysis of polygenic variation
- Its estimates are based on frequencies or ratios
- Hardy and Weinberg laid the foundation of population genetics in 1908.

What is meant by polygenic characters and write the significance of polygenes in crop improvement?

Polygenic Characters: The characters which are expressed in terms of degree rather than kind and show continuous variation from one extreme to the other are called polygenic characters or quantitative characters. Genes controlling them are called minor genes.

Features of polygenic characters:

- Continuous variation
- Effect of single gene is small and not easily detectable
- Transgressive segregation is possible
- Analysis is carried out from the mean, variances and covariance.
- Correlated response to selection is high for quantitative characters.
- Decline or regain of vigour by inbreeding and trait hybridization.
- Classify by observer not by character itself.
- Sensitive to environment.
- Balanced polygenic system.
- Polygenic have pleiotropic effects and can act both as modified or suppressors of other genes.

Significance of Polygenes:-

- Polygenes are of prime importance to the plant breeder for evolution of improved cultivars.
- Polygenes have great evolutionary significance. They provide variation for fine adjustment and are systems of smooth adaptive change and of speciation.
- The potential genetic variability is stored in the form of linked polygenic complexes. This variability is selected after intermating of such genotypes with other genotypes, due to segregation and recombination.

How does the distribution of quantitative character become continuous? Or How did the findings off Nilson-Ehle show that the quantitative characters are inherited according to Mendelian inheritance?

Multiple factor hypothesis:

- In 1906, Yule suggested that many genes with small and similar effect could produce continuous variation. He proposed that these genes were transmitted according to the laws of Mendel.
- In 1908, Nilsson–Ehle presented experimental evidence to support the hypothesis of Yule. He studied the inheritance of seed colour in wheat and oat.
- The F_2 generations from various crosses had red and white grains in the ratio of, 3:1, 15:1, 63:1, It indicates seed colour was governed by one, two and three genes respectively.
- Close study of seed colour revealed that the seeds classified as red differed in the intensity of colour
- Thus the 15:1 ratio was actually 1:4:6:4:1. It is obtained when seed colour is governed by two genes with similar small and additive effect.

Parents $R_1 R_1 R_2 R_2$ X $r_1 r_1 r_2 r_2$

(Dark red) ↓ (white)

F_1 $R_1 r_1 R_2 r_2$ (medium red)

In F_2 1: 4: 6: 4: 1

Dark red (1): medium dark (4): medium red (6): light red (4): white red (1)

Thus Nilson-Ehle was able to show that certain characters features are governed by genes that have small and cumulative effect. This is multiple factor hypothesis or polygenic inheritance.

Features of multiple factor hypotheses:-

- Each of the factors follows Mendelian inheritance.
- Each of them has similar and cumulative effect.
- Thus a phenotype expression is simply equal to the sum of the effect of a definite number of genic differences. Such a view has persisted since the earlier days of Mendalism.

Multiple factor hypothesis	Polygenic system
No coordinated gene	Integrated genetic system
Identification of genes i.e., each if the loci has potentially identifiable for its site and specific effect	No identification of genes concerned under polygenic system of metric system.

Discuss the parallelism between qualitative and quantitative characters is they follow basic principles of inheritance? Or What is contribution of East in understanding inheritance of quantitative characters?

- Nilson-Ehle demonstrated that genes possessed the necessary qualities i.e., small and cumulative effect, to explain continuous variation. But the evidence was not conclusive because the character studied exhibited discontinuous variation.
- East in 1916 presented conclusive evidence that quantitative characters are governed by many genes with small and cumulative effect. He studied inheritance of corolla length in *Nicotiana longiflora*. The two parents used in the cross had contrasting corolla lengths viz., 40 and 93mm. while the F_1 was intermediate with 63mm long corolla. The F_2 showed a continuous variation for corolla length, which could not be grouped in to separate classes. To analyze his data, East used statistical parameters such as, frequency distribution, mean and coefficient of variation. The essential features of his study are

Qualitative (Seed texture)	Quantitative (corolla length in tobacco)
P_1 X P_2 Round Wrinkled (Here no variation even environment not) ↓ F_1 ↓ F_2 Variability released here ↓ F_3 Here some plants shows within variability, Round (Rr) shows within variability Homozygous (RR) do not show variability.	P_1 X P_2 Long short (No genetic variation within P1 and P2 variation only due to environment) ↓ F_1 (Intermediate)- Variation due to environment and genotypically no variation. ↓ F_2 Shows continuous distribution and first time variability released in F2. ↓ F_3 Here study mean and variability Here mean value in F_3 coincide with actual mean value in F_2. ↓ Extreme plants shows minimum within family variability but intermediate plants shows high family variability.

- As goes on increasing selfing, homozygosity increased, so variability decreased. Variability in F_4 never more than F_3, but equal or less than F_3. So, $F_2 > F_3 > F_4 > F_5$. So the perfect parallelism is there between qualitative and quantitative characters.

Discuss the role of environment in quantitative genetics:

Quantitative characters, by definition, are considerably affected by environment. The main result of this effect is that the relationship between genotype and phenotype is partially or completely hidden, i.e., the phenotype does not reveal the genotype.

For example, if the environmental effect is zero, the phenotype would be produced by the genotype, thus there will be a complete correspondence between the phenotype and genotype alone.

- To estimate the effect of environment on a character, large number of strains/genotypes is grown in a replicated trial and the data is subjected to analysis of variance as per the experimental design used.
- The phenotype described according to a mathematical model to facilitate statistical analysis and interpretation. The phenotypic mean i.e., x, of a given genotype from the trial may be expressed as,

- An estimate of magnitude of contribution of genotype to the phenotype may be obtained by using the mean squares (variances) from analysis of variances table.

Source of variation	Degrees of freedom	Expectations of mean squares
Genotypes	G-1	$\sigma^2 e + R\,\sigma^2 g$
Replications	R-1	$\sigma^2 e + G\,\sigma^2 r$
Error	(R-1)(G-1)	$\sigma^2 e$
Total	RG-1	

Genotypic variance ($\sigma^2 g$) = (MS due to genotypes - MS due to error)/ R

$$= [(\sigma^2 e + R\,\sigma^2 g) - \sigma^2 e]/R$$

$$= R\,\sigma^2 g/R$$

$$= \sigma^2 g$$

- Similarly, phenotypic variance ($\sigma^2 p$) can be estimated by totaling the genotypic and environmental variances. Thus,

$$\sigma^2 p = \sigma^2 g + \sigma^2 e$$

An alternative method for estimating the genotypic variance consists of growing the two parents (P_1 and P_2), the F_1 and the F_2 of a cross in a trial. Variances are estimated for P_1, P_2, F_1 and F_2 generations separately. In case of P_1, P_2 and F_1 the variances are due to the environment only since all the plants with in each of them have the same genotype. The variance for F_2, on the other hand, has both genetic and environmental components. The environmental variance (V_e) may be estimated as

$$V_e = (V_{P1} + V_{P2} + V_{F1})/3$$

Therefore, the variance of F_2 may be written as follows

$$V_{F2} = V_g + V_e$$

$$V_g = V_{F2} - [(V_{P1} + V_{P2} + V_{F1})/3]$$

What is meant by population mean? And write a brief note on use of statistical parameters for inheritance of quantitative characters?

Population mean is nothing but mean of all members of a population.

Statistical parameters used for inheritance of quantitative characters:

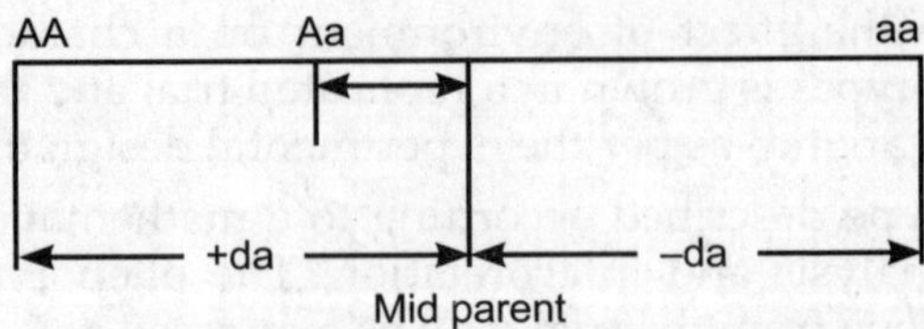

The deviation from mid parental value (d) is zero → additive gene action,

d=0 → additive gene action

d<0 → partial dominance

d=a → complete dominance

d>a → over dominance

Genotype	A_1A_1	A_1A_2	A_2A_2
Frequency of each genotype	p^2	2pq	q^2
Observed genotypic value	+a	d	-a

Population mean = $ap^2+2pqd-aq^2/(p^2+2pq+q^2)$

$= a(p^2-q^2) + 2pqd/1$

$= a(p-q)(p+q)+2pqd$

Mean of population of one of the loci (w. r. t one character) = a(p-q)+2pqd

Population mean considering all loci = Σa(p-q)+Σ2pqd

What are the components of genetic variance? And explain why additive component is fixable?

- Fisher in 1918 divided the genotypic variance in to three components that is additive, dominance and interaction or epistatic components. Later, Haymen and Mather partitioned the epistatic components into three types of interactions, viz, additive x additive, additive × dominance and dominance × dominance.
- Wright in 1935 divided the genotypic variance in to additive and non additive components.
- Mather in 1949 divided the genotypic variance in to heritable fixable and heritable non fixable components.

(1) Additive variance: It is the component arising from difference between the two homozygotes for a gene Eg. AA and aa.

Main features of Additive genetic variance are:

- Additive genes show lack of dominance i.e., intermediate expression
- It is associated with homozygosity and therefore it is expected to be maximum in self pollinated crops and minimum in cross pollinated crops.
- It is fixable and, therefore selection for traits governed by such variance is very effective. And this is the only variance which responds to selection
- It is required for estimation of heritability in narrow sense and response to selection is directly proportionate to narrow sense heritability.

- Breeding value of an individual is measured directly by the additive gene effects. The general combining ability (gca) effect of parent is a measure of additive gene effects.
- It gets depleted proportionate to the improvement made by selection.
- It is the chief cause of resemblance between relatives and progress by selection is directly proportional to the degree of resemblance between the parent and its progeny.
- Transgressive segregation is the result of additive gene action.

(2) Dominance variance: It is due to the deviation of heterozygote (Aa) phenotype from the average of phenotype values of the two homozygotes (AA and aa). Main features of dominance variance are

- It is associated with heterozygosity and, therefore, it is expected to be maximum in cross pollinated crops and minimum in self pollinated crops.
- It is not fixable and, therefore selection for traits controlled by such variance is not effective.
- It is the chief cause of heterosis or hybrid vigour.
- Specific combining ability variance is the measure of dominance variance in diallel and line × tester cross analysis.
- It gets depleted through selling or inbreeding

(3) Interaction or Epistatic variance: It results from an interaction between two or more genes. Main feature of epistatic variance are

- It includes both additive and non additive components. It is of three types, viz., additive x additive, additive × dominance and dominance × dominance.
    (*a*) Additive × Additive: It refers to interaction between two or more loci each exhibiting lack of dominance individually. It is fixable and it comes under additive gene action. So, selection is effective.
    (*b*) Additive × Dominance: It refers to interaction between two or more loci, one exhibiting lack of dominance and the other dominance individually and it is not fixable
    (*c*) Dominance × Dominance: It refers to interaction between two or more loci, each exhibiting lack of dominance and it is not fixable
- In generation mean analysis, the epistatic gene interactions are classified on the basis of sign of [h] and [l] in to two types only, viz., complementary and duplicate. When [h] and [l] have same sign, it is called complementary type and when [h] and [l] have opposite sign, it is termed as duplicate type of epistasis.

Additive component of variance is fixable:

By virtue of its very nature, the additive component is fixable i.e.,

- The intensity of trait expression controlled by genes acting additively does not change due to the segregation effect.
- If a trait controlled by additive genes, the selection of the carrier genotypes becomes reliable. That is homozygotes and heterozygotes distinguished confidently.
- Variation arising from additive geneaction narrows down rapidly as the selection proceeds.
- Elaboration of these three points hereunder.

(*a*) Additive effect helps in selection: If additive gene action be operative for a trait, the heterozygotes (Aa) in F_2 or advance segregating generations will be intermediate of the two parents (AA or aa), and hence could be easily identified and rejected. On the contrary, if dominance is operative, the distinction between a homozygote and heterozygote would not be pronounced, since Aa=AA. Thus, the pure breeding type (homozygotes) can easily be selected upon when only the additive type of gene action is consequential.

(*b*) Additive component is fixable: For genes acting additively, the F_1 between two homozygote equal to the parental mean, that is

$$\overline{F}_1 = \tfrac{1}{2}(\overline{P}_1 + \overline{P}_2)$$

Similarly, $F_2 = \tfrac{1}{4}(2\overline{F}_1 + \overline{P}_1 + \overline{P}_2)$

and $F_3 = \tfrac{1}{4}(2F_2 + \overline{P}_1 + \overline{P}_2)$

and $F_1 = \overline{F}_2 = \overline{F}_3$ and so on.

Then, such genes are fixable because the mean value of F_1 does not change from generation to generation. Thus additive gene effects are fixable. This is the reason why additive gene action is important in population improvement programme.

(*c*) Variation arising from additive effect narrows on selection: Selection tends to reduce the additive genetic variation, since it facilitates the identification of homozygotes. This implies that if in a segregating population, AA is selected at the cost of Aa and aa, the frequency of AA will increase enormously and that of aa will decrease. Thus, additive genetic variance that arises from the difference AA-aa, is exhausted rapidly in the absence of aa types. The consequence is freezing of genetic variability within the selected progenies.

Discuss how would you estimate the values of D and H components from variances of different generations?

The additive and dominance components of genetic variance may be estimated from a simple experiment. The experiment consists of two parents, P_1 and P_2, their F_1, and F_2, and two backcrosses, $B_1(F_1 \times P_1)$ and $B_2(F_1 \times P_2)$. The variances from these generations are used to obtain estimates of additive and dominance variances. These estimates are based on the assumption that epistasis is absent.

The theoretical basis for estimating the additive and dominance components of genetic variance may be explained by taking the example of single gene, 'A' with two alleles A and a. Three genotypes are possible for this gene which may be represents as follows.

The symbol 'da' denotes the additive effect of positive allele 'A', and '–da' is the additive effect of its negative allele 'a'. The positive and negative signs denote the direction of effects of the two alleles on the phenotype.

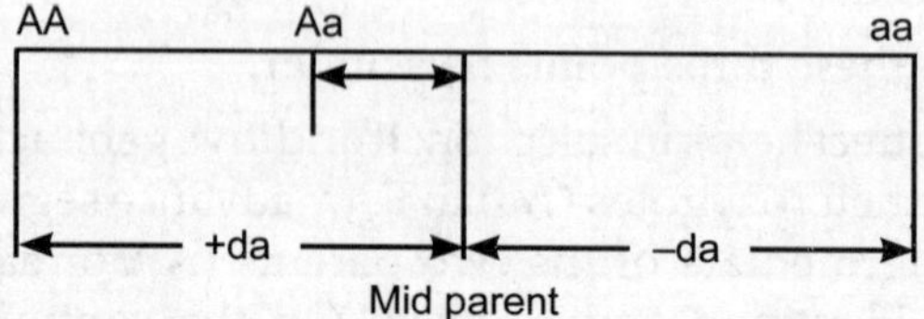

(Relationships between the two homozygote's (AA and aa) and heterozygote (Aa) at a single locus in respect of the phenotypic expression of a polygenic trait. Midparent is the average of the AA and aa phenotypic values.)

We may estimate the variances of the different segregating generations with respect to a single gene, Aa as follows.

Generation F_2:-	Genotype	AA	Aa	aa
	Phenoype	da	ha	-da
	Frequency	¼	½	¼

Thus, mean = ¼ da + ½ ha – ¼ da (since the population size is 1)

= 1/2 ha

Sum of squares = ¼ $(da)^2$ +1/2 $(ha)^2$ +1/4 $(-da)^2$

And sum of squares of deviations from F_2 mean

= [1/4 $(da)^2$ + ½ $(ha)^2$ + ¼ $(-da)^2$] – $(1/2\ ha)^2$

= [1/2 da^2 + ½ ha^2]-1/4 ha^2

=1/2 da^2 + ¼ ha^2

Since population size is equal to 1, the sum of squares is also the variance of F2. Since there will be more than one gene effecting the character, than the formula may be rewritten as

$$VF_2 \text{ (variance of } F_2) = 1/2\sum d^2 + ¼ \sum h^2$$

For simplicity, we may replace with D and with H

$$VF_2 = 1/2\ D + 1/4H + E$$

Similarly, variances may be estimated for the two backcross populations i.e., B_1 and B_2

Generation B_1	**Generation B_2**
Genotype AA Aa	Genotype AA Aa
Phenotype da ha	Phenotype ha –da
Frequency ½ ½	Frequency ½ ½

Thus, mean = ½ da +1/2 ha	Thus, mean = ½ ha +1/2 (-da) = ½ ha-1/2da
Sum of squares = $1/2 (da)^2 + ½ (ha)^2$	Sum of squares = $½ (ha)^2+1/2 (-da)^2$
Sum of squares of deviation from the B_1 mean $= [1/2(da)^2 +1/2(ha)^2]-[1/2da +1/2ha]^2$ $= (½ da^2+ 1/2ha^2)-(1/4 da^2+1/4 ha^2+1/2 da ha)$ $=1/4 da^2 +1/4 ha^2-1/2 da ha$	Sum of squares of deviation from the B_2 mean $= [½ (ha)^2+1/2 (-da)^2]-(1/2 ha- ½ da)^2$ $= [1/2 ha^2 +1/2 da^2]- [1/4 ha^2 +1/4 da^2-1/2da ha]$ $= 1/4 ha^2 +1/4 da^2 +1/2 da ha$
The above is also $VB_1=1/4 \sum d^2 +1/4 \sum h^2-1/2 \sum dh + E$ $=1/4 D +1/4 H- 1/2\sum dh +E$	The above also VB_2 $VB_2=1/4 \sum h^2+ ¼ \sum d^2 +1/2 \sum dh+ E$ $=1/4H +1/4D +1/2 \sum dh + E$

Where we add the variances for B_1 and B_2 generations, we get

$$VB_1 +VB_2 = (1/4 D+1/4H-1/2\sum dh +E) + (1/4D +1/4H +1/2 \sum dh +E)$$
$$= 1/2D +1/2H +2E$$

Thus the variances of the different segregating generations may be as follows

$$VF_2 = 1/2 D +1/4 H +E$$
$$VB_1+VB_2 = 1/2D +1/2 H+2E$$

From these equations, we can estimate 'D' as follows

$$2VF_2 -(VB_1+VB_2) = 1/2 D+2E -(1/2D +1/2H+2E)$$
$$2VF_2- (VB_1 +VB_2) = 1/2D$$

An estimate of the environmental variance can be obtained by averaging the variances of non-segregating populations

$$V_E = (VP_1+ VP_2+VF_1)/3$$

We can estimate H by subtracting the value of ½ D and E from VF_2

$$VF_2- ½ D-E = (1/2D+1/4H+E)-1/2D-E = 1/4H$$

Importance of plant breeding:

- If additive genetic variances is high, reliance should be placed on mass selection in self pollinated species and synthetic breeding in cross pollination species
- If the dominance variance is predominant development of hybrids in both self and cross pollinated crops
- If epistasis is high selection between families and lines is effective
- If all genetic components are of equal magnitude, either composite or population improvement programme should be taken up for the development of superior lines with several desirable genes.

What is meant by heritability? Explain the methods of estimation of heritability? And also discuss about the limits of use and generalities of heritability?

Heritability:

It refers to the measure of heritable portion of observed variability or measure of dependence of breeding value on phenotypic value or regression of breeding value on phenotypic value i.e. it is the ratio of genetic variance to total variance.

Heritability can be estimated by three different models

1. From analysis of variance table of a trials consisting of a large number of genotypes i.e., from simple trials

   $$H = V_g/V_p = V_g/(V_g+V_e)$$

   Where, V_g, V_p and V_e are the genotypic, phenotypic and environmental components of variance, respectively.
2. Estimation of V_g and V_e from the variances of F_2, P_1, P_2 and F_1 generations of cross i.e., from generation means analysis

   $$H = (VF_2 - VF_1)/VF_2 \times 100$$
3. Parent- offspring regression upon doubling provides an estimate of heritability. Thus H=2b, where b is the regression of progeny means on parental values.

Heritability estimated from the above three methods is known as broad sense heritability. Broad sense heritability estimates are valid for homozygous lines or populations. For segregating generations, it consists of additive and non additive component of variance. So the ratio of additive component of variance to the total phenotypic variance is more appropriate estimation of heritability. This is known as narrow sense heritability.

$$H_{ns} = 1/2D/V_{F2}$$

Where, D and V_{F2} are additive and phenotypic variances, respectively. Only 1/2D is used for the estimation of heritability since the phenotypic variance of F_2 has only 1/2D as its component.

Narrow sense heritability was estimated by following methods

1. Diallel analysis:

   Heritability = $[(1/2D+1/2H_1-1/2H_2-1/2F)/(1/2D+1/2H_1-1/4H_2-1/2F+E)] \times 100$

   Verhalen and Murray (1969) poposed the following formula for calculation of heritability in narrow sense from F_2 generation of a diallel cross

   Heritability = $[(1/4D)/(1/4D+1/16H_1-1/8F+E)] \times 100$
2. Generation mean analysis:

   Heritability as per Warner (1952) = $[(1/2D)/(VF_2)] \times 100$

   As per Mather (1949) = $[D/D+H+E] \times 100$

Where, D,H and E are additive, non additive and error variances respectively.

Limits of use and generalities of heritability: General considerations concerning estimates of heritability are

(*a*) h^2 is a property not only of a trait but also of the population. Since the value of h^2 is dependent on the magnitude of all the components of variance i.e., V_A, V_D and V_I.

(*b*) h^2 is greatly influenced by the development of population i.e. the selection for a trait, hence by the gene frequencies concerned. In particular, small populations maintained under domestication are expected to show low h^2 than the large population. Primitive land races, unselected, generally show high V_A. Hence high h^2. Advanced breeding lines which have undergone considerable selection show low V_A and hence low h^2.

(*c*) The h^2 estimate cannot be generalised to a crop since it is highly specific, and valid for the material involved in the experiment and for the environment of experimentation.

(*d*) Environmental variance is dependent upon the conditions of culture and management. The more variable conditions tend to lower down the h^2, while more uniform conditions increase it.

(*e*) The h^2 values enhance with generations. For instance, h^2 for yield may be relatively high when measured in early segregating generations of a population created by inter specific hybridization.

(*f*) Characters associated with reproductive fitness, such as yield, manifest low h_2, while those that are least important show high heritabilities.

(*g*) Heritability in a broad sense indicates only whether there is sufficient genetic variation present in population or not, which implies then whether a population will respond to selection pressure or in other words, whether a selection can be operative. But the success of selection in respect of actual gain judged by selection response (R) is determined by narrow sense heritability only.

$$R/S = h_2$$

Classification of heritability values (Johnson *et al.*, 1955)

Less than 30 per cent: Low

30-60 per cent: Moderate

More than 60 per cent: High

What is the formula for narrow sense heritability? Derive it?

$$\text{Narrow sense heritability } (H_{ns}) = V_A/V_P$$

Heritability is the measure of heritable portion of observed variability is additive effects or measure of dependence of breeding value on phenotypic value (b_{AP}) or regression of breeding value on phenotypic value.

$$h^2 = b_{AP} = \text{Cov}_{AP}/\sigma 2_P$$

$$= \text{Cov}_{\text{A with A}} + \text{Cov}_{\text{A with D}} + \text{Cov}_{\text{A with I}} / \sigma^2_P$$

Here, $\text{Cov}_{\text{A with D}}$ and $\text{Cov}_{\text{A with I}}$ will become zero

$$= \text{Cov}_{\text{A with A}} / \sigma^2_P \qquad \text{Since, } \text{Cov}_{xx} = V_x$$

$$= V_A/\sigma^2_P = h^2$$

When can broad sense heritability be used?

Broad sense heritability $b_{(GP)} = h^2_{(bs)} = V_G/V_P$

Broad sense heritability used in case of asexually propagated crops where entire genotypic value transmitted to the progeny. So in asexually propagated crops observed V_G transmitted to the progeny.

But in sexually propagated crops we have to look for narrow sense heritability.= $(V_A/V_P$. In sexually propagated crops only V_A is transmitted to the progeny.

The broad sense heritability is greater than narrow sense heritability. Numerator is always a part of denominator

$$h^2_{(bs)} > h^2_{(ns)}$$

The magnitude of heritability values, variance components are given for two situations. Give the implication of the following in plant breeding?

Heritability		V_P
Broad sense	Narrow sense	
98	18	High
94	90	Low

- In first case, variability is high, and broad sense heritability is 98 and narrow sense heritability is 18. That means V_D is predominant. So observed genotypic values not transmitted to the progeny. But in case of asexually propagated crops, the selection is effective because entire genotypic value transmitted to the progeny.
- In second case, both broad sense and narrow sense heritabilities are high, but variability is low. So for breeder to selection, first variability is required in the population. If variability is high than look for broad sense heritability and than for narrow sense heritability. So here selection is not effective.

Write short notes on coheritability?

Coheritability: Coheritability refers to the ratio of genetic covariance to the phenotypic covariance. Thus, it deals with the simultaneous inheritance of two characters.

Features of coheritability:

- It measures inheritance of two characters simultaneously.
- Estimation of Coheritability is based on genotypic and phenotypic covariances of the characters under consideration.
- Estimates of Coheritability are useful in the simultaneous improvement of two characters.
- Coheritability is estimated with the help of following formula.

Coheritability = GCOV x_1 x_2 / (PCOV x_1x_2) × 100

Where, GCOV and PCOV are the genotypic and phenotypic covariances of characters x_1 and x_2 respectively.

- Coheritability takes both genotypic as well as phenotypic covariances into account and helps in understanding changes taking place in pairs of polygenic characters.
- High values of Coheritability estimate suggest that increase in one polygenic trait will lead to simultaneous increase in another co heritable character.
- Thus Coheritability may form a more meaningful index for achieving the breeding objectives (Biswas and Sasmal, 1989).

What is meant by breeding value? What are the types of breeding value? Derive the breeding value of A_1A_1, A_1A_2 and A_2A_2 genotypes? Explain the role of breeding value in crop improvement programme?

Breeding value: (What portion of gene transmitted to progeny):

Sum of breeding potentiality of alleles containing in a genotype or Sum of average effects of alleles containing in a genotype or Portion of genotypic value transmitted to the progeny i.e. Additive gene action or Value of an individual as a parent, parents transfer a random sample of genes to their offspring.

Types of Breeding Value:

1. Theoretical breeding value 2. Practical breeding value

Theoretical breeding value: It is defined as the sum of average effects of genes over all loci. For single gene with two alleles, the breeding values of the genotypes are as follows.

- Breeding potentiality or average effect of A_1A_1 depends on A_1 allele and A_1A_2 depends on A_1 and A_2 alleles, and A_2A_2 depends on A_2 allele.
- Breeding potentiality of A1 and A2 alleles find out first because breeding potentiality of three genotypes depends on breeding potentiality of A_1 and A_2 alleles.

Breeding potentiality of A_1 allele:

Male / Female	A_1	A_2
A_1	A_1A_1	A_1A_2
Frequency	P	q
Genotypic value	A	d

Progeny derived = progeny row mean – population mean

= [pa+qd/1] – [a (p-q) + 2pqd]

= pa+qd-ap+aq-2pqd

= q [a+d – 2pd]

= q [a+d (1-2p)]

= q [a+d (1-p-p)] Simce, p + q = 1

= q [a+d (q-p)]

Breeding potentiality of A_1 allele = q [a+d (q-p)]

Breeding potentiality of A_2 allele:

Male / Female	A_1	A_2
A_2	A_1A_2	A_2A_2
Frequency	P	q
Genotypic value	D	-a

Deviation of progeny mean obtained from A_2 alleles subtracting from population mean.

Progeny derived = progeny row mean – population mean

= [pd-qa/1] – [a (p-q) + 2pqd]

= pd-qa-ap+aq-2pqd

= pd-ap-2pqd

= -p [a+d+2qd]

= -p [a+d (-1+2q)]

= -p[a+d (-1+q+q)]

= -p [a+d (q-(1-q))]

= -p [a+d (q-p)]

Breeding potentiality of A_2 allele = -p [a+d (q-p)]

So,

1. Breeding potentiality of A_1A_1 genotype

   = 2 × Breeding potentiality of A_1 allele

   = 2 × q[a+d (q-p)] [Assume, [a+d (q-p)] = α]

   = $2q\alpha$

2. Breeding potentiality of A_2A_2 genotype

   = 2 × Breeding potentiality of A_2 allele

   = 2 × -p[a+d (q-p)]

   = $-2p\alpha$

3. Breeding potentiality of A_1A_2 genotype = Breeding potentiality of A_1 allele + Breeding potentiality of A_2 allele

$= q\ [a+d\ (q\text{-}p)] + \text{-}p\ [a+d\ (q\text{-}p)]$

$= \alpha\ (q\text{-}p)$

Problem: Consider the pygmy gene and its effect on body weight for which a = 4 gms and d = 2 gms, if the frequency of (pg) is 0.1. Calculate the average effect of + gene or A_1 allele and pg gene or A_2 and also calculate the average effect of substituting + for pg.

Average effect of A_1 allele $= q\ [a + d\ (q - p)]$

$= 0.1[4 + 2(0.1 - 0.9)]$

$= 0.1[4 + 2 \times \text{-}0.8]$

$= 0.1[4 + (\text{-}1.6)]$

$= 0.1(2.4)$

$= 0.1 \times 2.4$

$= 0.24$

Average effect of A_2 allele $= \text{-}p\ [a + d\ (q - p)]$

$= \text{-}0.9\ [4 + 2\ (0.1 - 0.9)]$

$= \text{-}0.9\ [4 + 2 X (\text{-}\ 0.8)]$

$= \text{-}0.9\ [4\ (\text{-}1.6)]$

$= \text{-}2.16$

Average effect of gene substitution $\alpha = a + d\ (q - p)$.

$= 4+2(o.1\text{-}0.9)$

$= 4+ (\text{-}1.6)$

$= 2.4$

If frequency of pg is 0.4, Calculate A1, A2 and α

Average Effect	**q = 0.1**	**q = 0.4**
Average Effect of + or A_1	0.24	1.44
Average Effect of pg or A_2	-2.16	-2.16
Average Effect α	2.40	3.60

Note: The average effect of gene substitution is greater, if the frequency of recessive allele is greater.

Practical breeding value: When the breeding value of an individual is measured by the mean value of its progeny it is referred to as practical breeding value.

- If an individual is mated with a random sample of individuals from a population, its breeding value for a given trait is twice the mean deviation of the progeny from the population mean. The mean deviation has to be doubled because the progeny receive half the gene from the parent and other half genes come from the random population.
- That is the difference between half sib mean and overall mean is called GCA effect. It is also called as half of the breeding value.

Role of breeding value in crop improvement programme:

- In crop improvement programmes, breeding values are used in the selection of suitable parents in the hybridization programmes for developing superior cultivars.
- Breeding values are also used in the choice of breeding procedure for genetic improvemt of various polygenic characters.
- Breeding values of parents are useful in predicting the performance of their offspring.
- Breeding values are also useful in making decisions for the selection of advanced generation breeding population.
- In tree breeding (forest breeding) breeding values are useful in selection of superior stocks of timber trees.
- In animal breeding, breeding values are used for selection of superior sire and dam both for beef and milk production.

What is meant by dominance deviation? Derive the dominance deviation of A_1A_1 and A_2A_2 genotypes?

Dominance deviation: Portion of genotype not passed on to progeny.

Dominance deviation of A_1A_1 genotype: Generally, genotypic values expressed as deviations from mid parental value. And if we want genotypic values are expressed as deviations from population mean also.

= a- population mean

$= a - [a (p-q) + 2pqd]$

$= a - a (p-q) - 2pqd$

$= a (1-p+q) - 2pqd$

$= 2qa - 2pqd$

$= 2q [\alpha - d (q-p)] - 2pqd$

$= 2q\alpha - 2qd (q-p) - 2pqd$

$= 2q\alpha - 2qd (q-p+p)$

$= 2q\alpha - 2q^2d$

In above equation, '$2q\alpha$' is breeding value and '$-2q^2d$' is dominance deviation.

Dominance deviation of A_2A_2 genotype:

= -a- population mean

$= -a - a (p-q) + 2pqd$

$= -a (1+p-q) - 2pqd$

$= -2p [\alpha - d (q-p)] - 2pqd$

$= -2p\alpha + 2qd (q-p) - 2pqd$

$= -2p\alpha + 2qd (q-p-q)$

$= -2p\alpha - 2p^2d$

In above equation, '$-2p\alpha$' is breeding value and '$-2p^2d$' is dominance deviation.

Genotype	Frequency	Genotypic value	Breeding value	Dominance deviation
A_1A_1	p^2	A	$2q\alpha$	$-2q^2d$
A_1A_2	2pq	D	$(q-p)\alpha$	2pqd
A_2A_2	q^2	-a	$-2p\alpha$	$-2p^2d$

- Variance of breeding value is additive variance (V_A). And it is equal to half of the variance due to additive effect.

$$V_A = 2pq\alpha^2$$
$$= 2pq\,[a+d\,(q-p)]^2$$
$$= 2 \times \tfrac{1}{2} \times \tfrac{1}{2}\,[a+0(0)]^2$$
$$= \tfrac{1}{2}\,a^2$$
$$V_A = \tfrac{1}{2} \text{ variance due to additive effect}$$

- Variance of dominance deviation is dominance variance (V_D). And it is equal to one fourth variance due to dominance effects i.e. $V_D = \frac{1}{4}\,d^2$

$$V_D = (2pqd)^2$$

- Variance of interaction deviation is interaction variance (V_E)

Note: Presence of additive variance does not mean presence of additive gene action. It also possess some no or partial or complete or over dominance gene action. But dominance variance related to variance due to dominance effects only.

- Presence of additive gene action indicates presence of additive variance. Converse need not be true.
- Presence of dominance gene action indicates presence of dominance variance. Converse is true.

To estimate the genotypic variance from entire phenotypic variance first we will estimate the environmental variance. But it is not easy to find out the components of genotypic variance. So to find out V_A and V_D, We have to develop offspring and parent relationships or full sib relationships to find out genetic components of genotypic variance.

Write in detail about parent progeny relationships to find out genetic components of genotypic variance?

Parent progeny relationships: This can be expressed in two ways. 1. Between offspring and either of the parents, abbreviated as OP. 2. Between offspring and mid parent (i.e. average of the two parents), abbreviated as $\overline{OP}$.

Resemblance at genotypic level: The covariance of relationships tends to express in terms of components of genetic variance as follows. (When inbreeding depression (F) = 0)

$Cov_{(HS)}$ = Covariance of half sib = $\frac{1}{4}\,V_A$

$Cov_{(FS)}$ = Covariance of full sib = $\frac{1}{2}\,V_A + 1/4V_D$

$Cov_{(OP)}$ = Covariance of offspring on parent = $\frac{1}{2}\,V_A$

$Cov_{\overline{(OP)}}$ = Covariance of offspring on mid parental value $=1/2\ V_A$

Note that it is only the $Cov_{(FS)}$ that has dominance variance which seems to contribute to the resemblance. i.e. covariance. Further, the resemblance between different sorts of relatives is expressed as:

$$Cov_{\overline{(OP)}} = 2\ Cov_{(HS)} \text{ and}$$
$$Cov_{(FS)} = 2\ Cov_{(HS)} + \tfrac{1}{4}\ VD$$

Therefore,

$$\tfrac{1}{4}\ V_D = Cov_{(FS)} - 2\ Cov_{(HS)}$$

However, in practice this equivalence is absent chiefly due to the presence of non allelic interactions. Hence,

$$Cov_{(OP)} > 2\ Cov_{(HS)} \text{ and}$$
$$Cov_{(FS)} > 2\ Cov_{(HS)} + \tfrac{1}{4}\ V_D$$

Resemblance at phenotypic level: Phenotypic resemblance between the two major classes of relatives' sibs (HS and FS) and parent-offspring relationship (OP and $\overline{OP}$), can be ascertained by 1. Intra-class correlation and 2. Regression of offspring on parents respectively

Give the meaning of intra class correlation? How is it used to determine heritability?

Intra class correlation (t): Covariance of FS or HS divided by the respective total phenotypic variance (V_P) is 't'

$$t_{(FS)} = Cov_{(FS)}/V_P = (\tfrac{1}{2}\ V_A + 1/4 V_D)/\ Vp$$

And $$t_{(HS)} = Cov_{(HS)}/V_P = (1/4\ V_A)/\ V_p = \tfrac{1}{4}\ h^2$$

$t_{(FS)}$ is greater than $t_{(HS)}$ because $\sigma^2_{b/w}$ FS is higher than $\sigma^2_{b/w\ HS}$ and also $\sigma^2_{with\ in\ HS}$ is higher than σ^2 with in F_S because individuals within full sib are more likely to each other.

In order to derive't' a number of FS families are developed and evaluated in a field trial. The ANOVA runs as below.

Where, f = number of families, and n = number of samples observed. Now, estimates are obtained as under:

$$\sigma^2_f = (M_1 - M_2)/n$$
$$\sigma^2_w = M_2$$

Then, $$\sigma^2_P \text{ or } V_P = \sigma^2_f + \sigma^2_w$$
$$t_{(FS)} = \sigma^2_f / \sigma^2_P$$

Note that, σ^2_f is the variance of family means about the population mean and σ^2_w is the variance of individual values about the family mean. Degree of resemblance (i.e. t) is expressed in terms of the σ^2_f as a proportion of the total variance (i.e. V_P). In exactly the same way, the t values for HS families are also obtained.

Regression of offspring on parent: The covariance of P and $\overline{OP}$ divided by the variance of parents (V_P) gives the estimate of 'b'. Thus

$$\text{Cov}_{(OP)} = b_{OP} = (\tfrac{1}{2} V_A)/V_P$$
$$b_{OP1} = (\tfrac{1}{2} V_A)/V_P$$
$$b_{OP2} = (\tfrac{1}{2} V_A)/V_P$$

In a random mating population, the variance of parents and progeny is the same. The correlation and regression between parents and progenies are also the same. However, in case of mid parental value ($\overline{OP}$), the regression of offspring on mid parental value is Cov (OP) divided by ½ the variance of parents that is

$$\text{Cov}_{(\overline{OP})} = b_{(\overline{OP})} = (\tfrac{1}{2} V_A)/(\tfrac{1}{2} V_P) = V_A/V_P = h^2$$

Therefore, $\text{Cov}_{(\overline{OP})}$ gives the direct or robust or correct estimation of heritability because $\text{Cov}_{(\overline{OP})} = h^2$ (It self is heritability) but for $\text{Cov}_{(\overline{OP})} = \tfrac{1}{2} h^2$, so for we have to multiply by two and for t(HS) multiply by four.

Interms of intra-class correlation i.e. t and offspring parent regression b. represents the ratio of covariances (i.e. V_A and V_P). Then h^2 can be shown as follows

Full sib correlation	$t_{(FS)}$	$(\tfrac{1}{2} V_A + 1/4 V_D)/V_P$	$\tfrac{1}{2} h^2$
Half sib correlation	$t_{(HS)}$	$(1/4\, V_A)/V_P$	$\tfrac{1}{4} h^2$
Offspring and one parent regression	b_{OP}	$(\tfrac{1}{2} V_A)/V_P$	$\tfrac{1}{2} h^2$
Offspring and mid parent regression	$b_{\overline{OP}}$	V_A/V_P	h^2

Write the factors influencing genetic variance?

Factors influencing genetic variance: Several forces tend to influence the nature and magnitude of genetic variance. These are

1. Environment: the genes interact not only with other genes in the genetic system but also with the conditions of environment to which the genotypes are subjected. Therefore, non heritable variation is the rule with all metric traits in plants.
   (*a*) **With-in family environment (E_1):** Each individual within a family (FS or HS) may encounter somewhat different kind of environmental conditions with respect to cultural practices or inputs. Therefore E_1 is not common to all the constituent genotypes belonging to a family.
   (*b*) **Between family environment or common environment (E_2):** On the contrary, individuals related by descent, particularly of female parent, also enjoy a common environment. Mostly provided by maternal influence in addition to E_1 as above.

Therefore, the first requirement of any genetic experiment is to provide 1. An accurate measure of non-heritable variation, with E_1 and E_2 components measured separately, and 2. A means of separating this from the heritable component of phenotypic variation.

Direct estimation of V_{E2} can be obtained from genetically homogenous families, such as inbred parents and F_1 hybrids and that of VE_1 by using appropriate field designs with adequate randomization.

2. Linkage: Linkage is of major consequence in the following ways.
   (*a*) Effects on means: The degree of linkage determines the contribution of non-allelic interactions to the means of segregating generations (F_2, F_3, BC_1 and BC_2 etc). in the absence of interaction, linkage has no effect on the generation means of non segregating families (P_1, P_2 and F_1).
   (*b*) Effects on additive variance: The effect of linkage on V_A depends on the repulsion or coupling phase of linkage. V_A would be biased upward (positive) for gene pairs entering the crosses in the coupling phase and downward for gene pairs in the repulsion phase.

   Robinson *et al.* (1960) and Moll et al (1964) further demonstrated that the preponderance of the repulsion phase increases V_A in advanced generations, while the coupling phase decrease it.
   (*c*) Effects on dominance variance and level of dominance: Linkage disequilibrium is always a source of upward bias in the estimate of V_D. Consequently, the level of dominance, V_D/V_A, would be overestimated due to the presence of linkage. The effects of repulsion phase linkage of favourable dominant genes tend to mimic the over dominance effect, i.e. pseudo-over dominance which is indistinguishable from true over dominance in the absence of crossing over.
   (*d*) Effects on covariance between relatives: $Cov_{(HS)}$ and $Cov_{(FS)}$ are increased due to linkage, the closer the linkage, the greater is the increase. $Cov_{(OP)}$ is not effected by linkage.
   (*e*) Effects on heterosis: Linkage leads to an overestimation of the contribution of dominance, and an underestimation of that of interaction, to the magnitude of heterosis. In general, linkage always downgrades the significance of interaction as a cause of heterosis.
3. Maternal effect: Significant maternal effect tends to inflate the variance due to females, and hence the variance components. The σ^2_g (V_A) will be biased upward by additive maternal effect (σ^2_{mat}) and the non-additive variance component (σ^2_s) by maternal interaction as follows.

$$\sigma^2_g = (\sigma^2_f + \sigma^2_m)/2 = \sigma^2_g + \sigma^2_{mat}$$

$$\sigma^2_s = \sigma^2_{mf} = \sigma^2_g + \sigma^2_r$$

   The maternal contribution makes the progenies resemble more closely with maternal than with the paternal lines, and thus in turn causes reciprocal differences.
4. Mating system and gene frequency: The additive genetic variance (V_A) is basically an allelic property, where as V_D and V_I relate to the genotypic properties. Change in allelic frequency is the consequence of change in the mating system and hence a shift in the relative significance of variance components in the population.

In a random mating equilibrium population, V_A is 2pq times the square of the additive effect of the locus where p is the frequency of A and q of a. In case of over dominance, p = q, then additive variance would be zero and all the intra-allelic variance would be attributed to V_D.

5. Past history of selection: The nature of genetic variation in a population for a metric trait is greatly influence by the previous history of selection. Broadly speaking, the absence of selection, stabilizing selection tends to maintain the genetic variation of a largely additive nature while strong directional selection reduces the total genetic variation. This will tend to become composed of non-allelic interaction and unidirectional dominance effects.
6. Multiple allelism: Multiple alleles, if present, tend to cause a positive bias in the estimate of dominance variance, but not in those of additive variance.

How for inference can be drawn about gene action based on magnitude of different components of genotypic variance?

$$V_P = V_G + V_E$$

$$V_G = V_A + V_D + V_I$$

- If 'V_D' is higher, observed phenotypic value is not transmitted to the progeny.
- If 'V_E' is higher, observed phenotypic values are not reliable. So here no selection.
- In sexually propagated crops: 'V_A' is the reliable proportion of observed value because 'V_A' is the portion of genotypic value transmitted to the progeny.
- But in asexually propagated crops 'V_G' itself is the reliable portion of observed phenotypic value.

What is meant by combining ability? What are the types of combining ability?

Combining ability:

Combining ability refers to a capacity or ability of a genotype to transmit superior performance to its crosses

The concept of combining ability as a measure of gene action was proposed by Sprague and Tatam

Types of combining ability: General Combining Ability and Specific Combining Ability

GCA: The average performance of a strain or genotype in series of hybrid combinations with other genotypes is termed as General Combining Ability. It is estimated from half sub families

$$Cov_{(HS)} = 1/4\ V_A$$
$$Var_{(gca)} = 1/4\ V_A$$

SCA: It refers to the deviation of a particular cross from the general combining ability. It is estimated from full sib families

$$Var_{(sca)} = Cov_{(FS)} - 2\ Cov_{(HS)}$$
$$= 1/2\ V_A + 1/4\ V_D - 2 \times 1/4\ V_A = 1/4\ V_D$$
$$Var_{(sca)} = 1/4\ V_D$$

Tests for GCA: 1. Polycross test 2. Top cross test 3. Diallel cross test

Give the principle involved in determining sca effect?

Expected performance of combination (X_{AB}) = $\overline{X}_{00}$ (overall mean) + gca 'A' + gca 'B'

Observed performance of combination ($\overline{X}_{AB}$) = $\overline{X}_{00}$ (overall mean) + gca 'A' + gca 'B' + sca effect

For example:

Observed yield = $\overline{X}_{00}$ (overall mean) +effect of 'A' + effect of 'B' + interaction effect

Therefore, Expected performance of combination – observed performance of combination = sca effect

Sca effect refers to cross and gca effect refers to genotype or parent.

Note: Gca effect reflects the breeding value. Total of gca effects is zero because all are deviations from population mean. Significance in the positive direction of gce effect is important.

Consider the population of two contrast crops, maize and wheat, give the distinction between out breeders and inbreeders, and also give the main features of F_2 population considering 'N' independent-loci each segregating for positive and negative effects of alleles?

S.No.	Inbreeders	Outbreeders
1	Mating is selfing	Random mating
2	Individuals are homozygote's and have advantages of homozygosim	heterozygote's and heterozygosity
3	Variability distributed between compound lines	Over entire population
4	Deleterious recessive genes are eliminated	Masked by dominant gene
5	Tolerant to inbreeding	Intolerant to inbreeding
6	Have close adaptation are less finable	Wide adaptation high flexibility
7	New gene combinations are not possible	New gene combinations are possible

In case of following situations which breeding methods would be appropriate and why?

	Breeding procedure	Possible out come
Additive gene action		
Self pollinated crop	Pure line selection Mass selection Progeny selection	Pure line variety Mass selected variety New variety
Cross pollinated crop	Hybridization and selection Recurrent selection for gca Synthetic breeding Composite breeding	Superior segregant or new variety Population improvement Synthetic variety Composite variety
Dominance gene action		
Self pollinated crop	Heterosis breeding	Hybrid variety
Cross pollinated crop	Heterosis breeding Recurrent selection for sca	Hybrid variety Population improvement
Both additive and dominant gene action	Reciprocal recurrent selection	Population improvement

Reasons:

1. Additive genetic variance is prerequisite for genetic gain under selection, because this is the only genetic variance which responds to selection. i.e. genetic improvement through selection is achieved at the expense of additive genetic variance.
2. Non-additive gene action prerequisite for launching a heterosis breeding programme.
3. If both additive and non additive are equal magnitude, population improvement programme should be taken up for the development of superior lines with desirable genes.

Heritability:

Broad sense heritability:

- If high, selection may not be useful because it includes both fixable and non fixable variances
- If low, selection is difficult due to environment

Narrow sense heritability:

- If high, selection for improvement because of additive gene action
- If low, heterosis breeding because of non additive gene action

Genetic advance:

- If high, selection for improvement of characters because of additive gene action
- If low, heterosis breeding because of non additive gene action

Both Heritability and Genetic advance:

Heritability	Genetic advance	Result
High	High	Selection effective due to additive gene action
High	Low	No selection due to non additive gene action and high heritability due to favourable environmental influence
Low	High	Selection may be effective due to additive gene action low heritability due to high environmental effect
Low	Low	Selection is ineffective because of high environmental effect.

High heritability estimates along with genetic advance are more helpful in predicting the genetic gain under selection than heritability estimates alone. However, it is not necessary that a character showing high heritability will also exhibit high genetic advance.

Enlist the various systems of mating. Explain in brief the genetic consequences of these systems.

System of Mating:

1. Random mating:
   - Each female gamete has equal chance to unite with every male gamete.
   - The variability, homozygosity, heterozygosity, population mean and genetic correlation between relatives remains constant.
   - Random mating with selection helps in development of synthesis and composites
2. Genetic assortative mating:-
   - Mating between genetically similar individuals
   - Variability among the lines increased and within lines decreased
   - Homozygosity:- Increased due to fixation of genes.
   - Hetrozygosity:- Decrease, Population mean – decreases, Genetic correlation increases due to prepotency
   - Useful for development of inbreds.
3. Genetic dissortative mating:-
   - Mating between genetically dissimilar individuals
   - Variability: increased

- Homozygosity: decreases
- Heterozygosity: increases population mean: increase genetic correlation: decreases
- Useful for intra and inter specific gene transfer.

4. Phenotypic assortative mating:-
   - Mating between Phenotypically similar individual (AA × Aa and aa × aa)
   - Variability: increased if both extreme types are maintained in population
   - Homozygosity and genetic correlation reaches unity in single generation only
   - Useful in development of extreme phenotypes.
5. Phenotypic dissortative mating:
   - Mating between phenotypically dissimilar individuals (AA × aa).
   - Variability, heterozygosity and population mean remains constant
   - Useful in development of more stable genotypes.

Define Hardy-Weinberg equilibrium law. And explain the factors affecting Hardy-Weinberg equilibrium in relevance to crop improvement?

Hardy Weinberg equilibrium:-

In a large random mating population gene and genotypic frequencies are remains constant without migration, mutation, selection and random genetic drift.

Factors affecting this equilibrium:-

Factors affecting Hardy Weinberg equilibrium are migration, mutation, selection and random genetic drift. Among these migration, mutation and selection are systemic forces i.e. both magnitude and direction of change understood. While random genetic drift is a dispersive force i.e. only change in magnitude will understood but not direction

1. Migration: Migration refers to the movement of individuals into a population from a different population.
   - Gene flow or Migration can change frequencies of alleles in populations
   - If allelic frequencies of migrated and original population is same than there is no problem. Problem arised only when there is difference in allelic frequencies.

   Change in the gene frequency after migration

   $$\Delta q = m\,(q_m - q_o)$$

   Δq = Change in gene frequency

   m = population size

q_m = gene frequency of migrated population

q_o = gene frequency of original population

Impact Δq is more when m is more or $qm - q_o$ is more.

2. **Mutation:** It refers to the sudden heritable change in the features of an organism
   - Gene mutations are ultimate sources of new alleles and thus of genetic variability.
   - Mutations may be non recurrent (happening in one direction) or recurrent (happening in the both directions)
   - Mutations lead to the alternation of gene frequencies in a population

Change in gene frequency (Δq) $= p_o u - q_o v$

u = rate of forward mutation

v= rate of backward mutation

A point of equilibrium reached i.e $\Delta q=0$

$q_e = u/u+v$ = rate of forward mutation /u+v

$p_e = v/u+v$ = rate of backward mutation / u+v

3. **Selection:** Selection is defined as differential rates of reproduction of different genotypes in a population
   - Force of selection is described in terms of relative reduction in the contribution of genotype to the gametic pool
   - This relative reduction in contribution of genotype to the gametic pool is called as selection coefficient(s) or selection differential
4. **Fitness:** Relative contribution of genotype to genetic pool this relative value is called as fitness or selective value or adaptive value (W)

$$W = R_i/R_h$$

R_i = rate of reproduction of concerned genotype

R_h = rate of reproduction of that genotype, which has the highest reproduction rate .

Fitness may be any value between 0 and 1

Selection Co-efficient (s) = 1-w

Impact of selection:

- complete dominance $=\Delta q = -sq^2p_o/1-sq^2$
- Additive gene action $\Delta q = -spq/1-2sq$
- Selection favouring heterozygote (Δq) $=qp\,(s_1p-s_2q)/1-s_1p^2 - s_2q^2$
- Selection favouring homozygote's (Δq) = $spq(2q-1)\,/1-2spq$
- Selection against recessive phenotype (Δq) = $spq^2/\,1-sq^2$
- Selection against dominant phenotype (Δq) = $-spq^2\,/1-sp(1+q\,)$

5. **Random drift:** Random fluctuations in gene frequencies of a population due to chance or sampling error are called random genetic drift or genetic drift or random drift.
   - Random drift is greater in smaller populations than in a large one.
   - In this process of drifting, frequency of allele may be 0, these points are called as points of fixation.
   - The rate at which genes become fixed in a population is estimated by the formula, 1/2N is called random fixation.
   - Variance of gene frequency after t generations = $\sigma q^2 - p_o q_o (1\text{-}(1\text{-}1/2N)^t)$.

Other factors like inbreeding, unequal survival and population sub division are factors.

Explain the Proof of Hardey-Weinberg equilibrium:

Consequence of random mating of the genotypes in natural population (single gene model)

Random mating	Frequency of mating	Frequencies of progenies from random mating		
		AA	Aa	Aa
(p^2) AA × AA (p^2)	p^4	p^4	-	-
(p^2) AA × Aa (2pq)	$2(2p^3q)$	$2p^3q$	$2p^3q$	-
(p^2) AA × aa (q^2)	$2(p^2q^2)$	-	$2p^2q^2$	-
(2pq) Aa × Aa (2pq)	$4\,p^2q^2$	p^2q^2	$2\,p^2q^2$	p^2q^2
(2pq) Aa × aa (q^2)	$2(2pq^3)$	-	$2pq^3$	$2pq^3$
(q^2) aa × aa (q^2)	q4	-	-	q^4

- So total frequency of AA genotype after one generation of random mating

$$= p^4 + 2p^3q + p^2q^2$$
$$= p^2\,[p^2+2pq+q^2]$$
$$= p^2$$

So frequency of AA remains constant after one generation of random mating i.e. it is p^2.

- Frequency of Aa genotype after one generation of random mating

$$= 2p^3q + 2p^2q^2 + 2p^2q^2 + 2pq^3$$
$$= 2pq\,[p^2+pq+pq+q^2]$$
$$= 2pq$$

- Frequency of aa genotype after one generation of random mating

$$= p^2q^2 + 2pq^3 + q^4$$
$$= q^2\,[p^2+2pq+q^2]$$
$$= q^2$$

Whether Hardy-Weinberg law applies to X-linked genes and multiple alleles? Substantiate your answer? Or Write the Application of Hardy-Weinberg law in Genetics and Plant breeding.

Hardy-Weinberg law is applies to X-linked genes and multiple alleles.

X-linked genes:

Eg. Colour blindness, it is easily expressed in males compared to females.

X^CY – colour blindness

Suppose the frequency of this genotype (q^2) = 0.0001

Than $q = (q)^{1/2} = (0.0001)^{1/2} = 0.01$

So the frequency of 'p'= $1-q = 0.99$

In case of females, $X^CX^C = p^2$, $X^CX^c = 2pq$ and $X^cX^c = q^2$

Multiple alleles:

Eg. ABO blood group. (Frequency of 'A' blood group is 'p', for 'B' blood group is 'q' and for 'O' blood group is 'r')

Than $(p+q+r)^2$ (trinomial expansion)

$= p^2 + q^2 + r^2 + 2pq + 2pr + 2qr$

Give the relationship between allelic frequencies and digenic gametic frequencies and genotypic frequencies in an equilibrium population?

Allelic frequency: At the level of allele, the two alleles from locus unite in random manner than only it reaches the gametic phase equilibrium.

A_1	A_2	B_1	B_2
p (0.6)	q (0.4)	u (0.4)	v (0.6)

Gametic frequency: At the level of gametes, the gametes from male and female side unite in perfect random manner than only it reaches the zygotic phase equilibrium.

	uB_1 (0.4)	**uB_2 (0.6)**
pA_1 (0.6)	pu (0.24)	pv (0.36)
qA_2 (0.4)	qu (0.16)	qv (0.24)

At gametic frequency level, product of frequency of coupling phase gametes – product of frequency of repulsion phase gametes = 0

$$puqv - puqv = 0 \text{ i.e. } |d| = 0$$

Genotypic frequency: Already gametic phase equilibrium attained population, after one generation of random mating gives rise to zygotic phase equilibrium population.

Note: For reaching of gametic phase equilibrium it takes infinite number of generations, but after one generation (i.e. α+1) requires reaching zygotic phase equilibrium.

	B_1B_1 (u^2) (0.16)	B_1B_2 (2uv) (0.48)	B_2B_2 (v^2) (0.36)
A_1A_1 (p^2) (0.36)	p^2u^2 (0.0576)	$2p^2uv$ (0.1728)	p^2v^2 (0.1296)
A_1A_2 (2pq) (0.48)	$2pqu^2$ (0.0768)	4pquv (0.2304)	$2pqv^2$ (0.1728)
A_2A_2 (q^2) (0.16)	u^2q^2 (0.0256)	$2uvq^2$ (0.1728)	v^2q^2 (0.0576)

At what allelic frequency thus the homozygous recessive genotype (aa) becomes twice as frequent as heterozygous genotype (Aa) in a Hardy-Weinberg population.

$$q^2 = 2\,(2pq)$$
$$q^2 = 4pq$$
$$q^2 = 4\,[(1-q)q]$$
$$q^2 = 4q\,(1-q)$$
$$q^2 = 4q-4q^2$$
$$0 = 4q-4q^2-q^2$$
$$0 = 4q-5q^2$$
$$0 = q\,(4-5q)$$
$$0 = 4-5q$$
$$5q = 4$$
$$q = 0.8 \text{ and } p = 0.2$$

In a population pool the alleles A, a are at initial frequencies p and q respectively. Proove that gene frequencies, zygotic frequencies donot change from generation to generation.

	A(p)	a(q)
A(p)	AA (p^2)	Aa (pq)
a(q)	Aa (pq)	aa (q^2)

According to Hardy-Weinberg law, genes and genotypes does not change from generation to generation.

Frequencies of gametes containing 'A' allele = 'p' i.e., $p^2+pq = p(p+q) = p$

Frequencies of gametes containing 'a' allele = 'q' i.e., $q^2 + pq = q(p+q) = q$

'MN' blood group systems in human beings is governed by a codominant alleles l^m, l^n a sample of 208 people were tested for the presents of M and N antigens and found to contain 119 group M, 76 group MN, 13 N gene. If the frequency of lm = 0.3. How many individuals in the sample of size 500 could be expected belong to the group MN?

$$l^ml^m = 119$$
$$l^ml^n = 76$$
$$l^nl^n = 13$$

(*i*) $$l^m = [l^ml^m + \tfrac{1}{2}\, l^ml^n]/208$$
$$= [119 + \tfrac{1}{2} \times 76]/208 = 0.754$$

(*ii*) $l^n = [l^n l^n + ½ \, l^m l^n]/208$

$= [13 + ½ \times 76]/208 = 0.245$

$MN = 2 \, l^m l^n$

$= 2(0.3)(0.7) = 0.42 = 42\%$

For 100 → 42%

500 → ? 0.42 × 500 = 210

Let the frequencies of a pair of autosomal alleles (Aa) be represented by p^m and q^m in males and by p^f and q^f in females respectively. Given $p^f = 0.6$ and $q^m = 0.2$

(*a*) Determine the equilibrium gene frequencies in both success or after one generation of random mating

(*b*) Given the genotypic frequencies expected in the second generation of random mating.

$$p^m = 0.8 \text{ and } q^m = 0.2,\ p^f = 0.4 \text{ and } q^f = 0.6$$

First generation

	$p^m = 0.8$	$q^m = 0.2$
$p^f = 0.4$	0.32	0.08
$q^f = 0.6$	0.48	0.12

$$p = [AA + ½ \, Aa]/N = [0.32 + ½ \times 0.56] = 0.6$$

$$q = [aa + ½ \, Aa]/N = [0.12 + ½ \times 0.56] = 0.4$$

Second generation

	$p^m = 0.6$	$q^m = 0.4$
$p^f = 0.6$	0.36	0.24
$q^f = 0.4$	0.24	0.16

$$p = [0.36 + ½ \times 0.48]/1 = 0.6$$

$$q = [0.16 + ½ \times 0.48]/1 = 0.4$$

Therefore gene and gene frequencies do not change generation after generation.

CHAPTER - 7

Biometrical Genetics

Write short notes on the following.

1. D^2 statistics
2. Metroglyph analysis
3. Path coefficient analysis
4. Top cross analysis

D^2 statistics: The concept of D^2 statistics was originally developed by P.C. Mahalanobis in 1928. Rao in 1952 suggested the application of this technique for the assessment of genetic diversity in plant breeding. Main features of D^2 analysis are

- It is numerical approach which is used for measuring genetic divergence, in the germplasm collections.
- Estimates of D^2 analysis are based on second order statistics. And therefore, have lesser precision than metroglyph analysis.
- Analysis is more difficult than metroglyph analysis
- Analysis is possible from replicated data only
- Genetic diversity is depicted by cluster diagram.

Analysis of D^2 statistics consists of the following main steps.

1. Selection of genotypes
2. Evaluation of material
3. Biometrical analysis
4. Construction of cluster diagram

In the selection of parents on the basis of D^2 values three important points should be taken in to consideration. These are

- Relative contribution of each character to the total divergence.
- Choice of clusters with maximum genetic distance, and
- Selection of one or two genotypes from such clusters.

Metroglyph analysis

Anderson (1957) proposed this technique to study the pattern of morphological variation in crop species. It is a semi graphic method for assessing the variability in larger number of genotypes. The pattern of variability is depicted by glyph on the graph.

The analysis is based on mean value and following steps are involved in analysis are as follows:

1. Selection of genotypes
2. Evaluation of material
3. Assessment of variability.
   (*a*) Plotting of glyph on the graph
   (*b*) Depiction of variation
   (*c*) Construction of index score
   (*d*) Analysis of variations.

Mean value of each trait classified into three groups

1. Low index score
2. Medium
3. High.

Limitations

Inclusions of large number of genotypes sometimes lead to overlapping glyphs on the graph.

Path coefficient analysis: The concept of path analysis was originally developed by Wright in 1921, but the technique was first used for plant selection by Dewey and Lu in 1959.

- Path coefficient analysis is simply a standardized partial regression coefficient which splits the correlation coefficient into the measures of direct and indirect effects.
- In other words, it measures the direct and indirect contribution of various independent characters on a dependent character.
- It measures the cause of association between two variables.
- Analysis is based on the assumptions of linearity and additivity.
- It also estimates residual effects
- It helps in determining yield contributing characters and thus useful in indirect selection.

Top cross analysis:

- The top cross method deals with a series of single crosses developed between a single tester variety preferably open-pollinated with wide genetic base and a number of inbred lines.
- Since the tester parent with the common genetic background is variety, top cross is also reffered to as inbred-variety cross.
- Thus, with n inbreds and a tester, only n×1 = n number of crosses i.e. only one cross per parent, are developed.
- It was first proposed by Jenkins and Brunsen (1932) as amethod of testing inbred lines of maize in cross-bred combinations. Later, Tysdal and Grandall (1948) renamed it as "top cross".

- The inbred-variety crosses are in fact half-sib (HS) families. Since Cov (HS) = ¼ additive variance + small proportion of additive interactions.

Anova:

S.V	DF	SS	Mss
Replications (r)	r-1	rSS	rMS
Entries (g)	g-1	gSS	gMS
Parents (p)	p-1	pSS	pMS
Top crosses(c)	c-1	cSS	cMS
p vs c	1	pcSS	pcMS
Error (e)	(g-1)(r-1)	eSS	eMS
Total	(gr-1)	tSS	

Discuss in detail about diallele analysis?

Diallel cross analysis: Daiallel cross refers to mating of selected parents in all possible combinations and evaluation of set of diallel crosses is known as diallel analysis.

Diallel mating design is used to evaluate several inbred lines in terms of combining ability variances and effects.

Types of diallel cross:

- The diallel mating design involves three types of materials, namely, direct crosses or one way crosses, reciprocals and parents.
- Depending upon number of crosses to be made, diallel cross is of two types (simmonds, 1979 and Fehr, 1987).

S. No	Full diallel	Half diallel
1	Total number of crosses is equal to p(p-1), where is p is number of parents	Half crosses are required i.e. p(p-1)/2
2	Used when reciprocal differences are significant	Used when reciprocal differences are non significant
3	Used when parents do not have male sterility	Can be used when male sterility or self incompatability is present
4	Maternal efffects can be estimated	Maternal efffects cannot be estimated
5	Each parent is used as male and female parent in the mating	Each parent is used as male or female parent in the mating

Approaches of diallel analysis: There two approaches of diallel analysis

1. Hayman's graphical approach
2. Griffing's numerical approach

Hayman's graphical approach: Main features of approach are

1. This is graphical apporach which involves Vr-Wr graph
2. Analysis is based on estimation of components of variation
3. It provides information about six components, viz., D, H_1, H_2, E, F and h_2 where

   D = additive genetic variance

   H_1 = dominance variance

   $H_2 = H_1 [1-(u-v)^2]$, where u and v are the proportions of positive and negative genes, respectively in the parents

   E = expected environmental component of variance

   F = mean of Fr over the array, where Fr is the covariance of additive and dominant effects in a single array

   h^2 = dominance effects, as algebraic sum over all the loci in heterozygous phase in all crosses

The six components help in the estimation of various genetic ratios

- Average degree of dominance is estimated as $(H/D)^{1/2}$
- The ratio of dominant and recessive genes in the parents is estimated as follows

  $$[(4DH_1)^{1/2}+F]/[4DH_1)^{1/2} - F]$$

  =1 - dominance and recessive are in equal proportion

  <1 - excess of recessive genes

  >1 - excess of dominant genes

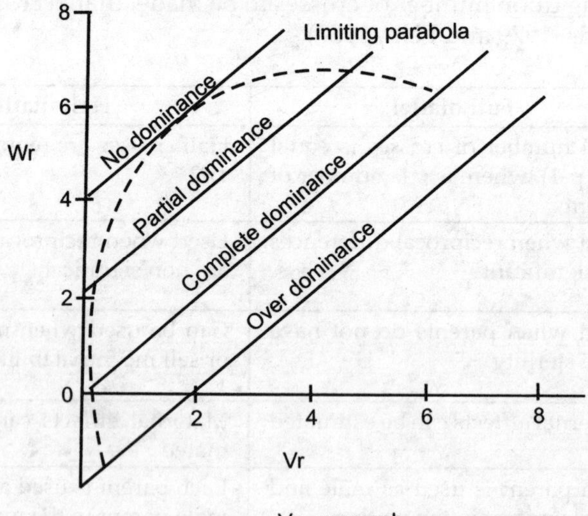

Vr - wr graph

- The number of gene groups which control the character and exhibit dominance is worked as h^2/H_2

- The proportion of genes with positive and negative effects in the parents is estimated as $H_2/4H_1$

Inferences from Vr-Wr graph:

The position of regression line on Vr-Wr graph provides information about the average of degree of dominance

- When the regression line passes through the origin, it indicates the complete dominance ($D=H_1$)
- When it passes above the origin cutting the Wr axis, it shows that there is a partial dominance($D>H_1$)
- When it passes above the origin cutting Wr axis and touching the limiting parabola, it suggests the absence of dominance
- When it passes the below the origin, cutting Vr axis it denotes the presence of over dominance

Griffing's numerical approach:

- This approach is based on the estimates of combining ability variances and effects
- Method 2 is the mostly commonly used method of combining ability analysis from diallel cross analysis.

Method	No of parents	Striaght crosses	Reciprocal crosses	Parents	Total
1	n	n(n-1)/2	n(n-1)/2	n	n^2
2	n	n(n-1)/2	-	n	n(n+1)/2
3	n	n(n-1)/2	n(n-1)/2	-	n(n-1)/2
4	n	n(n-1)/2	-	-	n(n-1)

Partition of variation and degrees of freedom in four methods of diallel analysis:

Source of variation	Method 1 F_1s + P + R	Method 2 F_1S + Parents	Method 3 F_1s + R	Method 4 F_1s only
Replications	r-1	r-1	r-1	r-1
Treatments	t-1	t-1	t-1	t-1
gca	P-1	P-1	P-1	P-1
sca	c	c	c-p	c-p
Replications	c	-	c	-
Error	(r-1)(t-1)	(r-1)(t-1)	(r-1)(t-1)	(r-1)(t-1)

Where r, t, p and c are number of replications, treatments, parents and all possible crosses, i.e. n(n-1)/2, respectively

Components of genetic variance:

- When the parents are inbred lines (F=1) the components of genetic variance are

  Cov HS = ½ VA + ¼ VAA + higher order of additive epistasis

  Cov FS = VA + VD + VAA + higher order of additive and dominant epistasis

  Cov FS – 2 Cov HS = (VA +VD) – 2(1/2 VA)
- When parents with F = 0

  Cov HS = ¼ VA + 1/6 VAA + higher order of additive epistasis

  Cov FS = ½ VA + ¼ VD + ¼ VAA + higher order of additive and dominant epistasis
- In Griffing's analysis, the additive genetic variance is equal to twice gca variance (2Vgca) and dominance variance is equal to Vsca

Explain in detail about partial diallele design.

Partial diallele design:

- Partial diallele design is a modified form of a diallele cross which utilises only a part of all possible crosses for analysis from a diallele mating (Proposed by Kemptherne in 1957 and was further elaborated by Kempthorne and Kurnow in 1961).
- In partial diallele, total number of crosses is equal to ns/2.
- Where n and s are number of parameters and sample crosses.
- S should be greater than or equal to n/2 and both n and s are neither be odd or even.
- For purpose of sampling, constant k is worked out as k = (n+1-s)/2
- Partial diallele provides information about gca and sca variance and gca effects and D and H components. But not sca effects.
- When F=1, $Cov_{(HS)} = 1/2\ V_A$ = Variance gca;

  $Cov_{(FS)} = V_A + V_D$
- Variance sca = Cov(FS)- 2 Cov(HS) = V_D
- Anova:

S.V	DF	SS	Mss	Expectations
Replications	r-1	rSS	rMS	
Crosses (c)	c1	cSS	cMS	
gca (n)	n-1	gSS	gMS	$\sigma^2_e + r\sigma^2_s + rs(n-2)/n-1\ \sigma^2_g$
sca (s)	c-n	sSS	sMS	$\sigma^2_e + r\sigma^2_s$
Error (e)	(c-1) (r-1)	eSS	eMS	σ^2_e
Total	(cr-1)	tSS		

Write short notes on Line X Tester analysis.

Line X Tester analysis:

- The Line x Tester design is basically extension of top cross analysis proposed by Kempthorne in 1957 and it is used when inbreed lines have self incompatibility and male sterility.
- L X T design is simple design which provides both FS and HS relatives simultaneously (Top cross poly cross provide only HS).That is why L X T is also referred as FS/HS analysis.
- L X T analysis involves the 'mf' crosses where m and f are number of males and females
- This technique can evaluate large number of germ plasm lines in terms of gca and sca effects and variances and D and H components.
- When F=1; $Cov_{(HS)f} = 1/2\, V_A$

  $V_D = Cov_{(FS)} - [Cov_{(HS)m} + (Cov_{(HS)f}]$
- When F=0; $Cov_{(HS)m}$ and $Cov_{(HS)f} = 1/4\, V_A$

  And $V_{gca} = Cov_{(HS)}$

  $V_{sca} = Cov_{FS} - 2Cov_{HS}$

  So additive genetic variance is equal to gca variance and dominance variance (V_D) is equal to sca variance.
- Anova:

S.V	D.F	SS	MSS	Expectation
Replications (r)	r-1	rSS	rMS	
Entries (e)	e-1	enSS	enMS	
Parents (p)	p-1	eSS	pMS	
Females (f)	f-1	fSS	fMS	
Males (m)	m-1	mSS	mMS	
Females vs Males	1	fmSS	fmMS	
Hybrids (c)	c-1	cSS	cMS	
Females in Hybrids (fh)	f-1	fhSS	fhMS	$\sigma^2_e + rCov_{(FS)} - 2Cov_{(HS)} + mrCov_{(HS)}$
Males in hybrids (mh)	m-1	mhSS	mhMS	$\sigma^2_e + rCov_{(FS)} - 2Cov_{(HS)} + frCov_{(HS)}$
fh x Mh (lxt)	(f - 1) (m-1)	fmhSS	fmhMS	$\sigma^2_e + rCov_{(FS)} - 2Cov_{(HS)}$
Parents Vs Hybrids	1	pcSS	pcMS	
error (e)	(e-1)(r-1)	eSS	eMS	σ^2_e
Total	(er-1)	tSS		

Characters of testers:-

- Broad genetic base
- Wider adaptability
- Low yielding potential
- Low performance for other traits.

What is meant by biparental mating designs? Explain the types of biparental mating designs?

Biparental mating designs: Crossing of randomly selected plants from a random mating population in F2 or subsequent generation of a cross between two pure lines in a definite fashion. (Comstock and Robinson, 1948)

North Carolina Design I (NCD-I): A polygamous design where one male is crossed with more than one female chosen randomly and each male with different group of females.

- This is also known as Nested design. And in this design, total number of crosses equal to 'ns', where 'n' is number of female plants per set and's' is number of sets.
- In this design the variance among single crosses is divided into two fractions viz.,
  1. Variance among males = ¼ V_A
  2. Variance due to females = ¼ V_A + ¼ V_D
- It is influenced by the presence of maternal effects.

ANOVA for North Carolina Design I

Source of Variation	D.f	SS	MSS	Expectation
Sets (Blocks)	s-1	bSS	bMS	
Replications s in sets (r)	s(r-1)	rSS	rMS	
males in sets (m)	s(m-1)	mSS	mMS	$\sigma^2_e + r\sigma^2_f + rn\sigma^2_m$
Females in males in sets (n)	sm(m-1)	fSS	fMS	$\sigma^2_e + r\sigma^2_f$
Remainder among plots	s(mn-1)(r-1)	eSS	eMS	σ^2_e
Total	(smnr-1)			

North Carolina Design II (NCD II): A polygamous mating design where one female is crossed with more than one male chosen randomly i.e., each male is crossed with each female of the same set.

- It is similar to line × tester analysis; it is also called Factorial design.
- Total number of crosses equal to 'mns'.
- Variance among single crosses divided into three fractions
  1. Variance due to males = ¼ V_A

2. Variance due to females = ¼ V_A
3. Variance due to males × females = ¼ V_D

- Influenced by presence of maternal effects.

ANOVA for North Carolina Design II

Source of Variation	D.f	SS	MSS	Expectation
Sets/Blocks (s)	(s-1)	bSS	bMS	
Replication in sets®	S(r-1)	rSS	rMS	
Males in sets (m)	S(m-1)	mSS	mMS	$\sigma^2_e + r\sigma^2_{mf} + rn\sigma^2_m$
Females in sets (n)	s(n-1)	fSS	fMS	$\sigma^2_e + r\sigma^2_{mf} + rm\sigma^2_f$
mxn in sets	S(m-1)(n-1)	fmSS	fmMS	$\sigma^2_e + r\sigma^2_{mf}$
Remainder among plots	S(mn-1)(r-1)	eSS	eMS	σ^2_e
Total	(Smnr-1)	tSS		

North Carolina Design III (NCD III): A F_2 backcross design where randomly selected genotypes in F_2 are used as males to backcross with each of the two parents involved in the cross.

- Total number of crosses equal to '2ns'.
- In this design variance among single crosses is separated in to two fractions
  1. Variance due to males = ¼ V_A
  2. Variance due to males × females = ¼ V_D
- Not influenced by presence of maternal effects.
- Estimation of linkages are dissipated by backcrossing in NCD III, therefore estimates of degree of dominance and σ^2A (hence, h^2_{NS}) are estimated more precisely in NCD III than NCD I or NCD II. The order of precision is NCD III> NCD II > NCDI.

ANOVA for North Carolina Design III

Source of variation	D.f	SS	MSS	Expectation
Sets(s)	s-1	sSS	sMS	
Replications in sets	r-1	rSS	rMS	
Inbred lines (females) in sets	s-1	fSS	fMS	
F_2 parents (males) in sets n (M)	s(m-1)	mSS	mSS	$\sigma^2_e + 2r\sigma^2_m$
F × M in sets (FM)	s(m-1)	fmSS	fmMS	$\sigma^2_e + r\sigma^2_{fm}$
Remainder among plots (e)	s(2m-1)(r-1)	eSS	eMS	σ^2_e
Total	2 sms-1	tSS		

Write short notes on the following?

1. **Generation mean analysis?**
2. **Grid design**
3. **Honey comb design**

Generation mean analysis: The concept of generation mean analysis was developed by Hayman (1958) and Jinks and Jones (1958) for the estimation of genetic components of variation.

The testing of epistasis is necessary before estimation of the components of genetic variation because it helps in deciding the method of analysis for the components of variation. In generation mean analysis, the test which determines presence or absence of epistasis and their type is known as scaling test. Mather (1949) has given four types of scaling tests viz., A, B, C and D.

$A = 2B_1 - P_1 - F_1$ $\qquad V_A = 4V_{B1} + V_{P1} + V_{F1}$

$B = 2B_2 - P_2 - F_1$ $\qquad V_B = 4V_{B2} + V_{P2} + V_{F1}$

$C = 4F_2 - 2F_1 - P_1 - P_2$ $\qquad V_C = 16V_{F2} + 4V_{F1} + V_{P1} + V_{P2}$

$D = 2F_2 - B_1 - B_2$ $\qquad V_D = 4V_{F2} - V_{B1} - V_{B2}$

When backcrosses are absent and F3 population is available, the D scale test is applied as follows:

$D = 4F_3 - 2F_2 - P_1 - P_2$ $\quad VD = 16V_{F3} + 4V_{F2} + V_{P1} + V_{P2}$

The type of epistasis is revealed by the significance of specific scale as follows:

- The significance of A and B scales indicates the presence of all the three types of non-allelic gene interactions, viz., additive × additive (i), additive × dominance (j) and dominance × dominance (l).
- The significance of C scale suggests dominance × dominance (l) type of non-allelic gene interactions.
- The significance of D scale reveals additive × additive (i) type of gene interaction, and significance of both C and D scales indicates additive × additive and dominance × dominance type of gene interactions.

Grid design:

Proposed by C.O. Gardener

The grid design refers to sub dividing the land into small blocks. The reason is that smaller blocks are likely to be more homogenous than large blocks plants are selected based on comparison among plants within each block only.

Advantages:

- Reduces the effects of soil heterogeneity.
- No precise arrangements are needed hence conventional planters can be used.
- Suitable for selection on the basis of plant traits with low heritability
- Easier or more effective to compare plants within the small group than in large group.
- Selection intensity can be varied by selecting more than one plant per block

Honey comb (hexagonal) design:-

Proposed by A. Fasoulas

Key Features

- Planting arrangement is such that each plant is equidistant from other in a hexagonal pattern.
- The spacing is determined to remove interplant competition
- Plants are selected only if they are superior to in hexagonal units.
- When a plant in the intermediate hexagon out yields its surroundings six plants it is selected by 14.3 selection intensity. If it is out yields the 18 plants within the second concentric hexagon it is selected by 5.3% selection intensity and so on
- Unit of selection at all stages in the breeding programme are single plants, not plots.

Disadvantages

- Conventional equipment cannot be used for planting.
- It is more complex to conduct.

Define gene action. Enlist the biometrical approaches which can be used to detect and determine the nature of gene action.

Gene action refers to the behaviour or mode of expression of genes in a genetic population. Gene action is measured in terms of components of genetic variance or combining ability variances and effects. Gene action can be studied with the help of various biometrical techniques. The choice of design is mainly depends on the type of genetic material to be evaluated for genetic variances.

S. No.	Designs	Type of material	Component of genetic variance obtained
1.	Diallel, Partial diallel and Line x Tester methods	Single crosses	Additive and dominance variances
2.	Triallel analysis	Three way crosses	Additive, dominance and epistatic variances
3.	Quadriallel analysis	Double crosses	Additive, dominance and epistatic variances
4.	Biparental mating designs NCDI and II designs	F_2 generation	Additive and dominance variances
	NCD III design	$P_1 P_2$ and F_2 generations	Additive and dominance variances
5.	Triple test cross analysis	$P_1 P_2$ and F_1 and F_2 generations	Additive and dominance presence and absence of epistasis
6.	Generation mean analysis	$P_1 P_2, B_1 B_2 F_1 F_2$ generations	Additive, dominance and epistatic variances

Use of combining ability as a measure of the type of gene action was suggested by Spragne and Tatum in 1942. The variation among the half sib families is an estimate of gca or additive gentic variance. The full- sib families provide an estimate of sca or non-additive gene action.

Mating design	Variances	Equivalent genetic component
Diallel (F=1)	V_g	½ V_A
	Vs	V_D
Partial diallel (F=1)	V_g	½ V_A
	V_s	V_D
Line × Tester (F=1)	$V_m = V_f$	½ V_A
	V_{mf}	V_D
NCD (F=0)		
NCD 1	V_m	¼ V_A
	V_f	¼ V_A + ¼ V_D
NCD 2	$V_m = V_f$	¼ V_A
	V_{mf}	¼ V_D
NCD 3	V_m	¼ V_A
	V_{mf}	½ V_D

Explain in detail about G× E interaction along with its significance in crop improvement?

Genotype × Environmental interaction: P=G+E+ (G×E) because, P≠G

Genotype:

Comprise all the crop varieties (cultivars), improved or unimproved, homogeneous or heterogeneous under domestication, and genetic stocks in breeder's nursery.

Environment (Comstock and Moll, 1963)

1. Micro-environment 2. Macro-environment

Environment may be predictable or unpredictable, rich or poor.

Therefore G×E interaction implies differential behavior of genotype under different environments.

Possible interaction $= L_{mn}/L_m L_n$

Where, m=genotypes, n= environments

When m=2, n=2, $L_4/L_2L_2 = 4\times3\times2/2\times2 = 6$ Significant interactions

Classification of genotype based on G×E interaction.

1. Specifically adapted genotypes:
   - To rich environment: B in fig .1 : Positive sensitivity
   - To poor environment: A in fig .1: Negative sensitivity

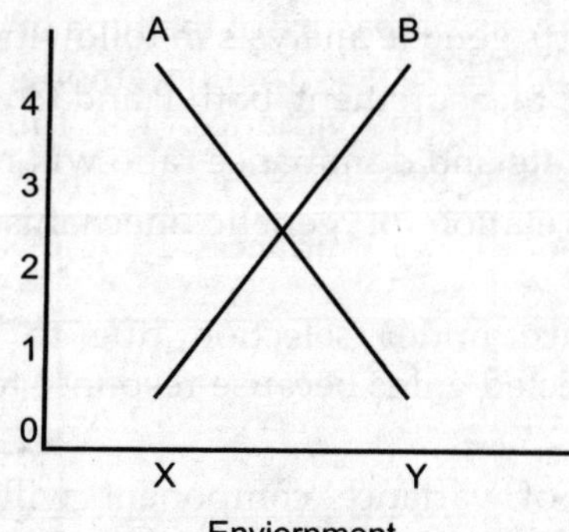

Where X = Poor environment and Y = Rich environment

2. Generally adapted genotypes:
   - Well adapted over all environments
   - Poorly adapted over all environments
   - Well adapted: Minimum G × E interaction, almost insensitive, consistently higher performance across environment.

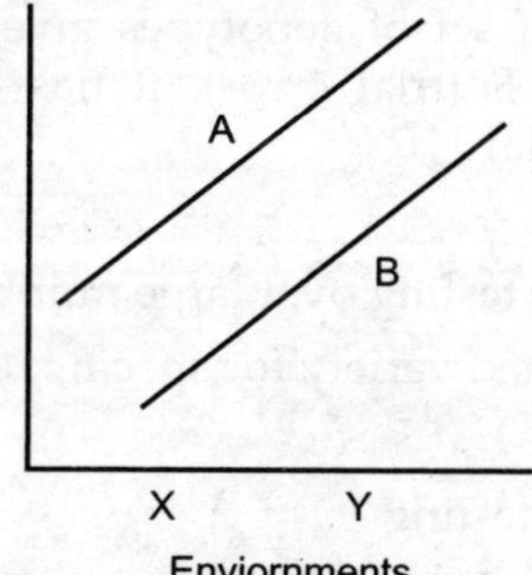

   - Poorly adapted: No change over environments, stable one but A>B Eg. genotypes, such genotype can't exploit favorable conditions so not suitable for intensive agriculture with higher inputs and better crop management.

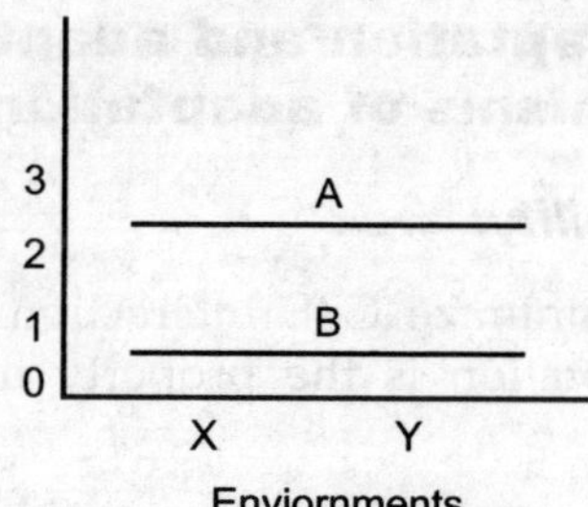

Significance of G × E interaction and adaptation: G × E interaction primarily interference with precise estimates of genetic parameters and stability of genotypic values.

G×E interaction distorts genetic analysis in following ways:

- In absence of G×E measurement, both P and G will be overestimated.
- Heritability estimate and dominance ratio will not be precise.
- No valid interpretation of genetic mechanism prediction of progeny performance.
- Limit the advance under selection due to the discrepancy between realized and expected gains because response to selection is function of h^2 (heritability).
- Biased estimate of variance component will lead to faulty breeding procedure.
- Loss of desirable genotype due to genetic slippage, as G×E interaction reduces correlation between genotype and phenotype.

G×E interaction limits the adaptive range:

- For specific area adaptation need large G×E interaction
- For wide area adaptation need small G×E interaction

Variance component and their importance:

Multilocation year trial of set of genotypes given different G×E interaction. (G×L), (G×Y), (G×L×Y) and E (trial error). It has clear implication in testing procedure of breeding material

1. Large (G×L) variance
   - Suggests to go for testing over large number of location.
   - Leads to most suited variety for specific region
2. Large (G×Y) variance
   - Over year season testing
   - For a final end product multilocation testing is best option than multi season testing
3. Large (GxLxY) variance

   Essential to test material over a number of different environments

What is meant by adaptation and adaptability? And explain its types and mechanisms of adaptation?

Adaptation and adaptability:

Adaptation: Capacity to minimize G×E interaction under different situations (Simmonds, 1962) or Adaptation is the property of genotypes which permit survival under selection.

Salient features:

- Capacity of genotypes to adjust in particular condition to reach a certain level of phenotype expression.

- Related to survival value
- Exploit micro environment and predictable component of environment.
- It is of two types- morphological and physiological.
- It is associated with environment where as fitness may be independent of environment.

Fitness: Fitness refers to ability of genotype to leave viable offspring, which can reproduce further and analogous to adaptation.

Adaptability: Capacity of genotype to overcome macro environmental factor in its favour in order to manifest a consistent performance overtime and space or ability of genotype, to react on average to change in environment.

Salient features:

- Phenotypic stability is main effect.
- Selection is the source of all types of adaptation hence adaptability is the capacity of genetic response to selection or genetic change in adaptation.
- Cultivar with narrow genetic base manifest close or specific adaptation.
- Cultivar with broad genetic base result in wide adaptation.
- Related to mean in statistical model.

Forms of adaptation: Based on type of genetic material (Simmonds, 1962)

Genotypic adaptation: It is associated with homogenous individual genotype, adapted to specific or range of environment.

Eg. Inbred (homozygous), Hybrid (heterozygous)

Specific genotypic adaptation: Adaptation of single line or clone to a limited environment, expected in few out breeders.

Eg. Vegetable crop suitable for specific areas and deep water rice with rapid internode early.

General genotype adaptation: Flexibility of adaptation of certain lines and clones to a variety of environments, also termed as genotypic adaptability or phenotypic flexibility, Adaptability is the feature of both heterozygosity and specific genotypes, hybrid perform better in stress and favourable condition

Eg. Semi dwarf wheat, rice etc.

Population adaptation: Heterogeneous population to a particular or diverse environment.

Specific population adaptation: It results due to a interaction between component rather than to adaptation of individual component, in limited environment, performance of mixture of genotype excel the means of component. Series of population at equilibrium suited to local needs are maintained.

Eg. Malwa cotton (Central asia), Mixture of *desi* cotton and American upland cotton.

General population adaptation: It is property of population, not of component also known as population adaptability.

Eg. Maize composite (Vikram, Jawahar and Kisan)

Factors affecting adaptability

1. Heterogeneity 2. Heterozygosity 3. Genetic polymorphism. 4. Mode of pollination

Mechanism of adaptation: Change in environment results in plastic response and altered physiology, generally governed by genetic mechanism.

Mechanism – Fairbaim (1970)

1. Invariant adaptation:
   - Rigidly programmed
   - Generally not modified significantly
   - It related to bio chemical unity
   - Eg. Genetic code.
2. Variant adaptation: Biochemical versatility or diversity, control and regulate synthesis of an array of characteristics protein

   Fairbaim further classified variant adaptation in to:
   - Epigenetic: which refers to orderly growth and development throughout lifecycle of organism, more rigidly programmed and
   - Expleitive: Genetic capacity to bring phenotypic change in response to an altered environment to achieve acclimatization. It depends on phenotypic flexibility. It can be achieved in two ways

     (*a*) Behavioural flexibility:- Reversible, non-heritable modification

     (*b*) Developmental flexibility:- Irreversible, heritable

   Phenotypic flexibility is complementary to genetic flexibility

Explain in brief about phenotypic stability?

Phenotypic stability: It is just opposite to phenotype flexibility and related to keeping morpho physiological processes in steady state also known as homeostasis and the term coined by (Canon 1932). It refers to totality of steady state maintained in an organism through the coordination of its physiological process to give near optimal phenotypic expression

Eg. Tolerance to inbreeding in tomato, facultative cross breeders, inter regulation of dominance and mutation rate, relationship between population six and mating system

It is of 2 types:

Developmental homeostasis:

- It is genetically controlled
- Responsible for preservation of genetic variability

- Also known as individual buffering
- Homogeneous population, narrow range of phenotype.

Genetic homeostasis: Also known as collective homeostasis or homeostasis of gene ratio. It works at population level and relates to gene frequency.

- Maintenance of gene frequency at equilibrium after disturbance by artificial selection.
- Heterogeneous population confers both individual and population buffering for stability.
- Population buffering is due to interaction among different consistent.
- Eg. Mixture of pureline give more stable yield performance than pureline.

Cannalisation and genetic assimilation:

Term coined by Waddington and its main features are:

- Related with genetic homeostasis.
- Selection of plastic genotypes.
- Genetic assimilation of acquired character.

Genetic assimilation is an adaptive mechanism of organism in all natural selection.

Components of Adaptive response:

Predictable response: Definite response as per environment richness or status in positive or negative direction

Unpredictable response: Non predictable

Phenotypic performance of genotypes over environment can be defined in terms of these three components

- Genotypic mean over environments
- Linear response
- Non-linear response
- Change in environment
- Determine stability

Explain the various approaches for measurement of G × E interaction?

Measurements of G x E interaction: (List various approaches for estimation of G×E interaction.

It can be done by four different statistical approaches

1. Partitioning of variance
2. Regression analysis
3. Multivariate techniques
4. Non-parametric statistics

Partioning of variance:

- ANOVA – multilocation /seasonal replicated trial of large population
  - Fixed effect model (model I)
  - Random effect model
- Stability factor – Lewis SF = $\frac{X_{HE}}{X_{LE}}$
  - If SF=1 means maximum phenotypic stability
- Ecovalance of genotypes developed by Wrickel (1962).

Regression analysis:

- Finally and Wilkinson (two stability parameter)
- Eberhart and Russell (three stability parameter)
- Fitting two – phase regression analysis

Multivariate approach:

Environments can also be grouped to decide the environment for screening and evaluation of genotype

(*a*) Clustering
(*b*) Principal component analysis
(*c*) Factor analysis
(*d*) Additive main effects and multiplicative interaction (AMMI)

Non parametric statistics:

Huehn proposed the concept of analyzing relative ranking of genotypes in different environments. A genotype is stable over environment if its ranks are the same over environment

The similarity of ranking in any two environments can be estimated by "spearman's rank correlation coefficient.

Define the term stability and explain briefly the various measures of stability of genotype used in plant breeding?

Stability refers to deviations from average response for which no straight forward mechanism can be envisioned or simply, suitability of variety for general cultivation over a wide range of environments.

In breeding terms, stability is thus closely related to the variance where as adaptability is related to mean in statistical models.

Main features:

- Based on multi location trial data.
- Used for both parental and segregating population.

- Assist in identification of suitable genotypes.
- Helps in predicting varietal responses under different environment.
- Based on first order statistics.

Measures of stability performance (Stability parameters):

1. Stability factor of Lewis (1954)- Employing only the mean is expressed as $SF = X_{H.E.}/X_{L.E.} = 1.0$

   $X_{H.E.}$ and $X_{L.E.}$ are the mean values under high and low yielding environments. The maximum phenotypic stability is reflected by SF=1. A deviation from unit value means low stability.

2. Ecovalance of genotypes developed by Wrickle (1962)- referred to the contribution of each genotype to the G×E sum of squares and is expressed as its percentage

   $$Wi = \left[\frac{\Sigma yij}{r} - \frac{y \cdot i}{n} - \frac{yj}{g} + \frac{y..}{gn}\right]$$

   Where 'Wi' is Ecovalance of genotype

   Y_{ij} is the performance of ith genotype in j^{th} environment

   Y_i and Y_j are totals of ith genotype over environment and total of jth environment over genotypes.

   Y… is the grand total of all genotypes over all environments.

   The variety with maximum value of Wi is considered to be more stable and vice versa.

3. Adaptability measurements of Finlay and Wilkinson model (1963):
   - It is two parameter model (μ and b), here b = regression, μ = mean performance over environment.
   - Performance in different environment over the respective environment mean and it gives idea about different level of stability when b=1 (Average stability), b>1 (below average stability), b<1 (above average stability) and b=0 (Absolute stability)
   - For ideal variety ,μ > x and b=1
   - Lack statistical perfection
   - Based on Yates and Cochran
   - Deviate from regression are not estimated

4. Eberhart and Russel model(1966):

   It is slight important over Finlay and Wilkinson model and it is 3 parameter model (μ, b and δ), here, δ is deviation from linear.
   - Ideal stable variety is one with μ > x, b=1 and no deviation from linearity (δ=0)
   - It does not provide independent estimation of mean performance and environmental index

- Total variance is divided into two components, i.e. Genotypes and environment plus interaction (E + G×E). The second component (E + G×E) is further sub divided into three components, viz., environment linear, genotype × environment (linear) and pooled deviations
- Degree of freedom for environment is 1.

5. Perkins and Jinks (1968) model :
   - Three parameter model
   - Does not provide the independent estimation of mean performance and environmental index.
   - Total variation is divided into three fractions, viz., genotypes, environments and G × E interaction. The G × E is sub divided into heterogeneity due to regression and S.S. due to remainder.
   - Degree of freedom for environment is 2
6. Freeman and Perkins model (1971):
   - Three parameter model
   - Provide the independent estimation of mean performance and environmental index.
   - Degrees of freedom for environment is 2
   - Total variation is divided into three fractions, viz., genotypes, environments and G × E interaction. The environmental S.S. is subdivided into two components, namely combined regression and residual 1. The interaction variance is also subdivided into two components namely, heterogeneity of regression and residual 2.

Eberhart and Russel model		**Perkin and Ginks**		**Freeman and Perkins model**	
S.V	**D.F**	**S.V**	**D.F**	**S.V**	**D.F**
Genotypes	g-1	Genotypes	g-1	Genotypes	g-1
Environmental(E)+ interaction (G×E)	g(e-1)	Environmental (E) + interaction (G × E)	e-1	Environments(E)	e-1
Environmental (linear)	1	Genotype x Environmental	(g-1) (e-1)	Combined regression	1
G×E (linear)	g-1	Heterozygosity Among regression	g-1	Residual (1)	e-2
Pooled deviatons	g(e-2)	Remainder	(g-1) (e-2)	Interaction	(g-1)(e-1)
Genotype 1	e-2	Error	ge(r-1)	Heterogeneity of regression	g-1
Genotype 2	e-2			Residual(2)	(g-1)(e-2)
Pooled error	ge(r-1)			Error (between replication)	ge(r-1)

Enlist the models for stability analysis. Explain about AAMI analysis?

Models for stability analysis:

A. Conventional models
   1. Stability factor model – Lewis (1954)
   2. Ecovalence model – Wrickle (1964)
   3. Stability variance model – Sukla (1972)
   4. Lin and Binns model (1988)

B. Regression coefficient models
   1. Finley and Wilkinson model (1963)
   2. Eberhart and Russell model (1966)
   3. Perkins and Jinks model (1968)
   4. Freeman and Perkins model (1971)
   5. Genotypic stability model (Tai, 1971)

C. Principal component analysis
   1. Perkins, 1972; Freeman and Dowker, 1975; Seif *et al.* 1979
   2. Additive main effects and multiplicative interaction effect – Gauch (1988), Freeman (1990) and Zobel (1990)
   3. Shifted multiplicative model (SHMM) – Cornelius et al. (1992)
   4. Redundancy analysis – VanEEuwijk, 1992
   5. Factor regression model – VanEEuwijk, 1992

D. Cluster analysis
   1. Grouping by cluster analysis – Westcot (1987)
   2. Webber and Wrickle (1990)

E. Pattern analysis
   1. Mungomery et al, 1974
   2. Delacy et al, 1990

F. Factor analysis
   1. Johnson and Wichern, 1982
   2. Calinski et al, 1987

Additive Main Effects and Multiplicative Interaction (AMMI) model:

The AMMI model is a combination of two earlier and simpler models. AMMI first calculates genotypes and environment additive (main) effects using analysis of variance (ANOVA) and then analyzes the residual from this model (namely the interaction) using principal components analysis (PCA).

AMMI is applicable and useful for data

(*i*) Structured in a two-way factorial, with at least three rows and three columns, replicated or not, (*ii*) Containing one kind of data, quantitative rather than categorical and

(*iii*) Fitted by the AMMI model reasonably well, as ordinarily happens when main effects and interaction are both significant.

The third requirement is fulfilled when all the three factors viz., genotype, environment and G×E interaction are significant in the pooled analysis of variance table.

The AMMI model mathematical equation is

$$Y_{ger} = \mu + \alpha_g + \beta_e + \Sigma_n \lambda_n \gamma_{gn} \delta_{en} + P_{ge} + \tau_{ger}$$

First, the additive part of the AAMI model uses ordinary ANOVA. The grand mean μ is estimated by $\overline{Y}$, the genotype deviation from the grand mean α_g by $\overline{Y}_g - \overline{Y}$, and the environment deviation β_e by $\overline{Y}_e - \overline{Y}$. ANOVA leaves a non-additive residual, the interaction.

Second, the multiplicative part of AMMI model uses PCA to decompose the interaction into PCA axes 1 to N, and a residual P_{ge} remains if not all axis are used. The multiplicative parameters are λ_n the singular value for PCA axis n, γ_{gn} the genotype eigenvector for axis n, and δ_{en} the environmental eigenvector. The eigenvectors are scaled as unit vectors, $\Sigma_g \gamma^2_g = \Sigma_e \delta^2_e = 1$. The eigenvector values γ_g and δ_e are unitless, whereas the singular value λ has the units of yield.

Anova: Environmental wise analysis and pooled analysis of variance are conducted as per the normal procedures. If genotypes, environments and G×E interaction are significant, the analysis may be proceeded further for AMMI analysis.

AMMI model						
S.V	Total	Treatment	Genotype	Environment	Interactions IPCA1, IPCA 2 and Residual	Blocks
D.F	ger-1	ge-1	g-1	e-1	(g-1)(e-1)	r-1

The IPCASS, MS and scores are computed as already explained in PCA analysis. If IPCAMS are significant and residual MS is non-significant, the steps may be continued for development of biplot.

The AMMI biplot is developed by placing both genotype and environment mean values on the X-axis and the respective IPCA axis eigenvectors on the Y-axis.

Interpretation of biplots:

Among points of the kind, displacements along the X-axis indicate differences in main (additive effects), whereas displacements along the ordinate (Y-axis) indicate differences in interaction effects. For points of different kinds, the AMMI model equation provides the expected values. The biplot has another important interpretation. The main effect for genotypes reflects breeding advances. Similarly, the main effect for environments reflects overall comparison of environments.

From the values of mean and IPCA 1, the genotypes are classified under four distinct classes.

Class I : Genotypes with high mean and positive IPCA 1

Class II : Genotypes with high mean and negative IPCA 1

Class III : Genotypes with low mean and negative IPCA 1

Class IV : Genotypes with low mean and positive IPCA 1

Nevertheless, sometimes there is in interest in more than just the first IPCA axis. Accordingly a different type of biplot, having IPCA 1on the X-axis and IPCA 2 on the ordinate may be developed. This biplot does not show the additive main effects; but by showing two IPCA axes, it is highly informative on interaction. When IPCA2 is sizable and significant, such a graph is a useful supplement to the usual biplot.

An experiment was conducted on ten (10) test candidates (G) of pea over eight (8) different environments (E). Give the d.f for {G,(E+(G+E)}, E (linear) , G×E (linear) and pooled deviation? G=10 and E=8

- G=g-1=10-1=9
- E+(G×E)=g(e–1=10(8-1)=70
- E(linear)=1
- G×E (linear)=g-1=10-1=9
- Pooled deviation=g(e-2)=10(8-2)=60

What is effect of linkage relationship of genes at different loci on possible gene recombination that obtains through hybridization? How different genetic phenomenon influence the planned hybridization.

Linkage among segregating loci:

- Linkage affects the frequency of neither the alleles of a gene (A:a) or (B:b) nor the three genetic classes in respect of each gene (say AA:Aa:aa) or (BB:Bb:bb) individually in segregating generations.
- Linkage only affects the joint frequencies with which various gene combinations occur without affecting the overall frequencies of alleles or genotypes at individual loci.
- It does not change the proportion of homozygosity or heterozygosity at individual loci but influences frequency of different combinations of homozygotes and heterozygotes at two or more loci.
- Linkage results in higher frequency of parental i.e. non-recombinant types than that of recombinants and the magnitude of this effect will depend upon the strength of linkage.

Genetic factors influencing planned hybridization are:

(*a*) Gene-character relationship

(*b*) Pleiotropy

(*c*) Penetrance and Expressivity.

CHAPTER - 8

Plant Breeding for Stress and Nutritional Quality

Write short notes on disease triangle.

Diseases Triangle: A plant becomes diseased only when it comes in contact with pathogen under environmental conditions that favour the disease.

- The interaction of three components of diseases has often been visualized as a triangle, generally referred to as the 'disease triangle'
- Each side of the triangle represents one of the three components.
- The length of each side is proportional to the sum total of the characteristics of each component that favour disease.
- The area of triangle would represent the amount of disease in a plant population.
- If any of the three components is zero, there can be no disease.

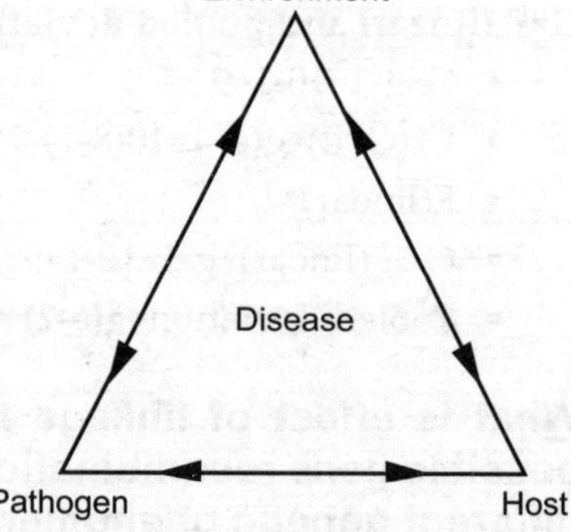

Explain in brief about the genetic basis of vertical and horizontal resistance.

Genetic basis of vertical resistance: When variety is more resistant to some races of a pathogen than it is to others, the resistance is called "vertical or perpendicular resistance".

- Vertical resistance is the true consequence of Flor's gene for gene hypothesis.
- It is commonly conditioned by monogenic or oligogenic differences.
- Influence of major genes is all the more pronounced on vertical resistance since the segregation of resistant and susceptible host is discontinuous, showing distinct phenotypic classes.
- Thus, simple inheritance 3:1 ratio is attendant upon vertical resistance.
- Eg. Yellow rust resistance in wheat, which is simply inherited recessive character observed in the early 1900's by Biffen (1907).

Genetic basis of Horizontal resistance: When the resistance is evenly spread against all races of the pathogen, it is called horizontal or lateral resistance.

- Terms polygenic resistance, minor gene resistance and quantitative resistance are all indicative of the fact that the HR is controlled by many genes with minor phenotypic consequences, showing a continuous pattern of segregation in segregating generations.
- Value of these genes is not basically dependent upon the pressure of the pathogenic race, and hence they do not exert selection pressure on the pathogen.
- Each resistance gene in the population contributes to the average degree of resistance of the population as a whole, and its contribution depends on the frequency of that gene in the population time it own effect.
- Since segregation of resistance and susceptible genotypes is not defined by class, and that it is continuous in nature, no distinct ratios can be standardised. Suitable metric scales, usually standardised disease score indicators, are employed to measure horizontal resistance.

Explain in brief about simple interest and compound interest diseases.

Robinson (1971) classified diseases according to whether the total inoculums at the beginning of the season is all that causes the disease (primary infection), or there is a multiplication of it on the host, i.e. whether the spores can reinfect the same or the other plants during the growing season (secondary infection).

- The diseases that are confined to the primary infection are called "simple interest" diseases (Eg. Loose smut of wheat) and
- Those spreading due to secondary infection are the "compound interest" diseases (Eg. Wheat rust and potato leaf blight).
- According to Robinson, vertical resistance is associated with simple interest diseases and horizontal resistance with compound interest diseases.
- Vander plank propounded that during the early stages of an epidemic, the disease (cereal rust), increases on the fashion of compound interest. He gave compound interest equation as

$$X_t = X_0 e^{rt}$$

Where X_t is the amount of disease at the time, t i.e. the amount after it has accumulated for t days, X_0 is initial amount of disease, 'r' is the infection rate and 'e' is 2.718, with the assumption that "interest" (i.e. amount of disease) is added as soon as it is earned (occurred).

Write short notes on the following:

1. Vertifolia effect
2. Boom and bust cycle
3. Differential host

1. **Vertifolia effect:** Vertifolia effect refers 'to an epidemic development in a variety carrying vertical resistance genes and low level of horizontal resistance genes, heading to heavy economic losses.

- Vander plank introduced the term vertifolia effect derived from the name of a German potato variety having the late blight resistance genes R_3 and $R_{4.}$
- The variety became susceptible when the appropriate pathogen (P_3, P_4) evolved in the pathogen leading to complete failure of crop
- The failure is mainly due to (i) level of horizontal resistance in the varieties carrying oligo genes for resistance is usually low and (ii) the pathogen is able to evolve the virulent phenotype.
- Parlevilet (1981) reported vertifolia effect is not universal, as Barley variety 'cebada capa' carrying P_7 (resgene-V_R) also having high level of polygenic partial resistance to leaf rust.
- To avoid vertifolia effect, one should try to reject susceptible one in place of selecting highly resistance plant. Kumar *et.al* (1977) suggested combining HR and VR and first selecting for HR, crossing with VR and selecting further in segregating generation.
- Gene for both HR and VR can be combined by alternating the type of inoculum.

2. **Boom and Bust cycle:**

   When a new variety carrying VR (vertical resistance) is in development it is resistant to the prevalent pathotype of the pathogen in question, therefore it becomes popular rapidly and occupy a large are under the concerned crop this period of boom continuous till the pathotype specific to the VR gene employed in the variety evolves and /or increase in frequency to cause on epidemic in the fields occupied by the variety.

   Now susceptible variety loss popularity and area under it declines sharply thus the variety is effectively busted by the virulent pathotype. This has been called as boom and bust cycle by Priestley (1970).

3. **Differential host or host Tester**: The concept of differential host is based on the ability of a variety to differentiate between races on the basis of disease reaction. It results in variety resistant to one but susceptible to other.

   - In case of two classes of expression i.e., resistance and susceptibility 'n' differentials can be used to differentiate 2^n races.

		Races			
		R_1	**R_2**	**R_3**	**R_4**
Varieties	**V_1**	R	R	S	S
	V_2	R	S	R	S

So, with three differential varieties 2^3 unique pattern of disease reaction are possible.

Varieties	Races							
	R_1	R_2	R_3	R_4	R_5	R_6	R_7	R_8
V_1	R	R	R	S	S	S	S	R
V_2	R	R	S	R	S	S	R	S
V_3	R	S	R	R	S	R	S	S

The ideal set of differential host (varieties) is one in which each variety carries gene for resistance to only one race. The pattern of disease reaction becomes a marker of a particular race. For more type of reaction as in stem rust of wheat, number of races identifiable will be r^n. Here 'r' is number of reaction type identifiable and 'n' is the number of differential hosts with a single unique resistant gene

What is meant by Gene for Gene hypothesis? Write the genetic basis, fundamental significance along with deviants of hypothesis?

Gene for Gene relationship: Proposed by Flor in linseeed rust (Melanspora lini).

According to gene for gene hypothesis, 'for every resistance gene in the host, the pathogen has a gene for virulence'. Susceptible reaction would result only when the pathogen is able to match all the resistance genes present in the host with appropriate virulence gene.

In most of the pathogens, virulence is recessive to avirulence

		Host	
		R-	**rr**
Pathogen	**V-**	R	S
	vv	S	S

		Host	
		$R_1R_1r_2r_2$	$r_1r_1R_2R_2$
Pathogen	$v_1v_1V_2V_2$	S	R
	$V_1V_1v_2v_2$	R	S

Genetic basis: Flor's gene for gene hypothesis specifies single-gene difference for resistance, i.e. resistance is simply inherited. The hybrid behaviour can be studied using two lines A and B and two parasite strains 1 and 2.

Parasite		Host lines		F_1.	F_2 (A×B)			
Strains	Genetic structure	A (RR)	B (rr)	(AB) (Rr)	RR	Rr	Rr	Rr
1	VV	-	+	-	-	-	-	+
2	vv	+	+	+	+	+	+	+

Here, three conclusions are simple and clear:

1. Resistance (-) is dominant over susceptibility (+), as F_1 between A and B inoculated with strain 1 shows all the resistant progenies and no susceptible plants.
2. F_2 segregation for resistance and susceptible plants in case of strain 1 is typical 3:1 ratio, indicating a simple inheritance of resistance. In case of strain 2, all the F_2 plants are susceptible (ratio of resistance to susceptibility being 0:4 or 0: All).
3. Simply inherited difference between two host lines are specified by strain 1 only and not by strain 2. That is, the specific parasitic strain used determines the ability to observe the presence or absence of R gene in the host.

Fundamental significance: Fundamental significance lies in three new vistas in the genetics of resistance which are now being actively pursued.

1. Functional allelism and fine structural analysis of resistance genes indicated that R genes frequently show allelism and likage, whereas V genes seldom do so.
2. Biochemical relationship of primary gene products in the host and parasite to the expression of resistant phenotypes.
3. Population dynamics and evolution of host-parasite systems.

Deviants of Flor's hypothesis: Two objections to the hypothesis are

1. Flor's hypothesis is purely a hypothesis of gene identities, in that; it identifies the resistant gene in the host and matching gene for the pathogenesity in the pathogen. But it fails to define the quality of genes with respect to their degree of resistance and pathogenesity. Vander plank, proposed a 'second gene for gene hypothesis' which defines the genes identity and its quality both.
2. The R gene of the monogenic model might be acting only as a switch for a complex system and the co-called monogenic resistance may be, as Eenink says "only the tip of the genetical iceberg" or more than one gene may be involved in the resistance mechanism.
3. Two explanations i.e., Vander plank's second gene for gene hypothesis and Switch mechanism (gene regulation) for disease reaction based on their theory of gene regulation, Britten and Davidson (1969).

Write the short notes on the following:

1. Second gene for gene hypothesis
2. Breeding methods for disease resistance
3. Durable resistance

Second gene for gene hypothesis:

Vander plank's second gene for gene hypothesis of gene identity and quality in contrast to Flor's hypothesis which deals with only gene identity (locus of gene).

- Vander plank's hypothesis (1968) deals with the identity and quality of gene both.
- By quality of gene, means the amount of protection that a resistance gene (identified by its locus) can confer upon its carrier against a disease.
- The quality of resistance gene in the host determines the fitness of the matching virulence gene in the parasite to survive when this virulence is unnecessary.
- When it is necessary, determines the quality of matching resistance gene, as judged by the protection it can give to the host.
- The hypothesis is based on observations on qualitative difference among five resistance gene, two in potato against potato blight (R_1 and R_4) and three in wheat against stem rust (Sr5, Sr6 and Sr9d).

Breeding for disease resistance:

Describe the back cross method of breeding if the disease governed by recessive gene.

Breeding methods:

1. Selection: Eg. Cotton (*G. Hirstum*) variety MCU1 (Madras Combodia Uganda 1) was selected from variety Coimbatore 4 (CO 4) for resistance to back arm.
2. Introduction: In wheat Kalinga sona and Sonalica introduced from Cimmyt, Mexico and were rust resistant.
3. Mutation
4. Hybridisation
   - Pedigree method - E.g. Kalyanasona and Sonalika
   - Back cross method
5. Somaclonal variation
6. Genetic engineering (programmed cell death)

Durable resistance:

1. The resistance that lasts for a longer period is referred to as durable resistance.
2. The procedure recommended for breeding for durable resistance is similar to horizontal resistance except that it starts with cultivars which have exhibited durable resistance.
3. Ex:- Cotton to bacteria blight and leaf curl virus and sorghum to Milo diseases
4. The durability of resistance mainly depends

- Formation of new races of pathogen
- Genetics of resistance
- Morphological features of host plant
- Biochemical substances associated with disease resistance

Explain in detail about the mechanisms of insect resistance.

Insect resistance is grouped into four categories: non preference, antibiosis, tolerance and avoidance. The first three types of resistance were recognised by Painter (1951), while the fourth, avoidance, was included subsequently.

Nonpreference: Host varieties exhibiting this type of resistance are unattractive or unsuitable for colonisation, oviposition or both by an insect pest.

- This type of resistance is also termed as nonacceptance or antixenosis.
- Nonacceptance appears to be more appropriate, since in this type of resistance, insects will not accept a resistant host variety even if there is no alternative source of food.

Antibiosis: It refers to an adverse effect of feeding on a resistant host plant on the development and or reproduction of the insect pest. In severe cases, it may even lead to the death of insect pest.

Tolerance: An insect tolerant variety is attacked by the insect pest to the same degree as a susceptible variety. But at the same level of infestation, a tolerant variety produces a larger yield than a susceptible variety.

Avoidance: Insect avoidance result from the host plants being at a much less susceptible developmental stage when the pest population is at its peak.

Eg. Early maturing cotton varieties escape pink bollworm infestation, which occurs late in the season.

Host crop	Insect pest	Cause of resistance	Mechanism of resistance
Wheat	Hessian fly Stem saw fly Green bugs	H_1 to H_8 genes Solid stem Nonpreference High content of benzyl alcohol Antibiosis	Antibiosis and antibiosis
Barley	Cereal leaf beetle	Hairiness and waxiness of leaves	Tolerance and antibiosis
Rice	Rice stem borer	High silica content Lignified stem More schlerenchymatous tissue in stem	Antibiosis Nonpreference Tolerance
	Brown plant hopper	Low asparagine content of leaves Red pericarp and purple stigma	Nonpreference Nonpreference

Maize	European corn borer	DIMBOA content in leaves	Antibiosis
	Maize stem borer	High aspartic acid, low nitrogen and sugar contents	Antibiosis
	Corn ear worm	Toughness of husk	Nonpreference
	Corn leaf aphid	High DIMBOA in leaves	Nonpreference and antibiosis
Cotton	Boll worms	High gossypol content	Antibiosis
		Smooth leaves	Nonpreference
		Nectariless	Nonpreference
		Earliness	Escape
	Boll weevil	Frego bract	Nonpreference
		Red plant body	Nonpreference
		Earliness	Escape
	Jassids	Hairiness of leaves	Nonpreference
Sugar beet	Aphids	Low free sugar	Nonpreference and antibiosis
Alfalfa	Spotted alfalfa aphid and pea aphid	High saponin content	Antibiosis
Brassica	Cabbage aphid	Low sinigrin content	Nonpreference and antibiosis
Potato	Aphid	Sticky trichome exudate	Antibiosis

Write the differences between allomones and kairomones (allelochemic factors) in plants and their effects on insects?

Allochemic factors	Effect on insect response
Allomones	**Give adaptive advantage to host plant**
Repellent	Orient insect away from the plant
Locomotor stimulants	Increase walking or flight activity
Suppressants	Inhibit biting or piercing
Oviposition inhibitors	Interrupt egg laying
Metabolic inhibitors	Disrupt normal physiological process
Kairomones	**Give adaptive advantage to insect pest**
Attractants	Orient insect towards plants
Excitants	Evoke biting or piercing
Feeding stimulants	Increase duration and rate of feeding
Oviposition inducers	Induce egg-laying
Metabolic regulators	Regulating normal life processes

Explain in brief about constraints in breeding for insect resistance in plants?

Breeding for insect resistance suffers from several problems like

1. Breeding for resistance to one insect leads to the susceptibility to another pest. This may be because the feature of host plant associated with resistance to one insect is associated with susceptibility to another insect.

   Eg. Glabrous (hairless) strains of cotton are resistant to bollworms, but they are susceptible to jassids.

2. Breeding for insect resistance reduces the quality of produce or may even make it unfit for consumption.

   Eg. High gossypol content is associated with resistance to bollworm in cotton.

3. Genes for insect resistance are available only in the related wild species. So problem with linkage drag i.e. linkage between the genes for resistance and some undesirable genes.
4. Screening for insect resistance is the most critical and difficult step in a breeding programme for insect resistance. This necessitates a close coordination between the breeder and the entomologist.
5. Breeding for insect resistance is a long term programme. Therefore, adequate financing over a long period of time with a follow-up action is essential.

What is meant by nutritional quality? And explain the important nutritional factors to be considered while breeding for improvement of quality in cereals, pulses and oil seeds?

Nutritional quality: The nutritional quality refers to the suitability or fitness of a plant product for human and animal consumption.

Important Nutritional factor consider while breeding for improvement of quality:

Cereals:

1. Improved lysine content is a major issue of protein quality modifications of cereals because in cereals prolamines constitute a major percentage of seed protein, but prolamine is deficient in rice.
   - Rice-8% of prolamine and wheat 45%.
   - Like all cereals, lysine is the first limiting amino acid in rice.
2. Apart from the nutritional quality, the processing quality in cereals depends much on seed proteins.
   - Eg. Gluten depends quality (elasticity and extensibility) is the major processing quality in wheat.

3. In Barley, hordein has a negative influence on malting quality but is also a source of hydrophobic peptides that are essential for foam stability and cling to the glass.

Pulses: 20-25% Protein and in soyabean (38% Protein)

1. Pulses are characterised by lysine rich but defined in sulphur containing amino acids like methionine and cysteine besides tryptophan.
2. These amino acids are limiting factors in pulse protein utilization

   Eg. Pigeon pea and black gram.
3. Biological value of pulse protein is also limited by several antinutritional factors or toxicants.
   - These are Trypsin inhibitors- Beans, peas, lentil, chick pea
   - Lectins- Beans, chick pea, green gram and black gram
   - Cyanogens- Lima beans and chick peas
   - Goitrogens- Soyabean and French beans.
     - Trypisin inhibitors activity reduces the protein efficiency ratio cyanogens are neutralised during cooking.
     - Lectins- Break down of red blood cells
     - Goitrogens- Enlargement of thyroid glands
     - Other less potent toxicants associated with pulse protein are flatulents, phytates, tannins and saponins.

Oilseeds:-

1. Albumins and globulins are major storage proteins these proteins are rich in amides and arginine but relatively poor in sulphur containing amino acids.
2. Lysine is especially low in groundnut, rape seed, sunflower, linseed and sesame.
3. Utilisation of oilseeds is limited by occurrence of several antinutritional factors.

   Eg;-Glucosinolates in rapeseed and mustard.
4. Occurrence of anti pyridoxine factors (linative) makes linseed meal toxic for chickens in the absence of pyridoxine in diet.
5. Brassica seeds also contain anti-nutritional compounds like sinapic acid testers (SAE)- limiting the use of high quality protein.
6. Trypsin inhibitor activity has also been reported in all low glucosinolate cultivars of Brassica and Groundnut.

Write short notes on the following?

1. Single zero and double zero varieties of mustard
2. Golden rice
3. Quality protein maize

Single 'zero' and double 'zero' varieties: (0-0) mustard:

1. Euric acid is undesirable in edible oil in Brassica. If the euric acid content is more than 2% in oil according to world standard it becomes harmful. So breeding the mustard and rape seeds varieties having low euric acid i.e., less than 2% euric acid in oil is called as zero varieties (or) single low (or) single zero varieties. These are referred as low euric acid rape seeds.
2. Gulcosinolate: - In the meal glucosinolate are hydrolyzed by the enzyme myrosinase to produce glucose and goitrogonic aglycones which have pungent odour and biting taste. This reduces the palatability of meals to animals. Therefore glucosinolate content need to be reduced by breeding.
   - International standard for glucosinolate is less than 30μmol/gm of defatted cake.
   - The varsities having low euric acid and low glucosinolate are called double zero or double low varieties in Canada.
   - First double low variety was 'candle' .This is yellow coloured seed coat is also low in fibre; this is referred to as triple low or triple zero.
3. The name canola was adopted in 1979 to apply in Canada to "double – low" cultivars. Therefore, by the most recent definition, 'canola' is referred as a rapeseed cultivar that contain less than 2% erucic acid in oil and less than 30μmol/gm of one or any combination of the four known aliphatic glucosinolate in defatted meal.

Golden rice: Golden rice is the lines of rice developed by Ingo potrykus of Switzerland and Peter Beyer of Germany utilising the transgenic technology.

- These lines contain the beta- carotene and carotenoids that are precursor of vitamin-A.
- Term 'golden rice' refers to golden colour of the rice grain due to pro-vitamin-A.
- Golden rice was developed by using two genes from Narcissus (Daffodils) (Phytoene *synthase-psy;lycopene β-cyclase=lcy)* and one gene from *Erwinia uredovora*(Phytoene desaturase-cttl)
- Golden rice is the best example for molecular breeding of biosynthesis pathway.

Steps involved:-

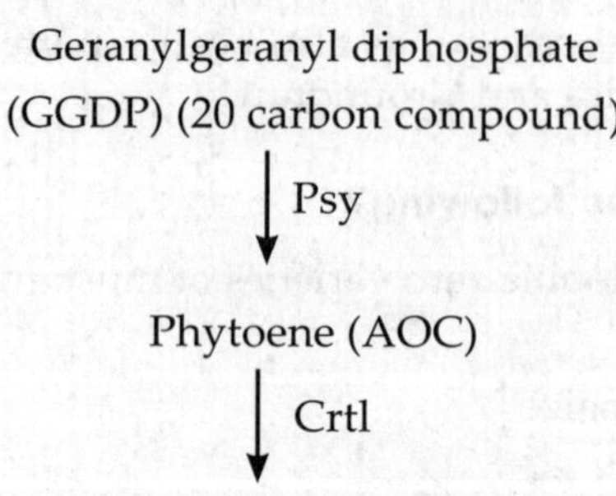

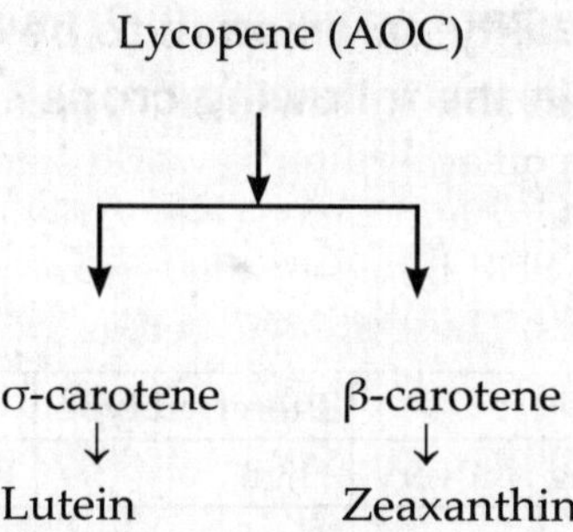

Genes utilized

1. Phytoene symthase gene from Narcissus
2. Phytoene desaturase gene from Erwinia
3. Lycopene β-cuclase gene from Narcissus

QPM (Quality Protein Maize)

1. QPM stands for high quality protein maize- lines which have high amount of lysine and tryptophan amino acids in the maize protein which are deficient in maize protein.
2. These lines of maize came to existence with discovery of opaque-2 (O_2) mutant line in 1963 at Purduew university U.S.A by Mertz, Batsland and Nelson, followed by discovery of other mutants like floury-2 ($fl_2 fl_7 fl_9$) that also has ability to alter endosperm nutritional quality.
3. But the mutant had some negative effects like
   - Soft endosperm
   - Increased susceptibility to pests and diseases
   - Inferior food processing
   - Reduced yield
4. Nutritional superiority and biological value of QPM
   - High lysine 3.3 to 4.0gm/100 gm of endosperm protein (twice improved)
   - Good quality protein and digestibility.
   - High Biological Value about 90% of milk.
   - Higher Niacin availability due to higher tryptophan and lower leucine content.
5. In India three composites namely Shakti, Rattan and Protina released in1971for commercial cultivation.
6. In 1997 a modified, nutritionally superior O_2 composite named as Shakti 1.
7. In 1998 under NATP, two QPM hybrids shaktiman-1 (triple cross) and shaktiman 2 (single cross) were released using CIMMYT QPM lines.

Write the important quality parameters to be considered for breeding for nutritional quality in the following crops?

1. Rice 2. Cotton 3. Wheat

Quality traits of rice:

S. No	Trait	Preferred type	Remarks
1.	Hulling and milling recovery	High	Range,70-79% (Total rice)
2.	Endosperm appearance	Transparent	Opaque appearance due to loose packing of starch, polygenic; low heritability
		Non-waxy	A single dominant gene
3.	Length and shape of grain	Long slender	Five groups: LB, LS, MS, ShB, Shs; polygene; high H

Cooking quality

S. No	Trait	Preferred Type	Remarks
1.	Amylase content	Intermediate	Groups: Low (up to 20%), intermediate (21-25%), high (>25%); single gene
2.	Grain elongation	Greater elongation	
3.	Volume expansion	More expansion	
4.	Water absorption	Intermediate	
5.	Gelatinization temperature	Intermediate	Range, 56-79^{0}C
6.	Grain elongation on cooking	Preferred	Special trait of 'basmati' rice
7.	Disintegration in 1-7% *Icdh* (milled rice grain)	Deep diffused network of proteins	Range, 6-8%, polygenic; good amino acid balance; lysine content 3.8-4%
8.	Aroma	Presence of aroma	Scented types fetch, higher price; oligogenic; recessive

L-long, S-slender, M-median, B-bold, Sh-short, H-heritability.

Quality traits of cotton:

S.No.	Trait	Preferred type	Remarks
1.	Fibre colour	White	Colours; cream light/ medium/ dark brown, light green, oligogenic.
2.	Fibre length		
3.	Halo length	Long	Length of fibre with attached seed

4.	Mean length	Long (24.5 to 26.00 mm)	Mean length of all the fibres present in a sample
5.	2.5% span length	Long (28 to 31.5 mm)	Length of the longest 2.5% of fibre
6.	Fibre strength	Good to very good (43.1 to> 47.5 g/tex)	Determines yarn strength
7.	Fibre fitness	Fine to very fine (< 3micronaire value)	Determines spinning capacity of cotton, uniformity and strength of yarn.
8.	Fibre maturity	To very high maturity (maturity coefficient > 68)	Index of the degree of fibre development; depends on secondary wall deposition.

Quality traits of wheat:

S. No.	Trait	Preferred type	Remarks
1.	Grain size	Bold , pump	
2.	Grain colour	Amber to white	
3.	Grain texture	Hard, vitreous, lustrous	
4.	Protein content	High	Atlas 66 and Naphal have over 19% protein
5.	Gluten content	High	(C 273 and UP302 have high gluten (13%)
6.	Polshanke value (Gluten strength)	Strong (>150 min) Medium strong (81-150 min) Weak (<80 min)	Suitable for bread making Suitable for chapathi making Suitable for biscuits and confectionery
7.	Sedimentary value	(20-40)	All purpose wheat; lower the value, weaker the gluten
8.	Alveogram	Strong (>300), medium strong (200-300)), weak (< 200)	Measures gluten strength
9.	Mixogram	-	Indicates baking quality and also glutens strength

What are problems encountered with respect to nutritional quality in the use of following crops? Briefly mention the strategies used to overcome these problems.

1. Lathyrus   2. Mustard.

Antinutritional factor	Crop	Effect on health	Remarks
BOAA (B-N-oxalyl-α-β-diamineo-propionic acid)	Khesari (L.sativus)	Lathyrism (paralysis of lower limbs)	Content reduced to 0.008-0.05% from the normal 0.55%
Euricic acid	Brassica spp	Reduced body weight	Zero varieties developed
Glucosinolates	Brassica sps	Goitre	Zero varieties developed
Gossypol	Cotton	Toxicity	Involved in insect response
Haemogglutinin	Broad bean (Vicia faba)	Favism	Due to consumption of raw braod beans
Oligopeptide	soyabean	Goiter	Destroyed by cooking/ heat treatment
Phenolic glycoside	Groundnut	Goiter	Destroyed by cooking/ heat treatment
Cyanogenic glycoside	Forage legumes	HCN toxicity	Removed by breedimg
Phytic acid	Pearl millet (Bajra)	Mineral deficiency	Occurs in seed coat and germ
Trypsin inhibitor	Soyabean		Destroyed by heat processing

Lathyrus: BOAA is a neurotoxin

(*i*) It is detoxified by steeping the dehusked seeds overnight followed by steaming for 30 min or by roasting at 150^{0}C for 70 min.

(*ii*) After extensive screening of Lathyrus accessions, some accessions Eg. P-24 having low levels of BOAA were identified.

(*iii*) Breeding efforts have been successful in flowering the BOAA content from normal 0.55 to as low as 0.03- 0.55%.

Brassica: Zero glucosinolate varieties have been developed in Brassica (see 0-0-mustard, previous pages).

What is meant by abiotic stress? And write its significance in crop improvement?

The various non biological factors of environment interferes with the complete expression of genotypic potential is called abiotic stress. Abiotic stresses result due to moisture, temperature, mineral, salinity, soil p^H, air pollution, etc.

Importance of abiotic stresses:

- Abiotic stresses not only determine the geographical and regional distribution of crops, but they also dictate if a potentially arable land piece can actually he used for cultivation.

- Temperature is the most effective factor affecting crop distribution/moisture and soil chemical properties are also important factors.
- Drought is the main abiotic factor as it affects 26 % of the arable area.

  It may be pointed out that the relative importance of different abiotic stresses varies greatly depending on the geographical region and the specific location within a given region.
- For example, the degree of moisture stress experienced by a crop will be greater under warmer conditions than that under relatively cooler environments, the soil moisture level and air humidity's being comparable under both the environments. In addition, the severity of the effects of a stress on a crop depends greatly on the duration of stress and on the stage of crop growth at which the stress has occurred.

Define drought. Explain the various methods used for screening of drought resistance/ tolerance? And also mention the scale used for scoring the drought resistance in rice?

Drought may be the "inadequacy of water availability, including precipitation and soil moisture storage capacity, in quantity and distribution during the life cycle of a crop to restrict the expression of its full genetic yield potential."

Methods for screening of drought resistance/tolerance:

I. Physiological methods

1. **Relative water content:** Water content in percentage on the basis of full turgidity

   RWC = (Fresh weight – dry weight/ Turgid weight – dry weight) × 100

   • High RWC indicates high drought tolerance.
2. **Chlorophyll stability index:**

   CSI = (Total chlorophyll of heated sample/Total chlorophyll of unheated sample) × 100

   • High CSI indicates high drought, temperature tolerance.
3. **Membrane injury index:**

   $MII = 1 – [(1-T_1/T_2)/ (1 – C_1/C_2)] \times 100$

   Here, T_1 and T_2 are Electrical conductivity of stress sample before and after autoclaving.

   And C_1 and C_2 are Electrical conductivity of controlled sample before and after autoclaving.

   • High MII indicates less temperature tolerance.
4. Membrane thermo stability and heat killing time:

   Injury (%) = E – I/F × 100

   Here, I = Initial conductance

E = Elevated temperature conductance

F = Final conductance after autoclaving

- High MTS indicates high temperature tolerance.
- High HKT indicates more heat tolerance.

5. Pollen assay technique: A quick and efficient technique for screening against temperature, salinity, herbicides, heavy metal stress.

- High pollen germination indicates more tolerant to salinity.

6. Total chlorophyll content:

$$TCC = [20.2\ (A644) + 8.02\ (A663)] \times [V/\ 100 \times W]$$

Here, A = Absorbance

V = Final volume of chlorophyll extract

W = Fresh weight of tissue extracted

- High TCC indicates high drought tolerance.

7. Chlorophyll fluorescence index (F_v/F_m ratio): A powerful screening technique to assess various abiotic stresses like heat, chilling, drought, water logging and salinity. Modulator flourometer is used for measurement of CFI.

$$CFI = F_v/F_m \qquad F_v = F_m - F_o$$

Here, F_v = Variable fluorescence

F_o = Minimal fluorescence

F_m = Maximum fluorescence

- High CFI indicates high drought tolerance.

8. SPAD Chlorophyll meter (SCMR): The SPAD meter is a simple hand hold and portable instrument, which provides information on the relative amount of chlorophyll.

II. Biochemical indices:

- Proline, ABA, Peroxide, Catalase and Lipid per oxidation are more in drought conditions.
- Super oxide dismutase, high under drought conditions.
- Amylase content, low amylase content indicates high tolerance
- Aldehyde content (submerged stress), is high under stress condition.
- Polyamines content (Cold tolerance), high accumulation.
- Spermamide more in wheat under drought condition than normal.

III. Agronomic indices:

1. **Stomatal diffusive resistance:**

   DMSI = (Total dry matter produced under stress/ Total dry matter produced under control) × 100

   High DMSI indicates high tolerance to drought.

2. **Yield stability ratio (YSR)**

   YSR = (Grain yield under stress/ Grain yield under control) × 100

   High YSR indicates high drought tolerance.

3. **Drought stability index:**

   DSI = $[1- Y_d/Y_p]/D$

   Here, Y_d = Grain yield under stress

   Y_p = Grain yield under control

   D = Drought intensity = 1 – Mean Y_d of all genotypes/ Mean Y_p of all genotypes.

4. **Drought tolerance index:**

   DTI = I_o/I_1

   I_o = Grain yield under moisture stress

   I_1 = Grain yield under control

Scoring of drought tolerance at vegetative and reproductive stages in Rice (IRRI, 1996)

Scale	Drought tolerance at vegetative stage	Spikelet fertility (%) at reproductive stage	Plant recovered (%)
0	No symptoms	-	-
1	Slight tip drying	>80	90-100
3	Tip drying extended up to ¼ length in most leaves	61-80	70-89
5	One fourth to ½ of all leaves fully dried	41-60	40-69
7	More than 2/3 of all leaves dried	11-40	20-39
9	All plants apparently dead	<11	0-19

Define mineral toxicity. Give the scale used for scoring of the salt and alkali injury, iron toxicity, phosphorous deficiency and zinc deficiency in rice

Mineral toxicity: The stress generated by the presence of toxic concentrations of available minerals in the root zone is called mineral toxicity.

Scoring of damage for both alkali and salt injury in field conditions in rice (IRRI, 1996)

Scale	Growth patterns
1	Growth and tillering nearly normal
3	Growth nearly normal but there is some reduction in tillering and some leaves discolored (alkali/whitish and rolled/salt)
5	Growth and tillering reduced, most leaves discoloured (alkali)/rolled (salt); only a few elongation
7	Growth completely ceases; most leaves dry; some plants drying
9	Almost all plants dead or drying

Scoring of iron toxicity in rice (IRRI, 1996)

Scale	Expression
1	Growth and tillering nearly normal
2	Growth and tillering nearly normal; reddish-brown spots or orange discoloration on older leaves
3	Growth and tillering nearly normal; older l;eaves reddish brown, purple or orange yellow
5	Growth and tillering retarded, most leaves discolored or dead
7	Growth and tillering ceases, most leaves discolored or dead
9	Almost all plants dead or drying

Scoring of phosphorus deficiency in rice (IRRI, 1996)

Scale	Relative tillers (%)
1	100-80
3	79-60
5	59-40
7	39-20
9	19-0

Scoring of zinc deficiency in rice (IRRI, 1996)

Scale	Expression
1	Growth and tillering nearly normal and healthy
2	Growth and tillering nearly normal, basal leaves slightly discoloured
3	Stunting slight, tillering decreased, some basal leaves brown or yellow
5	Growth and tillering severely retarded, about half of all leaves brown or yellow
7	Growth and tillering ceases, most leaves brown or yellow
9	All most all plants dead or drying

Define temperature stress and give its various types. And also provide the scale used for scoring the heat tolerance and cold tolerance in rice?

Each plant species has an optimum range of temperatures for its normal growth and development/the specific temperatures would depend not only on the genotype but also on the stage of growth and development of a given genotype. When temperature moves beyond this optimal range, it generates temperature stress, i.e., temperature interferes with the performance.

The various types of temperature stress:

Heat stress: The adverse effects of plants of temperatures higher than the optimal is considered as heat stress.

Chilling stress: In many environments, crop productivity is limited by low temperatures and when the temperatures remain above freezing, i.e., > 0 °C, it is called chilling stress.

Freezing stress: When the temperatures remain below freezing, i.e., < 0 °C, it is called freezing stress.

Scoring of heat tolerance under the field condition in Rice (IRRI, 1996)

Score	Spikelet fertility (%)
1	>80
3	61-80
5	41-60
7	11-40
9	<11

Scoring of cold tolerance at seedling and adult plant stages in rice (IRRI, 1996)

Seedling stage		Adult plant stage	
Scale	**Symptoms**	**Scale**	**Symptoms**
1	Seedlings dark green	1	Plant show a natural colour, rate of growth and flowering
3	Seedlings light green	3	Plants slightly stunted and growth slightly stunted and growth slightly delayed
5	Seedlings yellow	5	Plants moderately stunted, leaves pinkish and development delayed
7	Seedlings brown	7	Plants severely stunted, with leaves yellow, development delayed and panicles poorly exerted
9	Seedlings dead	9	Plants severely stunted, with leaves brown, development much delayed and panicles not exerted

CHAPTER - 9

Genetic Engineering and Biotechnological Tools in Plant Breeding

Define biotechnology? And explain the role of biotechnology in crop improvement?

Biotechnology: The term Biotechnology was coined by Karl Ereky a Hungarian engineer in 1919. This term is derived from a fusion of Biology and Technology

Biotechnology is the application of scientific and engineering principles to the processing of materials by biological agents to provide goods and services. This was given by OECD – the organisation for economic cooperation and development in 1981.

Role of biotechnology in crop improvement:

1. Crop improvement through plant tissue culture
    - Anther/ ovule/ pollen culture – To obtain haploid plants and there by development of dihaploids on chromosome doubling
    - Micropropagation – Mass production of plants
    - Meristem culture - Virus free plants
    - Production of secondary metabolites – Eg. Caffine from *Coffea arabica* and Nicotine from *Nicotiana rustica*
    - Somaclonal variation – Eg. 'ono' variety of sugarcane
    - Embryo culture – To overcome pre and post fertilisation barriers
    - Ovary culture – To obtain haploid plants from egg cell
    - Endosperm culture – To obtain triploid plants
2. Use of protoplasm fusion and somatic hybrids:
    - Transfer of CMS from *N. tabacum* to *N. sylvestris*
    - Transfer of diseases resistance from Solanum bravidens to S. tuberosum.
3. Genome mapping and MAS:
    - MAS for QTLs and other agronomic character
    - MAS in prediction of heterosis and identification of heterotic pool.
    - For locating, identification and isolation of gene.
    - Plant variety protection
    - Gene pyramid and QTL introgression.

4. Genetic engineering and transgenic plant:
   - Bollgaurd - cotton variety against American bollworm
   - Flavr savr – Tomato improved for quality
   - Endless summer – tomato variety
   - Male sterility – Barnase – Barstar in Brassica
5. Edible vaccines

Define somaclonal variations. Enlist the causes of somaclonal variations. How somaclonal variations are induced and selections made? Discuss the significance of somaclonal variations in crop improvement?

Somaclonal variations is the variation among the tissues or plant derived from the *in vitro* somatic cell culture i.e., callus or suspension cultures.

1. Somaclonal variations may be genetic or epigenetic out of which the genetic variation is heritable whereas the epigenetic variation caused by cultural conditions is not heritable and hence of no significance in the sexually propagated plants.
2. Genetic variation may result from the following causes:
   - Chromosomal changes: Eg. Numerical and structural chromosomal aberrations
     - Aneuploidy observed in oats, ryegrass, triticale and potato.
     - Deletions, inversions, translocations and duplications observed in barley, wheat, potato and maize.
   - Mitotic crossing over: M.C.O may account for some of the genetic variation that leads to recovery of homozygous recessive single gene mutations in some regenerated plants.
3. Apparent 'point' mutations
4. Cytoplasmic genetic changes

   E.g. Sensitivity to host specific toxin of *Drechlera maydis* race T, the causal agent of southern corn leaf blight is associated with all genotypes containing Texas male sterile (CMS-T) cytoplasm.
5. De-amplification and amplifications:
   - Deficiencies in ribosomal DNA i.e., rDNA deamplification observed at molecular level in flax, triticale and potato.
   - Gene amplifications observed in Nicotiana and tomato.
6. Transposable element activation: Eg. Nicotiana, alflfa and maize.
7. Virus elimination: Eg. Prior infection with Barley yellow dwarf virus causes susceptibility to powdery mildew in oats.
8. Methylation/deamination of DNA: Genome activity i.e. transcription, replication, rearrangements and the structural organisation of chromatin seems to be related to DNA methylation.

9. Altered expression of multigene families.

   E.g. Heritable somaclonal variation has been obtained for *gliadina storage* protein and β-amylases in wheat.

Induction and selection of somaclonal variation:

- Callus cultures are established from suitable explants and multiplied through periodic sub culturing.
- Plants are regenerated usually from long-term maintained callus/cell suspension cultures and transferred to soil and screened for variation in glass house or field.
- *In vitro* selection at cellular level can be carried out for some traits by growing cells from cell suspensions and calli on a medium supplemented with elevated levels of various biotic and abiotic stress factors.
- Moreover, it reduces the chances of diplonitic selection but it requires high level of correspondence between the trait selected in vitro and expressed in vivo.

Significance in crop management:

- Somaclonal variation appears to be an important alternative for creation of genetic variability in crops where tissue culture plant regeneration systems have been established.
- Recovery of novel variants which either do not exist or are rare in the natural gene pool such as atrazine resistance in maize, glyphosphate resistance in tobacco, improved lysine and methionine content in cereals, increased seedling vigour in lettuce are of much significance.
- Genetic and cytogenetic evidences have now been provided for increased recombinant frequency through cell culture.
- This will be important for breaking undesirable linkages and obtaining introgression from alien sources through tissue culture of wide hybrids.
- New varieties have been developed through somaclonal variation in tomato, sugarcane, celery, brassica and sorghum.
- This is simple and cost effective technique possess greater potential for the improvement of apomictic and vegetatively propagated crops and for seed propagated crops with narrow genetic base.
- In India, a somaclonal variant of a medicinal plant citronella java has been released as a commercial variety 'Bio 13' which gives higher yield and oil content.
- 'Pusa jai kisan' is another variety of B. juncea released as somaclonal variant of 'Varuma' variety.

Define totipotency? And discuss the significance of regenerating plants from single cell or protoplast under asceptic conditions?

Totipotency: Totipotency is the ability of plant cell to perform all the functions of development which are characteristic of zygote i.e., Ability of single cell to develop into complete plant.

- Haberlandt first demonstrated it in 1902 by culturing palisade cells in sucrose media. He only showed the accumulation of starch but failed to show division finally it was demonstrated by R.J. Gauthert and White.
- This character is too much vital because whole tissue culture is based on it. Later on embryo culture, pollen culture, protoplast cultures have been demonstrated.

Getting whole plant through it has several advantages:

1. Single cells culture or protoplast culture offer the possibility for large scale clonal propogation. Eg. Banana or orchids.
2. Somatic embryos from cell suspension can prove useful for long term storage in germplasm bank. Eg. Jack fruit and mango etc.
3. Secondary metabolite production – secondary metabolite production through cell culture have certain advantages over normal cultivation of whole plant. Eg. Nicotine –*Nicotiana tobaccum*. Vinoristrin – *Vinca rosea.* The reserpine – *Rawalphia serpentine*.
4. Pollen culture and ovule culture are also a specialised cell culture and very useful in haploid production and too much helpful in development of hybrid as well as mapping of gene.
5. One can harness somaclonal variation and by this one can produce so many novel germplasm.
6. Chemical mutagen can be effectively applied through cell culture.
7. Protoplast culture is the first step in production of somatic hybrids. This is especially useful in vegetatively propagated crop and crops where sexual hybrid production is difficult.

   Eg. CMS cytoplasm transfer
8. Production of artificial seed i.e. somatic embryo encapsulated by calcium alginate coating.
9. In genetic transformation the transformation of cell through transfer of foreign gene into cell. By this way transgenic plants can be produced.

What is meant by Somatic embryo genesis? And explain in brief about synthetic seed?

Somatic embryo genesis: Where an embryo is obtained from somatic cell rather than from zygote. Such somatic embryos can be encapsulated in a suitable matrix along with nutrients, growth regulators, antibiotics needed for development of complete plant to make what are known as synthetic or artificial seed

Synthetic seed:

The concept of synthetic seed was first conceived by Murashiqe (1978). However Kitto and Janick (1982) used it for first time in carrot.

- Synthetic seed involved the production of tissue culture derived somatic embryos encapsed in protective coating.
- It may be encapsulated somatic embryos, shoot buds, cell aggregates and any other tissue that can be used for sowing as a seed and that possess the ability to convert into plant under *invivo* or *invitro* condition and returns this potential after storage also.
- Somatic embryo is bipolar structure with both apical and meristamatic regions, which are capable of forming, shoot and root respectively. Somatic embryos are structurally same as zygotic embryos found in seed and possess many of their useful features including the ability to grow into a complete plant.
- Artificial seeds may be produced by one of the two following ways 1) Dessicated and 2) Hydrated systems
- In the dessicated system the somatic embryos are first hardened to withstand dessication and then are encapsulated in a suitable coating material to yield desiccated artificial seeds.
- In the Hydrated systems, somatic embryos are enclosed in gels which remain hydrated. Of the many gels evaluated "calcium alginate" is the most suitable.

Problems with synthetic seed or artificial seed:

- Synthetic seeds require procedure for making the embryo quiescent.
- Need to protect from desiccation.
- Very low recovery of plants.
- Embryo must be protected from micro organisms.

Give in detail the steps involved in micro propagation of plants with the help of flow diagram? And also mention the applications and problems of micro propagation?

Micro propagation: The multiplication of genetically identical copies of a cultivar by asexual means is called clonal propagation.

- A genetically uniform assembly of individual, derived originally from single individual by asexual propagation constitutes a clone.
- *In vitro* clonal propagation is called micro propagation.

Techniques of micro propagation:

The process of micro propagation involves 4 distinct stages

1. Selection of suitable explants, their sterilisation and transfer to nutrient medium for establishment / initiation of a sterile culture explant.

2. Proliferation of shoots from the explant on medium
3. Transfer of shoots to a rooting medium
4. Transfer of plants to soil / normal environment

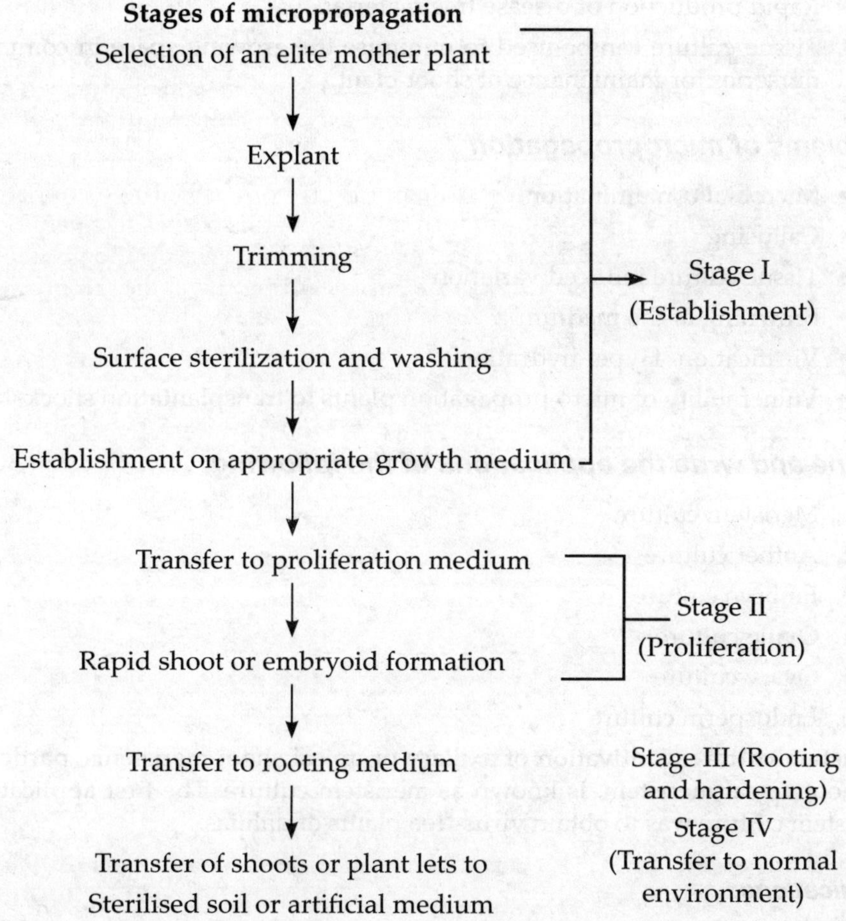

Applications of micro propagation:

1. Micro propagation of a hybrid has the greatest multiplication advantage since it can be result in large number of elite plants from a very small tissue clump taken from the hybrid plant.
2. Maintenance of inbred lines for producing F_1 hybrids.
3. Maintenance of male sterile genotypes of wheat and onion are useful in hybridisation.
4. Selective propagation of dioecious plants Eg. female plants of papaya, male plants of Asparagus.

5. Multiplication of particular heterozygous superior genotype with increased productivity Eg. Oil palm.
6. Shoot cultures of some species are maintained as slow growth culture for germ plasm conservation.
7. Rapid production of disease free material.
8. Tissue culture can be used to minimise the growing space in commercial nurseries for maintenance of shoot plant.

Problems of micropropagation

- Microbial contamination
- Callusing
- Tissue culture induced variation
- Browning of the medium
- Vitrification/ Hyper hydration
- Vulnerability of micro propagation plants to transplantation shocks

Define and write the applications of the following.

1. Meristem culture
2. Anther culture
3. Embryo culture
4. Ovule culture
5. Ovary culture
6. Endosperm culture

Meristem culture: Cultivation of axillary or apical shoot meristems, particularly of shoot apical meristem, is known as meristem culture. The first application of meristem culture was to obtain virus-free plants of dahlias.

Applications:

- Virus elimination generally improves the yield by 20-90% over infected controls.
- Virus-free plants serve as excellent experimental materials for evaluating the detrimental effects of infections by various viruses.
- The virus free bulbs grew more rapidly, plants were more vigorous, and they produced a greater number of larger flowers that had richer colour than the virus infected stock.
- The virus -free plants are deliberately infected by known viruses, and effects of the infection on performance of the host are assayed.
- Meristem culture can also help eliminate other pathogens, Eg., Mycoplasmas, bacteria and fungi. Bacteria and fungi present in explants

show up when they are cultured *in vitro* since tissue culture media provide excellent nutrition for the microbes.

- Meristem culture has been used to eliminate systemic bacteria form Diffenbachia and Pelargonium, and *Fusareum roseus* from carnations.

Anther/pollen culture: Anther culture is a means of obtaining haploids from the pollen grains. Haploid production through anther culture is known as "Androgenesis"

Applications:

- Dihaploid production through bulbosum technique is successful in barley.
- Through the induction of genetic variability (gametoclonal variation) haploids (or) plants with various ploidy levels and mutants are obtained.
- Selection of mutants resistant to disease. Haploid provides relatively easiest system for the induction of mutations. Therefore, they can be employed in rapid selection of mutants having traits for disease resistance. Eg. Tobacco mutants resistant to Black shank disease and wheat lines resistant to scar. *Fusarium graminorium*
- Developing asexual lines of trees (or) perennial species.
- Pollen derived rubber tree is taller by 6m which could be multiplied by asexual propagation
- Transfer of desired alien genes / genetic transformation.
- Chromosomal instability in haploids makes them potential tool for introduction of alien chromosomes (or) genes during wide crossing programme. Haploids are used for the development of substitutional and additional lines.
- Double haploids are used for QTL mapping and construction of genetic linkage maps etc.

Embryo culture: The term embryo culture means excision of embryos regardless of age, size and developmental stage from their natural environment and growing them under artificial environmental conditions.

Applications:

1. Production of rare hybrids from intergeneric and interspecific crosses.

   Cereals: *Hordeum vulgare x secale cereale*

   Triticum durrum x secale cereale

   Legumes: *Arachis hypogaea x Arachis monticola*

   Phaseolus vulgaris x Phaseolus. Vitensis
2. Development of disease resistant plants.
3. Embryo culture has been useful in evolving disease resistant plants.

Eg. Tomato resistance to virus, fungi and nematode. Lycopersican esculentum x L peruvianum.

4. Production of haploids: Chromosome elimination technique for production of haploids in cereals / Bulbosum technique (barley and wheat).
5. Overcoming seed dormancy: Embryo culture technique is applied to break seed dormancy which can be caused by numerous factors including endogenous inhibitors, specific light requirements, low temperature, dry storage requirements and embryo immaturity.
6. Shortening of Breeding cycle
7. Propagation of rare plants
8. Propagation of orchids
9. Prevention of embryo abortion with early ripening stone fruits
10. Clonal micro propogation
11. Rapid seed viability testing

Ovule culture: It is the technique by which ovules are aseptically isolated from ovary and are grown on chemically defined nutrient medium under controlled conditions.

Maheshwari (1958) isolated ovules of Papaver somniferum and cultured them on nutrient medium.

Applications:

1. To obtain interspecific and intergeneric hybrids Eg. Cotton, the hybrid embryo between tetraploid and diploid was rescued and obtained plants. *Gossypium barbadense* × *G. Arborium* and *Gossypium hirsutum* × *G. herbaceum.*
2. Orchid seed germinate only in association with fungus, but the cultured fertilised ovules germinate even in the absence of fungus
3. Test tube pollination and fertilization is possible through ovule culture
4. It helps in the development of several embryos in Citrus and other crop plants
5. Culture of unfertilised ovules helps in the formation of haploid callus.

Ovary culture: Culture of unfertilised ovaries to obtain haploid plants from egg cells or other haploid cell of the embryo sac is called ovary culture and this process of haploid production is termed as gynogenesis

Applications:

1. Interspecific hybrids have been successfully obtained in several genera including Brassica species (*B. compestries* × *B,oleracea*) seeds with well developed embryos have been obtained when ovaries are cultured 4 days after pollination in white's medium contain casein hydrolysate. For inter specific and inter generic crosses ovaries are excised at zygote stage (or) at two celled pro embryo stage.

2. It is used for the study of understanding of physiology of fruit development
3. The effect of phytoharmones on the development of parthenogenic fruit or haploids from the cultures of un pollinated flower can be studied
4. It is successful in inducing poly embryony.

Endosperm culture: Production of triploid plants from culturing the endosperm cells is called endosperm culture.

Applications of Endosperm culture in crop improvement

1. Triploid cells of endosperm have totipotency. So endosperm culture technique is applied to economically important cultivars for raising superior triploid plants.
2. Triploid plants are seed sterile and undersirable for plants where seeds are commercially important. Eg. Apple, Banana, Mulberry, Sugarbeet, Peach, Watermelon etc, which are commercially important for their edible parts.
3. In some cases triploids are superior in quality than diploid. Eg. Triploids of populus have better quality pulpwood.
4. To exploit in the biosynthesis of some natural products Eg. Cultured endosperm of coffee synthesizes caffeine the level of this alkaloid in callus is synthesized by three times after two weeks by and 6 times after 4-5 weeks.
5. Various trisomics developed from triploids may also be useful in gene mapping for cytogenetic studies.
6. Endosperm can be used as a nurse tissue for raising hybrid embryos. Eg. Using Hardeum endosperm as a nurse tissue the young embryos of hybrid between *Hardeum* × *Triticum*, *Hardeum* × *Secale* and *Hardeum* × *Agropyron*
7. These can be induced to germinate and form normal hybrid plants
8. The multiplication of triploids should not be a serious problem with the available technique of micro propagation

What is meant by protoplast culture? And explain in brief about somatic hybridization technique along with applications?

Protoplast culture:

Protoplast can be isolated either by mechanical (or) enzymatic methods.

1. Mechanical method:

The cells are kept in a suitable plasmolyticum and protoplasts are isolated by cutting the plasmolysed tissues with a sharp blade so that protoplasts are released from cells through the cell wall when the tissue is again deplasmolysed. This method is suitable for isolation of protoplast from highly vacuolated cells of storage tissues,

Eg. Onion bulb, scales, radish roots etc.

2. Enzymatic method:

Cocking (1960) demonstrated the possibility of enzymatic isolation of protoplast from higher plants. He used concentrated solution of cellulase to degrade the cell walls.

The enzymatic isolation of protoplasts can be performed in different ways

Two step (or) Sequential method: The tissue is first treated with macerozyme or pectinase enzyme which separates the cell by degrading the middle lamella. These free cells are then treated with cellulase which releases the protoplasts, cellulase enzyme digest the cellulose in plant cell walls while pectinase enzyme breakdown the pectin holding cells together

One step (or) Simultaneous method:

The tissue is subjected to a mixture of enzyme in first step reaction which includes both macerozyme and cellulose. The one step method is generally used because it is less labour intensive while the yield of protoplast isolated by two step method are better for culture studies.

Somatic hybridisation:

The technique of hybrid production through the fusion of protoplast from different genetic backgrounds is known as somatic hybridisation or parasexual hybridisation and protoplast fusion.

The technique of somatic hybridization is of special significance for the improvement of vegetatively propagated plants such as banana, cassava, potato, sweet potato, and sugarcane.

The production of somatic hybridisation involve a number of steps

1. Collection of explants
2. Protoplast isolation
3. Protoplast fusion
4. Selection of hybrid cells
5. Regeneration of plants from hybrid tissue
6. Characterisation of hybrid and cybrid plants

Applications of somatic hybridisation

1. Production of novel interspecific and intergeneric crosses between the plants that are difficult (or) impossible to hybridise conventionally.

   E.g. Tomato x Potato = Pomato
2. Transfer of desirable genes for disease resistance. Eg. Potato protoplast have been fused with those of Solanum brevidens and Solanum phuraja and *Lycopersicon* the resulting somatic hybrids are resistant to potato leaf roll virus, potato virus and Erwinia soft rot. Somatic hybrids resistant to *Phomalingam* (Black log disease) were produced by the fusion of protoplast of *Brassica napus* with those of *Brassica nigra* (resistant to pomalingam)

3. Transfer of desirable genes for Abiotic stress resistance, Atrazine (resistance) has been transferred from wild species of tomato Lycopersicon peruvianum into cultivated species (*Lycopersicon esculentum*)
4. Transfer of desirable genes for quality characters: Somatic hybrids produced between *Brassica napus* and *Eruca sativa* were fertile and had low concentration of Erucic acid.
5. Transfer of cytoplasmic male sterility: Male sterility can be induced by alloplasmic association. It results in interaction between nucleus of one species and cytoplasm of another species. Male sterile lines were developed by fusing protoplast of *N. tabacum* with X-ray irradiated protoplasm of *N. africana*
6. Identification of crop cultivar, somatic hybrids and hybrids in IPR issue and protection.

What is Meant by Recombinant DNA Technology? And Explain the Various Enzymes used for DNA Manipulation?

Recombinant DNA technology:

Recombinant DNA technology is the revolutionary technology which enables to transfer genes from any organism to another. This is feasible technology to transfer within or between species or taxa gene and even unrelated taxa and hence widened the scope of this technology is capable to use gene pool-4 and overcome the barrier of cross compatibility.

Recombinant DNA molecule used as a vector (plasmid, phage and virus) into which the desired DNA fragment has been inserted to enable its cloning in an appropriate host.

Recombinant DNA is product for following purpose.

- To obtain a large number of copies of specific DNA fragments
- To obtain large quantities of protein related to gene of interest.
- To integrate gene into chromosome

A cultivar developed by the rDNA procedure is called transgenic cultivar (or) genetically modified (GM) cultivar. The basic steps of all rDNA technologists are as follows.

- DNA of interest extracted from source organism through restriction enzyme.
- Desired DNA fragment is inserted into cloning vector.
- rDNA transferred and maintained in host through transformation.
- Host identified with cloned transgene.
- Cloned transgene can be manipulated to give protein product (expression of transgene).

Enzymes for DNA manipulation: The enzymes available for DNA manipulation fall into four broad categories.

1. **DNA polymerases:** Which are enzymes that synthesize new polynucleotides complementary to an existing DNA or RNA template i.e. template dependent DNA polymerase.

Important features of template dependent DNA polymerase:

(*i*) To initiate DNA synthesis, these must be a short, double stranded region to provide a 3′end on to which the enzyme will add new nucleotide. i.e. short, synthetic olegonucleotide, usually about 20 nucleotides in length, which acts as a primer for DNA synthesis.

(*ii*) Template – dependent DNA polymerase is multifunctional, being able to degrade DNA molecules as well as synthesize them.

(*a*) A 3′→5′ exonucleae activity enables the enzyme to remove nucleotides from the 3′ end of the strand that has just synthesized i.e. proof reading activity.

(*b*) A 5′→3′ exonuclease activity is less common

Types of DNA polymerases:

Polymerase	**Description**	**Main use**
DNA polymerase I	Unmodified *E. coli* enzyme	DNA labelling
Klenow polymerase	Modified version of *E. coli* DNA polymerase I	DNA labelling
Sequenase	Modified version of phage T7 DNA polymerase I	DNA sequencing
Taq polymerase	*Thermus aquaticus* DNA polymerase I	PCR
Reverse transcriptase	RNA-dependent DNA polymerase, obtained from various retroviruses	cDNA synthesis

Nucleases: Which degrade DNA molecules by breaking the phophodiester bonds that link one nucleotide to the next nucleotide.

- Nucleases have a broad range of activities but most are either exonucleases, removing nucleotides from the ends of DNA/RNA molecules or endonucleases, making cuts at internal phosphodiester bonds.
- Nucleases are specific for DNA and some for RNA, some work only on double-stranded DNA and others only on single-stranded DNA etc.

Nucleases	**Description**	Main use
Restriction endonucleases	Sequence-specific DNA endonucleases, from many sources	Restriction – many applications
Nuclease SI	Endonuclease specific for single-stranded DNA and RNA, from the fungus Aspergillus oryzae	Transcript mapping
Deoxyribonuclease I	Endonuclease specific for double-stranded DNA and RNA from E. Coli	Nuclease footprinting

- Restriction endonucleases, which play a central role in all aspects of recombinant DNA technology.

Restriction endonucleases: A, Restriction endonuclease is an enzyme that binds to a DNA molecule at a specific sequence and makes a double-stranded cut at or near that sequence.

There are three types of restriction restriction endonucleases (Type I, II and III).

- With Type I and III, there is no strict control over the position of the cut relative to the specific sequence in the DNA molecule that is recognised by the enzyme.
- Type II enzymes do not suffer from this disadvantage because the cut is always at the same place, either within the recognition sequence or very close to it. Eg. EcoRI cuts DNA only at the hexanucleotide sequence 5′-GAATTA-3′.
- Digestion of DNA with a Type II enzyme therefore gives a reproducible set of fragments whose sequences are predictable if the sequence of the target DNA molecule is known.
- Restriction enzymes cut DNA in two different ways. Many make a simple double-stranded cut, giving a blunt or flush ends. Eg. Alu1 and BSR B1 enzyme.
- Many enzymes cut the two DNA strands at different position, usually two or four nucleotides apart, so that the resulting DNA fragments have short-stranded overhangs at each end. These are called sticky or cohesive ends because base pairing between them can stick the DNA molecule back together again. Eg. Sau 3A1, EcoR1 and PstI enzymes.

3. **DNA ligases:** Which join DNA molecules together by synthesizing phosphodiester bonds between nucleotides at the ends of two different molecules, or at the two ends of a single molecule.
   - The most widely used DNA ligase is obtained from E. coli cells infected with T4 bacteriophage. Its natural role is to synthesize phosphodiester bonds between unlinked nucleotides present in one polynucleotide of a double-stranded molecule.
   - If the two molecules have complementary sticky ends, and the ends come together by random diffusion events in the ligation mixture, then transient base pairs might form between the two over-hangs.
   - If molecules blunt ended, then they cannot base pair to one another, not even temporarily, and ligation is a much less efficient process, even when the DNA concentration is high and pairs of ends are in relatively close proximity.

Conversion of blunt ends to sticky ends:

1. Short, double stranded molecules called linkers or adapters are attached to the blunt ends. They contain a recognition sequence for a restriction endonuclease and so produce a sticky end after treatment with the appropriate enzyme.
2. Creation of sticky ends is by homopolymers tailing, in which nucleotides are added one after the other to the 3′ terminus at a blunt end. The enzyme involved is called "terminal deoxy nucleotidyl transferase". Eg. Poly (G) tail, which would enable the molecule to pair to other molecules that carry poly (C) tails.

4. End-modification enzymes:

1. Terminal deoxynucleotidyl transferase is a template-independent DNA polymerase, because it is able to synthesize a new DNA polynucleotide without base pairing of the incoming nucleotides to an existing strand of DNA or RNA. Its main role in recombinant DNA technology is in homopolymer tailing.
2. Two other end-modification enzymes are alakaline phosphatase and T4 polynucleotide kinase

- Alkaline phosphatase, which removes phosphate groups from the 5′ ends of DNA molecules, which T4 polynucleotide kinase performs the reverse reaction to alkaline phosphatase i.e. adding phosphates to 5′ ends. These enzymes are used during complicated experiments, but its main application is in the end-labelling of DNA molecules.

What is meant by vector? And mention the properties of good vector?

Vector: A vector is a DNA molecule that has the ability to replicate in an appropriate host cell, and into which the DNA insert is integrated for cloning. Therefore, a vector must have a origin of DNA replication (Denoted as ori) that functions efficiently in the concerned host cell.

The vector is a vehicle (or) carrier which is used for cloning foreign DNA in bacteria. The cloning vehicles are called vectors. Any extra-chromosomal small genome, Eg. Plasmid, phage and virus may be used as vector.

Properties of a good vector:

- It should be able to replicate autonomously that is independent of the replication of host chromosome
- It should be easy to isolate and purify
- It should be easily introduced into the host cells.
- The vector should have suitable marker genes that allow easy detection or / and selection of the transformed the host cell. Eg. Genes for ampicillin and Tetracycline resistance.
- The cells transform with recombination DNA should be indentifiable (or) selectable from those transformed by the unaltered vector.

- A vector should contain unique target sites for as many restriction enzymes as possible into which the DNA insert can be integrated. When expression of the DNA insert is desired, the vector should contain suitable regulatory elements like promoter, operator, and ribosome binding sites.

Explain in detail about the different types of vectors used in cloning?

The most commonly used vectors for cloning are plasmid vectors, cosmid vectors, phagemid vectors, shuttle vectors, yeast vectors, bacterial artificial chromosome vectors (BAC).

Plasmid vectors:

Many different E. coli plasmids are used as vectors. Plasmids are self replicating double standard circular DNA molecules which exist in cells as extra chromosomal units in the bacteria (or) A plasmid is a DNA molecules other than bacterial chromosome that is capable of independent replication and transmission. Plasmids multiply at the same rate as that of bacterial cell multiply at a rate independent of that of host cell. Such that even upto 1000 copies per cell are obtained using the plasmid as vectors. The length of the DNA segments to be cloned is 10-15 base pairs. Any segments of DNA to be cloned is called as DNA insert

1. pBR – 322 – Ideal plasmids vector

The name pBR denotes 'p' signifies plasmid BR is from Boliver Rodriguez (1977) the two initials of the scientist who developed pBR-322 which is the most widely used plasmids

Features

- It contains 4362 base pair (4.3kb: 10 base pairs) of double stranded DNA, its entire base sequence is known.
- It contains col-E1 replication origin with relaxed replication control.
- It has two selectable markers i.e., two antibiotic resistance genes, tetr for tetracycline and ampr for ampicillin.
- It has several unique recognition sites for 12 different restriction endonuclease enzymes located within the tetr and ampr genes.
- These plasmids can take up a DNA insert of 10 kb (kilo base pairs) length.
- The presence of restriction sites within the markers tetr and ampr permits an easy selection for cells transformed with the recombinant pBR-322.
- Insertion of the DNA fragment into the plasmid using restriction enzyme (*PstI* or *PvuI*) places the DNA insert within the gene ampr, this makes ampr non functional.
- Bacterial cells containing such a recombinant PBR 322 will be unable to grow in the presence of ampicillin, but will grow on tetracycline.

When restriction enzyme *Bam HI* and sal *I* is used, the DNA insert is placed within the gene tetr making it non – functional. Bacterial cells possessing such a recombinant PBR – 322 will therefore, grow on ampicillin but not on tetracycline. This allows an early selection of a single bacterial cell having recombinant p PBR 322 from among other types of cells.

2. pUC vector / plasmid

The name pUC is derived because it was developed in the University of California (UC) by Messings and his colleagues. This vector is a derivative of pBR322 and is much smaller 2.7kb. It has all the essential parts of pBR322

Eg. The ampicillin resistant genes and Col E ori of replication.

The second scorable marker is due to E.coli gene lac Z encoding the fragment of β-galactosidase, which aplits lactose into glucose and galactose. When E.coil cells are transformed by these pUC vectors there will be

- Cells which are transformed but non recombinant these cells are ampr and able to synthesize β-galactosidase when an inducer of lac operon IPTG and X-gal (5 Bromo 4 chloro 3 Indolyl β-D galactopyranoside) a substrate for ß-galactosidase was added in the media, it will be broken down by the enzymes to give deep blue colonies.
- Cells with recombinant pUC vectors – these cells are ampr but unable to synthesize β-galactosidase. When IPTG and X-gal are put in the media it will give white colonies because the substrate cannot be broken down. Recombinants can therefore be selected by the colour of the colonies and there will be no need of replica plating as is done incase of pBR322.

  Cells plated + agar + ampicillin + IPTG + X-gal = Non recombinant -Blue colonies

  Cells plated + agar + ampicillin + IPTG (Isopropyl thio-galactoside) + X-gal = Recombinant-white colonies

3. Ti plasmid and Ri plasmid

The Ti plasmid and Ri plasmid is a large conjugative plasmid or megaplasmid of about 200 kb. These are commonly used plasmids for transformation of plant cells

- *A. tumefaciens* has the Ti plasmid (ca 200 kb) while *A. rhizogenes* has the Ri plasmid have similar general features and can be interchanged between the two species. These are plant pathogenic gram negative soil bacteria to cause crown gall (*A. tumefaciens*) and hairy root diseases of dicot plants. They infect plant cells near wounds, usually at the crown of roots at the soil surface.
- These plasmids naturally transfer a part of their DNA, the T – DNA into the host plant genome, which makes *Agrobactrium* a natural genetic engineer. The T – region (ca 23 kb) which contain genes for opine metabolism and phytohormone independence, this region is transferred into host cells

and is integrated into their genome. They have another region, called 'vir' region, which produces, an endonuclease essential for the excision transfer of T-region into plant cells. The genes to be transferred are placed within T DNA and contain eukaryotic regulatory sequences. As a result these genes are expresses only in plant cells. They are not expressed in the Agrobacterium.

Bacteriophage vectors

Bacteriophages are viruses that infect bacteria. These are usually called phages. The phages are constructed from two basic components.

(*i*) Capsid (or) protein coat within which the nucleic acid genome is enclosed.

(*ii*) Nucleic acid genome which is packaged with in the capsid phages can be both DNA and RNA phages (MS2)

The DNA phages can be double strandard (T2 T4 T6), (or) single standard (X 174 and M-13)

RNA virus – MS2

Several bacteria phages are used as cloning vectors, the most commonly used bacteria phage vectors being (lambda) phage, M-13 phages.

Phage Vectors Present Two Advantages Over Plasmid Vectors

1. They are more efficient than plasmids for cloning of large DNA fragments. The length of the DNA segments is to be cloned is upto 25 kb.
2. It is easier to screen a large number of phage plaques than bacterial colonies for the identification of recombinant pla ques / clones.

Bacteriophage vectors (Lambda)

The genome (48.502 base pairs) contains an origin of replication and genes for head and tail proteins and enzymes of DNA replication, lysis and lysogeny, and single stranded protruding cohesive ends of 12 bases at its 51 ends. These two cohesive ends are referred to as cos sites (sticky (or) cohesive ends) these cohesive ends enable DNA to form a circular molecule when it is injected into the *E. coli* cells. The central part (Nonessential segment) of the chromosome may be excised with restriction enzyme and replace with foreign DNA.

Advantages

- Large size DNA fragment upto 25 kb can be cloned as compared to plasmid vector which can take up 10kb
- Large number of phage plaques can be easily screened
- They are more efficient than plasmids for cloning of large DNA fragments

Phage M -13 vectors: They are derived from the 6.4 kb genome of the E. Coli filamentous bacteriophage M13. This phage has a single stranded linear DNA genome in phase particles which converts into double stranded circular DNA molecule in the host cells. It contains origin of replication and a scorable marker gene lac-Z that compliments the gal host giving blue colonies. On transformation only white (or) clear plaques are obtained, thus permitting easy selection of recombinant plaques.

Cosmid vectors: Cosmids are essentially plasmids that contain a minimum of 250bp of DNA including *cos* site and sequences needed for binding of and cleavage by terminase

A typical cosmid has

- Origin of replication
- Selectable markers from a plasmid ampr and tetr
- Cloning sites (unique restriction sites) and Cos site (The sequence yielding cohensive ends) of which are essential for efficient packing of -DNA into virus particle which infect host cells. Packaged cosmids infect host cells like -particle but inside the host they replicate and propagate like plasmids.

The typical features of cosmids are as follows

(*i*) Cosmids can be used to clone DNA inserts upto 45kb

(*ii*) They can be packaged into -particles which infect host cells, which is many fold more efficient than plasmid transformation.

(*iii*) Selection of recombinant DNA is based on the procedure applicable to the plasmid making up the cosmid.

(*iv*) Finally, these vectors are amplified and maintained in the same manner as the contributing plasmid

Eg. Cosmid PHC-79 and cosmid PJB-B

Phagemid vectors: - Plasmids + Bacteriophage, a plasmid vector that contains the original from the page, in addition to that of plasmid.

These are artificially constructed vectors containing the characters of plasmids and bacteriophages. These consist of socrable and selectable markers for antibiotic resistance.

Eg. PUC-118, PUC-119

Yeast vector: Yeast is a Eucaryotic which can grow as a single cell and produce colonies on agar plate. The vectors used in yeast are

1. Plasmid vectors
2. ARS (Autonomously Replicating sequences) vectors.
3. Minichromosome vectors
4. Yeast artificial chromosomes (YAC). Among these the last type of vectors are most commonly used

Yeast artificial chromosome vectors: developed byt David Burke *et al* (1987). These are linear vectors that behave like a Yeast chromosome hence they are called yeast artificial chromosome vectors.

A typical YAC contains the following functional elements from yeast. E.g. pYAC 3

(*i*) An ARS sequence for replication

(*ii*) A CEN4 sequence – centromeric sequence function

(*iii*) A TEL sequence (Telomeric sequence) at two ends for protection from exonuclease action.

(*iv*) One (or) two selectable marker genes Eg. TRP 1 and URA 3

(*v*) SUP4, a selectable marker into which the DNA insert is integrated.

(*vi*) And necessary sequences from E.coli plasmid for selection and propagation in E.coli YAC's are used for cloning very large (1000-2000kb) DNA segments used for mapping of complex Eukaryotic chromosomes.

The problem faced with YAC's is the recombination between the copies of inserts and more particularly deletions in DNA vectors.

Bacterial Artificial Chromosome (BAC):

- Artificial chromosomes are circular or linear vectors that are stably maintained in, 1 to 2 copies per cell.
- Bacterial artificial chromosome is another cloning vector system in *E.coli* developed by Melsimon and his colleagues have the origin of replication 'Ori'S' of *E.coli* 'F' factor which allows a strict copy number control at 1 / 2 copies per cell. The low copy number helps to maintain the DNA inserts without any change arising from recombination between the copies of DNA inserts and avoids any counter selection that may arise due to over expression of cloned genes.
- These vectors are used to clone the DNA inserts up to 300kb.
- These vectors are able to maintain in stable state and extensively used in analysis of large genomes but the main disadvantage of BAC vectors is somewhat laborious construction of BAC libraries.

Shuttle vectors:

- These are designed to replicate in cells of two different species. (These are plasmids capable of propagating and transferring (Shuttling) genes between two different organisms).
- Therefore they contain two origins of replication, one specific for each host species, as well as those genes necessary for their replication and not provided by the host cells.
- These vectors are created by recombinant DNA technique. Some of them can be grown in two different prokaryotic species, while others can propagate in prokaryotic species (E.coli) and a Eukaryotic one (yeast, plants and animals). Since these vectors can be grown in one host and then moved into another without any extra manipulation they are called shuttle vectors.

- Shuttle vectors have been designed to specifically satisfy the need i.e., for the initial cloning of DNA inserts in E.coli and sub-sequent functional tests in the species to which the DNA inserts belong. Most of the Eukaryotic vectors are infacts shuttle vectors.

Phasmid vectors:

- These vectors are shortened linear genomes containing DNA replication and lytic functions plus the cohesive ends of the phage.
- Their middle non-essential segement is replaced by a linerised plasmid with intact replication nodule.
- A phasemid vector contains several tandem copies of the plasmid to make it longer than 38 kb, the minimum size needed for packaging in particles

What is meant by gene cloning? And write the steps involved in gene cloning

Gene cloning: Gene cloning (or) DNA cloning is defined as insertion of fragment of DNA representing a gene into a cloning vector and subsequent propagation of DNA molecule in a host organism. It is defined as isolation and amplification of an individual gene sequence by immersion of that sequence into a bacterium where it can be replicated. The basic aim of cloning genes to produce multiple copies of DNA to isolate and characterize specific genes for their use in basic and applied research. Basic events involved in gene cloning are

- Isolation and synthesis of gene of interest
- Insertion of isolated gene DNA fragment into a suitable vector to produce rDNA.
- Introduction (of) rDNA (vector) into a suitable organism / cell called host. This process is known as transformation. Generally, E. coli is used for initial cloning
- Selection of transformed host cell by rDNA
- Multiplication of rDNA within the host cell to produce a number of identical copies of cloned gene
- A rDNA molecule is produced by joining together two (or) more DNA segments usually originating from different organisms. More specially a rDNA mol is a vector (Eg. A plasmid phage (or) virus) into which the desirable DNA fragment has been inserted to enable its cloning in an appropriate host. This is achieved by using specific enzymes (restriction enzymes) for cutting of DNA into suitable fragments and then for joining (ligases) together the appropriate fragments.
- The rDNA molecule produced which contains the coding region from one organism joined to regulatory sequence from another organism and such a gene is called chimeric gene. The capability to produce rDNA molecules has given human beings the power and opportunity to create novel gene functions to suit specific needs

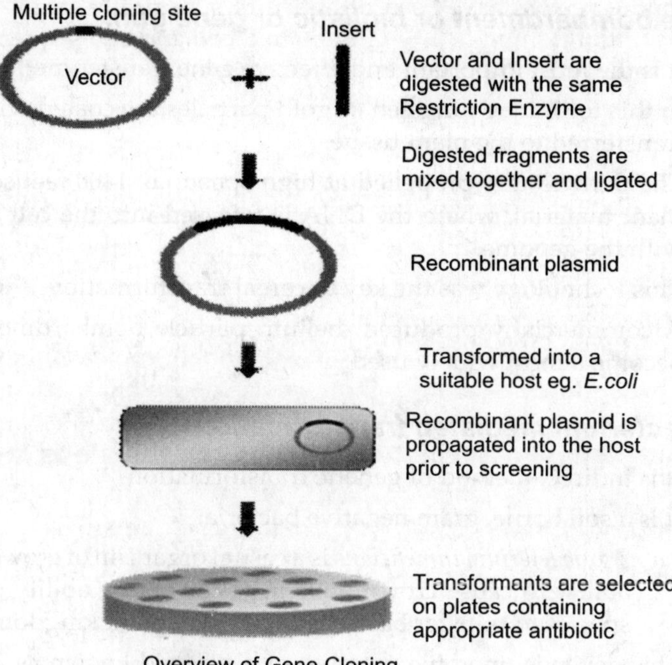

Overview of Gene Cloning

Discuss the Methods of Genetic Transformation in Plants.

Genetic transformation is nothing but genetic improvement of a living organism by way of integrating a gene or introduction of foreign DNA into plant followed by stable transmission through germ line.

Methods of gene transfer:

Direct transfer: Introduction of DNA into plant cells without the involvement of biological agents such as *Agrobacterium* and leading to stable transformation is called direct gene transfer.

- Particle bombardment method
- Electroporation'
- Laser induced transformation
- DNA uptake into protoplast (PEG) mediated transformation
- Micro injection and Macro injection
- Liposome mediated DNA transfer.
- Use of silicon carbon papers
- Genotype independent transformation
  - (*i*) Shoot apical meristem
  - (*ii*) Pollen tube pathway
  - (*iii*) Flower dip method

Particle bombardment or biolistic or gene gun:

- It is the most important and effective gene transfer method in regular use.
- In this technique tungsten or gold particles are coated with DNA i.e., to be transferred to the plant tissue.
- The particles are propelled at high speed i.e. 1400 feet/sec. into the target plant material, where the DNA is released into the cell and can integrate with the genome.
- This technology was the key to cereal transformation.
- A commercially produced, helium particle bombardment apparatus has become a most widely used.

Agrobacterium mediated transformation:

- It is indirect method of genetic transformation
- It is a soil borne, gram negative bacteria.
    - o *Agrobacterium tumeficians* is a casual organism of crown gall disease. The ability to cause crown galls depends on the ability of agro bacterium species to transfer bacterial genes into plant kingdom.
- This depends upon the presence of a plasmid known as Ti plasmid. (The T DNA region is actually transferred from the bacteria and into the plant cell)
- Where it becomes integrated into the genome of the host plant. It involves
    - o Disarming of Ti plasmid
    - o Incorporation of desired DNA into Ti plasmid
    - o Transformation of host cell.

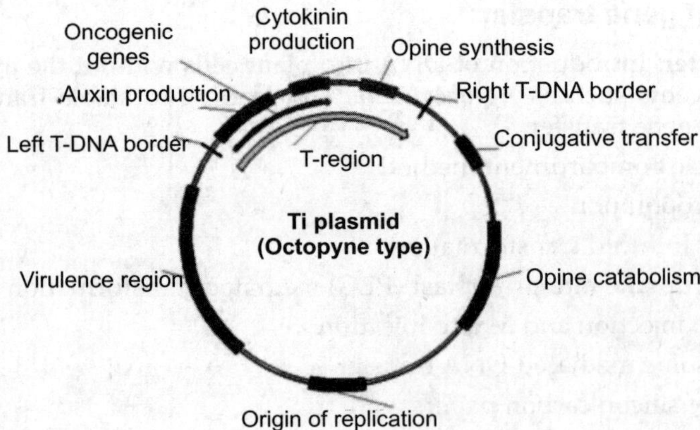

What is meant by PCR? And discuss the steps involved in PCR

Polymerase chain reaction:

PCR technique is developed by Kary mullis 1985. It is one of the most powerful molecular biology technique used to multiply minute (or) trace amounts (µg microgram quantities) of DNA copies of the desired DNA by multiple cycles of cooling and heating in a reaction catalysed by a heat stable DNA polymerase enzyme. (The PCR is carried out *in vitro*). PCR is based on the features of semi conservative DNA replication carried out by DNA polymerase in prokaryotic and eukaryotic cells.

Procedure of PCR: Amplification of DNA is achieved by a repetitive series of cycles involving 3 steps.

1. **Denaturation:** The DNA double helix separated into two complimentary single strands by heating reaction mixture to temperature between 90-98°C that ensures DNA denaturation. The duration of this step in the first cycle of PCR is usually 2 min at 94°C, but in subsequent cycles it is of only 1min duration.

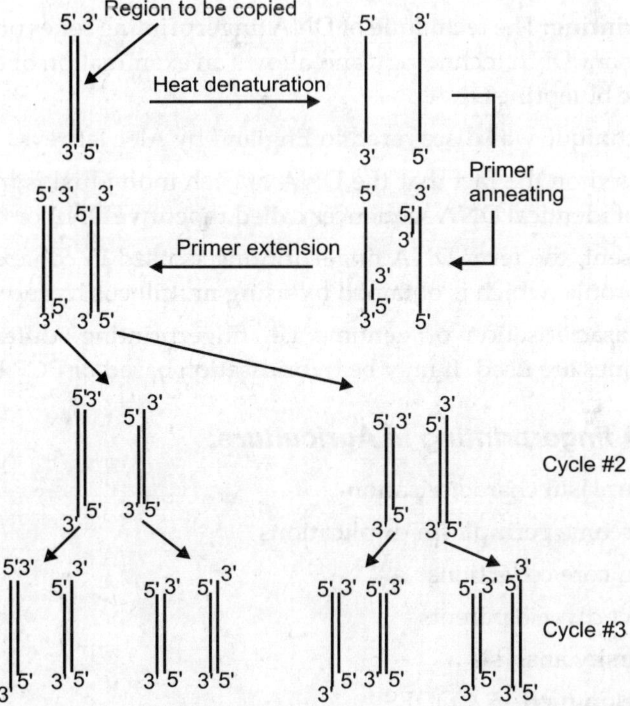

2. **Annealing:** The mixture is now cooled to a temperature of generally 40-60°C that permits annealing of primer to the complimentary sequences in the DNA. The duration of annealing step is usually 1 min during the first as well as the subsequent cycles of PCR. The annealed primers are then extended (i.e. synthesis of DNA) with Taq DNA polymerase.

3. Primer Extension (or) synthesis: It involves heating the mixture to 72°C at which a special polymerase synthesizes the new DNA strand by using the original strand as the template starting by the primer utilizing their 3′OH free ends and continuing in the 3′ direction.

Application:

- DNA cloning for sequencing
- DNA based phylogenetic studies.
- Functional analysis of gene
- Diagnosis of hereditary diseases
- Identification of genetic finger prints used in forensics and paternity studies
- The detection and Diagnosis of infectious diseases
- It can be used to determine the sex of embryos

Explain in brief about DNA Fingerprinting.

DNA fingerprinting: The technique of DNA fingerprinting relies on developments from recombinant DNA technology and allows an examination of each individual unique genetic blueprint-DNA.

- The technique was discovered in England by Alec Jeffreys.
- It is based on the fact that the DNA of each individual is interrupted by a series of identical DNA sequences called repetitive DNA or tandem repeats.
- At present, the term DNA fingerprinting is used in context of multilocus DNA profile, which is obtained by using multilocus hypervariable probes.
- For characterisation of genome i.e. fingerprinting, different molecular techniques are used. It may be hybridisation based or PCR based.

Role of DNA fingerprinting in Agriculture:

- In germplasm characterisation
- To overcome germplasm duplications
- To form core collections
- To select diverse parents
- In diversity analysis
- In developing BAR CODES
- As an supplementary tool in patenting regime
- In identifying EDV's
- It can be used to know the exact genetic diversity at molecular level rather than at morphological level.

Define molecular marker? And discuss the molecular markers, their principle, advantages and disadvantages with diagram

Molecular marker is a DNA segment that is readily detected and whose inheritance can easily monitored. The use of molecular markers is based on the naturally occurring DNA polymorphism.

A DNA marker is a small region of DNA showing sequence polymorphism in different individuals with in a species or among different species.

The DNA markers are most widely used type of markers predominantly due to their abundance; the first such DNA markers to be used for the fragments produced by the digestion with restriction endonucleases that means restriction fragment length polymorphism (RFLP) based genetic markers. A wide range of molecular techniques is now available to detect the polymorphism at DNA level. These have been grouped in to the following.

1. Non PCR based approaches: *Eg.* RFLP
2. PCR based approaches: E.g. RAPD, AFLP, SSR and SNP etc

Restriction Fragment Length Polymorphism (RFLP):

- First molecular marker developed in 1980 by Botestein and is based on southern hybridization technique.
- RFLP means variation in homologous fragment length of DNA on being digested with restriction endonucleases.
- In this, low copy/single copy number DNA sequence is used as probe. So the size of the DNA fragments homologous to the probe DNA depends upon the location of target sites.
- Base changes can change the sequence of target site and deletion and insertions can also change the location of target sites.

Diagram:

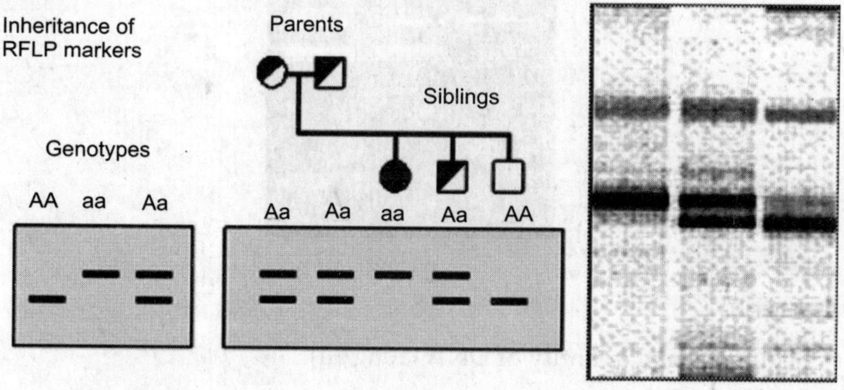

Advantages:

- Transferable across the population
- Have good repeatability, highly reproducible and very robust
- Useful in comparative genome mapping
- Co dominant marker.

Disadvantages:

- The assay is tedious and time consuming
- Requires large quantity of DNA (20µl of DNA)
- Extremely limited utility in marker assisted selection due to very low assay efficiency.

Random Amplified Polymorphic DNA (RAPD):

- It is dominant marker based on PCR technique
- It employs a decamer primer of arbitrary sequence which is annealed to template DNA at 37°C
- Variation in RAPD profile in the form of presence or absence of bands results from variations in primer binding sites.
- The number of primer binding sites or number of fragments expected to be amplified = $2fN/16^n$, where N = Genome size (bp), n = primer size (bp) and 2f = probability of number primer binding sites including both the strands

Diagram:

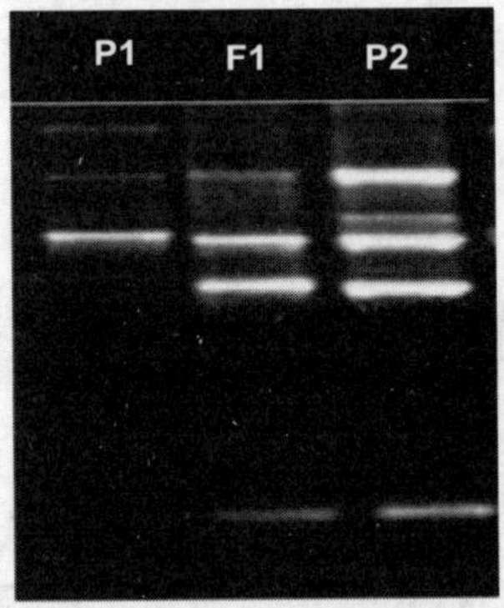

Advantages:

- Requires small quantity of DNA (10ng/µl)
- Quality of DNA is not important requirement
- Needs limited investment in time and training.

Disadvantages:

- Non reproducible due to low annealing temperature. For this RAPD marker will convert in to SCAR marker by sequencing RAPD marker termini and design a longer (eg. 24 nt) primers for specific amplification through PCR.
  - ❖ SCAR marker is Co dominant marker.

S. No	RAPD	SCAR
1	Primer size is 10 nucleotides (decamer)	Up to 24 nucleotides
2	Results are not consistent	Results are consistent
3	Single primer	Both forward and reverse primers are different
4	Size of fragment is small	Size of fragment is large
5	Number of fragments more	Only single fragment
6	Non reproducible and less robust	Reproducible and robust

Sequence tagged sites (STS):

- STS markers generated by unmodified PCR markers are designed based on cDNA sequence, random genomic DNA and ends of large genomic DNA fragment cloned in cosmid, λ and artificial vectors.

Advantages:

- Has all advantages of PCR
- Very robust in nature

Disadvantages:

- Considerable effort and time required to generate DNA sequence for designing primers
- Fragments amplified by primers may not show length variation i.e. monomophic. So convert into CAPS markers using restriction endonuclease enzymes.
  - ❖ CAPS will not work if there is no restriction site in sequence
  - ❖ CAPS marker is Co dominant marker.

Amplified Fragment Length Polymorphism (AFLP):

- AFLP are based on both RFLP and PCR
- AFLP markers are generated by selective amplification of DNA fragments obtained by restriction enzyme digestion.
- High molecular weight DNA is digested with two restriction enzymes one hexa cutter (EcoRI) and one tetra cutter (Mse I)

- It provides the highest number of amplified products among all the DNA profiling systems.
- Dominant marker, because of many fragments, we just cannot find out its homologous band, so it is almost like dominant marker

Double stranded DNA

↓

Restriction digestion with EcoRI and MseI

↓

Ligation of EcorI and MseI adaptors

↓

Annealing of primers with one additional random nucleotide at 3′ end (primer + 1)

↓

Pre selective amplification with primer + 1

↓

Annealing of primers with the additional nucleotides at 3′ end (primer + 3)

↓

Selective amplification with primer + 3

↓

Amplified restriction fragment

↓

Amplified product separated on DNA sequencing gel

Advantages:

- No prior sequence information needed
- Plenty in number
- Polymorphic information content is high
- Can be used in unsequenced crops
- Provide raw material for STS derivation.

Disadvantages:

- Lengthier and costlier than other PCR based markers
- Requires good quality DNA for complete digestion by enzyme, because partial digestion of DNA results in non reproducible variation in DNA profile.

Variable Number of Tandem Repeats (VNTR)/Simple sequence Repeats (SSR):

- Variable number of tandem repeats of mono or di or tri or tetra or penta nucleotides dispersed throughout the genome
- Variation due to number of repeats in different individuals at a given locus and these variation constitutes the alleles of VNTR loci.
- Classification:

Feature	Microsatellites	Minisatellites
DNA fragment	100-200 bp	0.2-2 kb
Tandem repeat sequence	2-9 bp	15-16 bp
Number of repeat units per locus	100 times	1000 times

  ❖ In microsatellite marker, the fragment length of amplification is usually bigger than the expected size.

- Co dominant marker.

Advantages:

- Transferable between populations
- Abundant and uniformly distributed in the genome
- Hyper variable (large number of alleles per locus)
- Highly reliable and robustable.

Disadvantages:

- Expensive and time consuming to detect SSR loci
- Not available in all crop plants

S .No	SSR	ISSR
1	Simple sequence repeats	Inter simple sequence repeats
2	Amplified region contains repeats	Contains region between repeats i.e., non repeat sequence
3	Both forward and reverse primers different sequence	Only one primer acts as forward and reverse primers
4	Genetic basis of primer is only unique flanking region of repeats	Genetic basis of primer is repeat perse
5	Genetic basis of polymorphism is variation in number of repeats at particular locus.	Variation in inter region between repeats
6	Amplification profile is single locus	Amplification profile is multilocus because non repeat region may vary
7	Genetic nature of polymorphism is Co dominant	Dominant marker

Single Nucleotide Polymorphism (SNP):

- Polymorphism observed between individuals may arise due to deletion or addition of multiple bases or due to single nucleotide substitution
- SNPs are biallelic.

Advantages:

- Co dominant marker
- Abundant in number
- Well suited for automated high through put genotyping
- Most useful to development of haplotype tags
- Serve to integrate physical and genetic maps

Disadvantages:

- Requirement of sequence information
- High development or start up cost

Why SNPs have only two alleles? And explain the SNP detection methods with suitable diagram?

Any four nucleotides could be present in any position in the genome, so it might be imagined that SNP have four nucleotides. Theoretically it is possible but practically most SNPs exist just as two variants. This is because the way in which SNP arise and spread in population.

SNP arises when a point mutation occurs in a genome converting one nucleotide to another nucleotide. If this mutation occurs in reproductive cells of an individual, one or more offspring inhert this mutation and after many generations, this SNP become established in the population.

But there are two alleles i.e. original and mutated version. For third allele to arise a new mutation must occur at the same position in the genome in another individual and his or her offspring must reproduce in such a way that the new allele becomes established. This scenario is not impossible, but it is unlikely. Consequently most SNPs are biallelic.

SNP detection: SNPs are detected by Oligonucleotide hybridisation, DNA chip technology and Solution hybridisation technique

Oligonucleotide hybridisation:

An oligonucleotide is a short, single stranded DNA molecule, usually less than 50 nucleotides in length that is synthesised in a test tube. If the conditions are just right, then the oligonucleotie will hybridize with another DNA molecule only if the oligonucleotide forms a completely base paired structure with the second molecule. If there is a single mismatch a single position within a oligonucleotide that does not form a base pair, then hybridisation does not occur. Oligonucleotide hybridisation can therefore discriminate between the two alleles of an SNP.

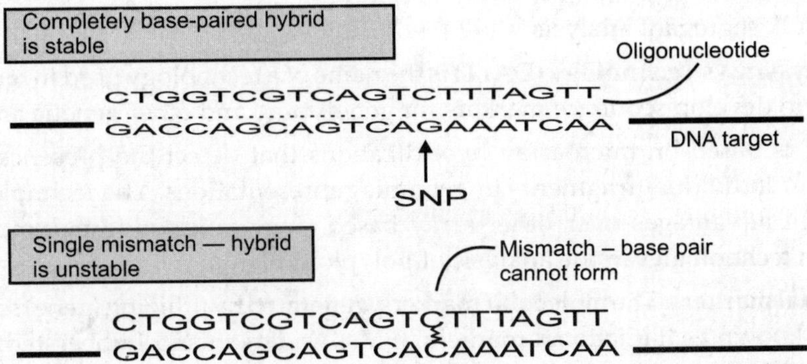

DNA chip and solution hybridisation techniques also used to detect the SNPs. **Solution hybridisation techniques** are carried out in the wells of microtitre tray, each well containing a different oligonucleotide, using detection system that can discriminate between unhybridised, single stranded DNA and the double-stranded product that results when an oligonucleoide hybridizes to a test DNA. Several systems have been developed, one of which makes use of a pair of labels comprising a fluorescent dye and a compound that quenches the fluorescent signal where brought into close proximity with the dye. The dye is attached to one end of an oligonucleotide and the quenching compound to the other end. Normally there is no fluorescence because the oligonucleotide is designed in such a way that the two ends base pair to one another placing the quencher next to the dye. Hybridisation between oligonucleotide and the test DNA disrupts this base pairing, the quenches away from the dye and enabling the fluorescent signal to be generated.

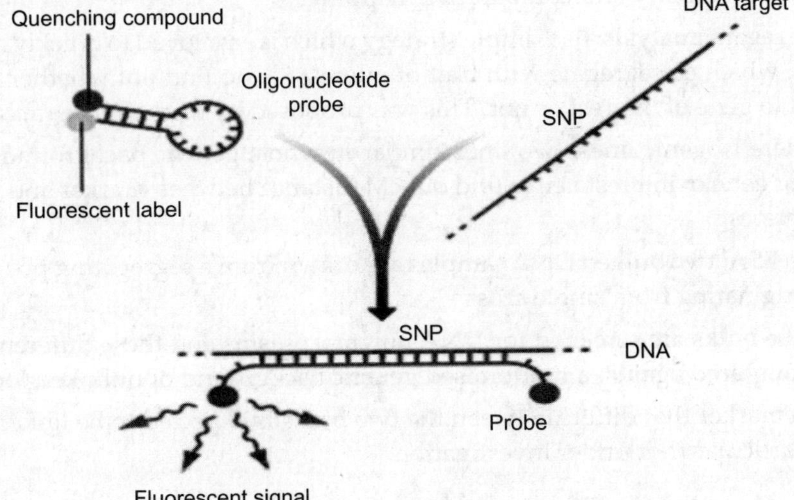

Write short notes on the following:

1. Diversity array technology marker (DArT)
2. Functional markers
3. Bulk segregant analysis

Diversity Arrays Technology (DArT) is the name of a technology used in molecular genetics to develop sequence markers for genotyping and other genetic analysis.

DArT is based on microarray hybridizations that detect the presence versus absence of individual fragments in genomic representations. The technology has significant advantages over other array based single-nucleotide polymorphism detection technologies in the analysis of polyploid plants.

Functional markers: The molecular markers, generated by utilizing (gene) sequence data are known as 'Functional markers' (FMs, Anderson and Lubberstedt, 2003). These markers are superior to random DNA markers (derived from genomic DNA) owing to complete linkage with potential trait locus alleles.

The FMs has advantages over genomic DNA markers as the later were designed from genomic DNA sequences such markers sample genetic variation in the genome more or less randomly and are sometimes referred to 'neutral' or random markers (RMs).

Applications:

- Enhanced breeding efficiency with precision
- Marker trait association: FMs are perfect markers; the polymorphism may yield insights into gene function.
- Functional diversity
- Transferability and comparative mapping.

Bulk segregant analysis: It is simple strategy which is employed to quickly identify markers, which co segregate with trait of interest i.e., to find out whether marker is linked to gene of interest or not. This was proposed by Michelmore *et al.* 1991.

- Here isogenic lines, two lines similar in whole genetic background except for gene of interest i.e., to find out cM distance between marker and gene of interest.
- In BSA, two bulked DNA samples are drawn from a segregating population originating from single cross.
- The bulks are screened for DNA polymorphisms and these differences are compared against a randomised genetic background of unlinked loci.
- A marker that differs between the two bulks is expected to be linked to the particular trait under investigation.

Explain briefly the role of molecule markers in crop management?

Role of molecular markers in crop improvement are

1. Construction of high density (saturated) genetic maps of crops to provide assessment of each segment of genome for economic use, on the basis of map position of the gene the DNA molecule rather than through its phenotypic effect.
2. Estimation of genetic diversity, DNA finger printing of crop varieties and genetic collections help to characterize genetic diversity for effective management and utilisation of germplasm resources.
3. Comparative gene mapping in different plant species to determine parallelism in the gene order (synteny) and evolutionary relationship.
4. Marker aided early generation selection of transgressive segregants to increase the speed and efficiency of developing new varieties.
5. Introgression of genes from wild germplasm.
6. Mapping of poly genes: QTLs.
7. Gene pyramiding: Assembly of many genes and alleles with similar effect from different loci is practically difficult through screening for molecular markers associated with each of such genes. The genes for different characters can also be assembled in to genotype with help of markers.
8. Map based cloning of genes.

Three specific applications of molecular markers with specific examples:

1. **Gene introgression:** Wild relatives of crop plants constitute a rich source of genetic variation which can be efficiently utilised for improvement of both qualitative and quantitative traits.
   - Eg. Genes resistant to *Meloidogyne incognita* and tobacco mosaic virus have been introduced from related species *Lycopersicon peruvianum* in tomato.
   - Problem of linkage drag during gene introgression by backcrossing i.e., undesirable linked genes is often faced by breeders. Here molecular linkage maps provide a method to reduce the problem of linkage drag by allowing selection of individual containg recombinant chromosome which breaks linkage drag.
   - It has been estimated that the use of molecular markers can reduce linkage drag at least 10 fold in a fraction of the time needed by traditional breeding.
2. **Gene pyramiding:** Pyramiding of Xa4, Xa5, Xa13 and Xa21 genes conferring resistance to four different races of BLB pathogen has been achieved in rice by using molecular markers at IRRI, Philippines. Two of these genes

(Xa13 and Xa21) have been recently been combined with basmati quality characters in rice in IARI, New Delhi.

3. **Construction of heterotic hybrids:** In maize, through marker aided transfer of yield QTLs from inbreeds lines like TX303 to B73 and Oh43 to Mo17, Improved inbreeds were generated, which are called as enhanced lines. 15 enhanced B73 lines crossed with 18 enhanced Mo17 lines to produce 93 hybrids at North Carolina, USA. Six of hybrids exceed national check hybrids by two standard deviation and two hybrids gave 15% more yield than checks.

Define MAS. Explain the advantages of marker assisted selection over conventional breeding.

Marker Assisted Selection (MAS)

Marker assisted selection involves selection of plants carrying genomic regions that are involved in the expression of traits of interest through molecular markers.

With the development and availability of an array of molecular markers and dense molecular genetic maps in crop plants, MAS has become possible for traits both governed by major genes as well as quantitative trait loci (QTLs). In general, the success of a marker based breeding system depends on three main factors:

- A genetic map with an adequate number of uniformly-spaced polymorphic markers to accurately locate desired QTLs or major gene(s)
- Close linkage between the QTL or a major gene of interest and adjacent markers
- Adequate recombination between the markers and rest of the genome and
- Ability to analyse a large number of plants in a time and cost effective manner.

Advantages of Marker Assisted Selection over Conventional Breeding:

- The segregants can be scored at seedling stage for traits that are expressed late in plant development. This includes traits such as grain quality, male sterility and photoperiod sensitivity.
- It is possible to screen for traits that are extremely difficult, expensive or time consuming to score or measure such as tolerance to drought, salt, mineral deficiencies and toxicity, root morphology, resistance to nematodes or to specific races or biotypes of diseases or insects.
- Selection can be practiced for several traits simultaneously which is difficult or even impossible by conventional means.
- With the use of co-dominant markers, heterozygotes are easily identified and distinguished from the either homozygotes without resorting to progeny testing. This saves time and effort.

- Single plants can be selected using conventional screening methods for many traits, plant families or plants are grown because single plant selection is unreliable due to environmental factors. With MAS, individual plants can be selected based on their genotype.
- Markers can also be used to replacement for phenotyping, which allows selection in off season nurseries making it more cost effective to grow more generations per year.
- With MAS, total number of lines to be tested can be reduced and also selection of traits with low heritability.

Advantage of molecular markers is in terms of enabling "precise selection" and "unambiguous selection" even when the trait is unable to screened for. The inheritance is same as any other trait with the difference being one is able to see the genotype segregating rather than phenotype segregating. Thus the same approach can be adopted for.

- o Pedigree breeding whether the selections can be specifically effected right from F_2 to minimize the number by picking up maximum number of homozygous loci through co dominant markers.
- o Backcross breeding where both recurrent parent genome backgrounds can be assembled with minimum number of backcrosses looking for homozygous loci distributed over all chromosomes by opting for markers detectable spread over the entire chromosome on each arm.

While the target gene can easily be identified for the homozygous form in the BC_1F_2 generation through its tagged marker.

- o Recurrent selection where markers tagged to desire alleles or loci can be tracked for picking up the recombinants possessing most of the desired alleles with the minimum number of intermatings.
- o Hybrid breeding where markers can be employed to pick up heterotic divergent parents and confirmed hybrid progenies in large number of crosses with minimum field evaluations in initial stages.

Explain in detail about marker assisted backcross breeding:

Marker Assisted Backcross Breeding:

In general backcrossing, the recovery of recurrent parent gene is delayed still BC_4 and BC_5 generations. Backcrossing has been a widely used technique in plant breeding for almost a century. In most cases, the parent used for backcrossing has a large number of desirable attributes but is deficient in only a few characteristics (Allard 1999). The method was first described in 1922 and was widely used between the 1930s and 1960s. In general backcrossing, the recovery of recurrent parent gene is delayed still BC_4 and BC_5 generations. Where the average recovery of genome is 96.87% and 98.43% respectively. It takes more time and resources for generating these populations. But it is observed that BC_1 itself some plants have recurrent genotype up to 90%. These types of plants can be selected by polymorphic markers between parents. By employing background selection plants with minimum recurrent genome are identified and use for crop improvement. Thus the use of DNA markers in backcrossing greatly increases the efficiency of selection. Three general levels of marker-assisted backcrossing (MAB) can be described.

- In the first level, markers can be used in combination with or to replace screening for the target gene or QTL. This is referred to as 'foreground selection'. This may be particularly useful for traits that have laborious or time-consuming phenotypic screening procedures. It can also be used to select for reproductive-stage traits in the seedling stage, allowing the best plants to be identified for backcrossing. Furthermore, recessive alleles can be selected, which is difficult to do using conventional methods.
- The second level involves selecting BC progeny with the target gene and recombination events between the target loci and linked flanking markers. This is referred to as 'recombinant selection'. The purpose of recombinant selection is to reduce the size of the donor chromosome segment containing the target locus (i.e. size of the introgression). This is important because the rate of decrease of this donor fragment is slower than for unlinked regions and many undesirable genes that negatively affect crop performance may be linked to the target gene from the donor parent and this is referred to as 'linkage drag'.
- Third level involves the markers used for selection of recessive parent genome and this is referred to as 'Background selection'.

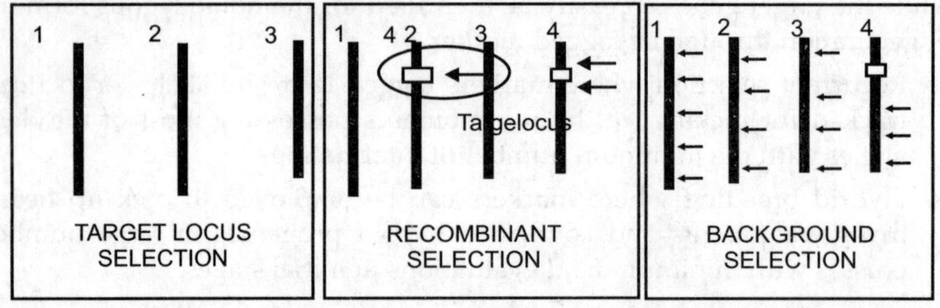

Recurrent parent × Donar parent

↓

F_1 × Recurrent parent

↓

BC_1F_1

↓ Foreground selection

BC_1F_1

↓ Background selection

BC_1F_1 × Recurrent parent

↓

BC_2F_1
↓Foreground selection
BC_2F_1
↓Background selection
BC_2F_2
↓ If we are sure about foreground and background markers than we will not go up to BC_2F_5,
↓ we will stop at BC_2F_2 and go for multiplication.
↓
BC_2F_5

The use of background selection during marker assisted breeding to accelerate the development of a recurrent parent with an additional gene is referred to as complete conversion (Ribaut *et al*, 2002).

Explain in Detail About Mapping Populations with Advantages and Limitations?

Types of mapping populations:

1. F_2 population
2. F_2 derived F_3 (F_2:F_3) population
3. Recombinant inbred lines (RILs)
4. Back cross populations
5. Doubled haploids (DHs)
6. Near – isogenic lines (NILs)

F_2 population:

- Best population for preliminary mapping
- Requires less time for development
- Can be developed with minimum efforts

But…

- It is not an immortal population
- F_2 population are of limited use for fine mapping
- Quantitative traits cannot be precisely mapped

F_2 derived F_3 (F_2:F_3) population:

- Developed by selfing F_2 population
- Suitable for Quantitative trait loci mapping
- Mapping recessive genes

But...

- Not immortal population

Recombinant inbred lines (RILs):

RILs are produced by continuous selfing or sib mating the progeny of individual members of an F_2 population until complete homozygosity is achieved.

Single seed descent (SSD) method is best suited for developing RILs. Bulk and Pedigree methods without selection can also be used for developing RILs.

- Once homozygosity is achieved, RILs can be propagated indefinitely without further segregation. This becomes an immortal mapping population.
- Since RILs are immortal population, they can be replicated over locations and years and therefore are of immense value in mapping QTLs.
- RILs being after several cycles of meiosis are very useful in identifying the tightly linked markers.
- Can become a public mapping tool.

But...

- Requires many seasons/generations for development
- Developing RILs is relatively difficult in crops with high inbreeding depression.

Back cross populations: Backcross population generated by backcrossing the F1 hybrid with any of the parent. Quick and easy to generate but have poor resolution.

Doubled haploids (DHs): chromosomal doubling of anther culture derived haploid plants from F1 generates Double haploids.

- DHs are products of one meiotic cycle, and hence comparable to F_2 in terms of recombinant information
- DHs are permanent mapping population and hence can be replicated over years and locations and maintained without any genotypic changes
- Useful for mapping both qualitative and quantitative characters
- Instant production of homozygous lines and thus saving a time.

But...

- Recombination from the male side alone is accounted

- Since it involves *invitro* techniques, relatively more technical skills are required in comparison with the development of other mapping populations.
- Often suitable culturing methods/haploid production methods are not available for number of crops and different crops differ significantly for their tissue culture response
- Further, anther culture induced variability should be taken care of.

Near – isogenic lines (NILs):

NILs are generated either by repeated selfing or backcrossing F_1 plants to the recurrent parent.

- NILs developed through backcrossing are similar to recurrent parent but for the gene of interest, while NILs develop through selfing are similar in pair but for the gene of interest.
- Like DHs and RILs, NILs are also immortal mapping population
- NILs are quite useful in functional genomics.

But...

- Require many generations for development
- Directly useful only for molecular tagging of the gene concerned, but not for linkage mapping.
- Linkage drag is a potential problem in constructing NILs, which has to be taken care of.

Indicate the genetic segregation ratios for following marker – population pair and explain, why do you expect the respective segregation ratios?

Genetic segregation ratio at marker loci in different mapping populations

Mapping population	Co dominant loci	Dominant loci
F_2 population	1:2:1	3:1
Back cross population	1:1	1:1
RILs	1:1	1:1
DHs	1:1	1:1
NILs	1:1	1:1

Mapping populations such as RILs and DHs equalise (neutralise) the type of marker because fixation of parental alleles at marker locus in homozygous condition. So these populations result in 1:1 segregation ratio at marker locus irrespective of genetic nature of marker. While F_2 population segregates in 1:2:1 for co dominant marker and 3:1 ratio for dominant marker i.e. F_2 population does not neutralise the marker type.

What is Meant by QTL. Explain the procedure involved in the mapping of QTLs?

QTL (Quantitative Trait Loci): A region of genome that is associated with effect on a quantitative trait.

- Concept given by Sax and term coined by Gelderman

Principle: QTL analysis is based on the principle of detecting an association between phenotype and genotype of markers. And QTL analysis is depends on "linkage disequilibrium".

Conceptual framework:

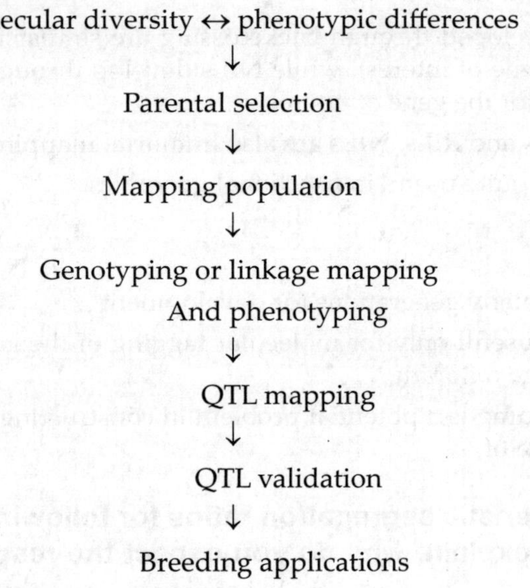

Factors affecting QTL mapping:

- Number of genes controlling the trait
- Heritability of trait
- Distribution of genetic effects and existence of genetic interactions
- Number of genes segregating in mapping populations
- Type and size of mapping populations
- Density of linkage map
- Statistical methodology employed
- Significance level used for QTL mapping

Approaches:

1. Single marker analysis

2. Simple interval mapping
3. Composite interval mapping
4. Multiple interval mapping
5. Bayesian interval mapping.

Single marker analysis: Association of trait mean with marker locus.

- It involve comparison of different class for each marker locus
- Marker class as factor and individual values within class as error term.
- Determined by simple t-test, ANOVA, Linear regression and maximum likelihood estimation.

Advantage: Does not require gene order and complete linkage map.

Disadvantage: unqualified presumption of tight linkage of QTL with marker locus.

- No idea about precise QTL position
- No estimation of its effects, no recombination (r) value.

Simple interval mapping: Proposed by Lander and Botstein

- It involves use of two marker on chromosome, which bracket the putative QTL
- Assuming only one QTL on chromosome
- Use of complete linkage map
- Provide value of r, so better estimation of QTL effect is possible, independent of its position.
- More probable location of QTL can be decided.

Advantage: Better estimation of location and effect of QTL, calculate r between marker and QTL

Disadvantages:

- Assumption of only single segregating QTL is unrealistic
- Resolution of exact number of QTLs not possible
- Exact position of QTL can't resolve when other QTLs are segregating in the population.

Complete interval mapping: Combine interval mapping for a single QTL with multiple regression analysis

Advantage: Powerful, because of elimination of variance caused by other QTLs.

Disadvantages:

- Highly dependent on background information
- Genetic marker incorporated a co factor in the frame work of simple interval mapping.

Multiple interval mapping:

- Use of marker as co factor to reduce genetic background noise
- Based on regressing the phenotype on a putative QTL in the given interval and at the same time on a number of markers located in the genome
- Need large mapping population
- Gives idea about number and position of QTL
- Interaction of significant QTLs and their contribution to the genetic variance
- Based on Cochran's model for interpreting genetic parameters

Bayesian interval mapping:

- Provide information regarding number, position and effect of QTLs
- Information about posterior estimate of marker QTL marker generated
- It can estimate QTL effect and position separately.

Types of QTLs: Based on presence and absence of epistasis

1. Main effect QTL: Defined as single mendalian factor at which effect on a given phenotype arises from allelic substitution.
   - Detected by marker trait associations using single factor ANOVA or interval mapping models
   - Includes two groups of genes

   (*a*) Major genes with very large effects on highly heritable traits.
   - Detected with very large LOD scores of >10.0
   - Each explains a large portion of total trait variation in mapping population
   - Eg. Sd – 1 for semi dwarf nature and Xa4 for BLB resistance.

   (*b*) Represents most QTLs reported to date and have relatively small effects.
2. Epistatic QTL: Defined as loci at which trait values are determined by interactions between alleles at two or more loci.
   - Detected by associations between trait values and multilocus marker genotypes using epistatic models.
   - Three types.

   (*a*) Type I: Two main effect QTLs are involved in epistasis and affect the same phenotypes.

   (*b*) Type II: Involves interactions between alleles at a main effect QTL and a modifying locus.

   (*c*) Type III: Represents epistasis between two complementary loci that donot have detectable main effects.

Compare the strengths and limitations of $F_{2:3}$ and RIL mapping populations in relation to QTL mapping in crop plants?

$F_{2:3}$ Population	RIL Population
Strengths: • Ability to measure the effects of additive and dominance gene action at specific loci. • Effect of heterozygosity estimated	**Strengths:** • Estimates only additive gene action. So multilocation trials are possible • Immortal population, so replicated over locations and years
Limitations: • Non immortal population. So not possible to grow over locations	**Limitations:** • Take more time for developing • Inbreeding depression on cross pollinated crops • Effect of heterozygosity cannot be estimated

Define Linkage disequilibrium and discuss the causes of Linkage disequilibrium?

Linkage disequilibrium (LD): Non random association of alleles at different loci in gametic phase is called gametic phase disequilibrium (GPD). Gametic phase disequilibrium is more accurate term for Linkage disequilibrium.

Causes of LD: 1. Linkage 2. Multi locus selection due to epistasis 3. Random drift in small population 4. Bottle necks in population size 5. Migration 6. Mutation

- LD increase by inbreeding, small population size, genetic isolation between lineages, population subdivision, low recombination rate, population admixture, natural and artificial selection and balancing selection etc.
- LD decrease by out crossing, high recombination rate and high mutation rate
- LD decaying: LD decrease/decaying with distance, it is called as LD decaying
- Incomplete LD due to crossing over and gene conversion
- Some genes have high LD while some genes have low LD because
  (*i*) Possibility of recombination- Genes which are nearer to centromere and telomere have low chance of recombination
  (*ii*) Mating system: out crossing and selfing. Eg. Rice has more LD than pearl millet and maize.
- Linkage disequilibrium represented by [D] and r^2 i.e [D] = 1 and $r^2 = 1$ But, in case of linkage equilibrium [D] = 0 and $r^2 = 0$.

Define association mapping. Explain the significance of association mapping in plant genetics?

Association mapping: Association mapping (genetics), also known as "linkage disequilibrium mapping», is a method of mapping quantitative trait loci (QTLs) that takes advantage of historic linkage disequilibrium to link phenotypes (observable characteristics) to genotypes (the genetic constitution of organisms). i.e. Genotyping of naturally available genotypes.

- No production of mapping population
- No development of linkage map
- Just genotype the naturally available genotypes

Significance of association mapping in plant genetics: In traditional QTL mapping approach generates linkage disequilibrium between genetic markers and QTL through crossing of different genotypes and creation of segregant population. But central problem with the conventional mapping approaches is limited number of meiosis have been occurred.

In association mapping, recombination's over many generations has broken up the linkage disequilibrium that exists initially between marker and novel QTL allele.

Limitation: Trait of interest is segregating over generations, so it is not possible to identify and integration of allele from wild species to cultivated species. But it is achieved by Advance Back cross-QTL analysis (AB-QTL). Linkage disequilibrium mapping is best when there is strong selection pressure on trait of interest.

Write short notes on the following.

1. AB QTL analysis
2. Nested association mapping
3. Allele mining

AB QTL analysis: AB-QTL strands for Advanced Backcross QTL analysis, proposed by Tanksley and Nelson (1996), in which the QTL mapping is delayed to BC_2 or BC_3 so that the contribution of epistatic interactions is reduced as compared with selfing generations.

The lines derived from backcross generations will be more close to the recurrent parent as a result of which QTL-NILs can be obtained with less additional efforts. In the AB-QTL analysis the elite inbred lines of cultivated species are crossed to exotic stocks and the resulting F_1s again crossed to IL to generate nearly 100 BC_1 progeny. After rejecting plants with obviously undesirable features of the wild stocks, the others are again backcrossed to elite line to produce a population of BC_2 generations of about 200 individuals. The BC_2 plants are assayed for the polymorphic DNA markers. The selfed seed of these plants i.e. BC_2S_1 and BC_2S_2 can be evaluated and compared with the parental elite line in replicated trail to identify BC_2 individuals with desirable QTLs from the wild donar stock.

The molecular data from BC_2 plants can be used to search for QTL association markers with performance in BC_2S_1 or BC_2S_2 families to identify useful QTL.

In this way the alleles from wild species can be evaluated for their agronomic worth as well as association with markers to be used as index of such alleles for their transfer to elite line.

AB-QTL analysis is a method for the simultaneous discovery and transfer of valuable QTLs from unadapted germplasm into elite breeding lines. It is a method of QTL mapping.

Nested Association Mapping: Structured sub population nested within unstructured population.

- High power
- High resolution
- Analysis of many alleles
- Moderate marker density

Allele mining: Allele mining aims to identify the allelic variation present in genetic resource collection

- For identifying genes of known function and basic DNA sequence, collections can be screened for allelic variation

  Eg. Tilling strategy using DNA chip technology
- New point mutations in relatively large DNA fragments, can be detected
- Once allelic variants of interest have been identified, the approach can be optimised by focusing on target set of polymorphism
- Eg. Using SNP detection methods.

What is meant by Candidate gene? How it is advantageous over other molecular approaches?

Candidate gene approach: Candidate genes either genes with molecular polymorphism genetically linked to major loci or QTLs or genes with molecular polymorphism statistically associated with variation of the trait being studied.

- CG approach is based on the hypothesis that known functional genes could correspond to the loci controlling trait of interest.
- CGs refers 1. Either to cloned genes presumed to effect the given trait (Functional CGs) 2. or to genes suggested by their close proximity on linkage maps to loci controlling the trait (Positional CGs)

Advantage over other approaches :

Main objective of molecular genetics is to identify and isolate a gene; methods such as positional cloning and insertional mutagenesis have been used with success to identify major genes. However, these methods are limited by genome size and or by lack of transposons in the species in the species being studied.

In this regard CG approach is an alternative strategy.

- CG approach provides markers of gene itself.
- Validated CGs for QTLs provides very efficient molecular marker because recombination between marker and QTL would be absent.

Define reverse genetics? Enlist the different approaches of reverse genetics? And explain in detail about the Anti sense RNA technology?

Reverse genetics: Reverse genetics is an approach to discover the function of a gene by analysing the phenotypic effects of specific gene sequences obtained by DNA sequencing. It is needed to elucidate gene and protein function in context of whole organism.

The reverse genetic strategy broadly fall into two general categories depends on whether mutagenesis is targeted specifically to the locus of interest or it's performed throughout the genome.

1. Targeting specific loci: It is attractive when only a few genes of interest exist. For this post transcriptional gene silencing (PTGS) increasingly popular. Here strategies include are
   (*i*) Antisense RNA suppression
   (*ii*) RNA_i based PTGS
   (*iii*) Homologous recombination
   (*iv*) Chimeric RNA/DNA oligonucleotides
2. Genome wide mutagenesis and screening: These are inherently suitable for reverse genetics on a large scale. These are
   (*i*) Insertional mutagenesis (T-DNA and Transposable elements)
   (*ii*) Fast neutron mutagenesis and size screening
   (*iii*) Tilling

Anti sense RNA technology: Anti sense RNA technology is one of the approaches to reduce endogenous gene expression.

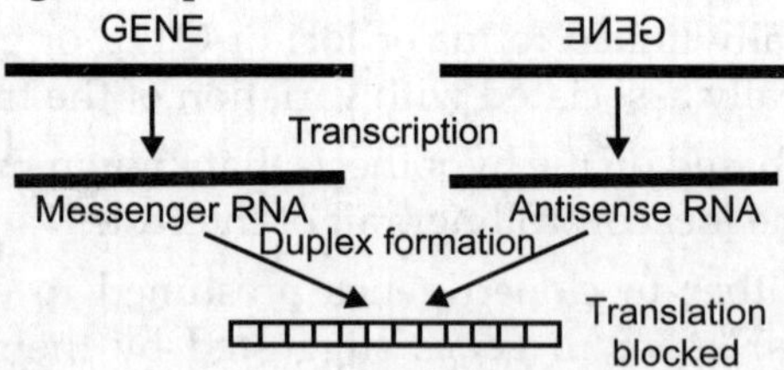

An anti sense RNA is produced by reversing the orientation of protein encoding region of gene in relation to its promoter. The RNA produced by this gene is known as antisense RNA. When an antisense RNA is present in same nucleus along with normal endogenous gene, it causes pairing of antisense RNA and sense RNA finally double stranded RNA formed. The implication of double stranded RNA as follows:

- mRNA unavailable for translation

- dS RNA degraded by dS specific RNAse
- Methylation of promoter and coding region leads to silencing of endogenous gene. So, Antisense RNA technology used for gene silencing and used for various purposes. Like

1. **Slow fruit softening tomato:**

   (*d*) Flavr savr-Polygalacturonase gene used. Improved flavour and, shelf life.

   (*e*) Endless summer – ethylene synthesis gene used (ACC synthase) ethylene <1%

2. **Change in fattyacid composition of *Brassica* oil:** Antisense gene construct of *Brassica* compestris stearyl-ACP desaturase gene.

   (*a*) Increase in level of steric acid (up to 40%)

   (*b*) Decrease in oleic acid content

3. **Delayed senescense in carnation:**

   (*a*) Gene aco suppressed using this thechnology

   (*b*) Vase life increases by 200%

4. **Male sterility:**

   (*a*) Flavonoid is essential for normal pollen development. Chalcone synthase (CHS) key enzyme of flavonoid synthesis. Use of antisense construct of CHS gene used to create male sterility

   (*b*) Apply flavonol during pollination of CHS antisense MS line to obtain 100% male sterile progeny

   (*c*) Also used to restore fertility-MS induced by role gene of *A. rhizogenes*

Explain in brief about 'Flavr Savr' Tomato?

'Flavr Savr' Tomato:

- The first approval for commercial scale of food product was a transgenic tomato 'Flavr Savr' with delayed ripening, developed by (calgene USA in 1994.)
- Calgene used antisense RNA against polygalacturonase (PG) enzyme encoding gene.
- On the basic of PG gene sequence, an antisense PG gene was constructed and tomato plants were transformed.
- The tomato plants produced both sense and antisense mRNA for the PG gene resulting in RNA-RNA pairing this would result in non production of PG gene product, and thus preventing the attack of PG gene upon pectin in the cell wall of ripening fruit thereby preventing softening of fruit.
- Calgene has given the brand name Mac Gregor to its transgenic tomato; it can stay on the market shelf for approximately two weeks longer without softening.

Define TILLING. Explain the Steps Involved in TILLING Process? And also Mention the Advantages of TILLING?

TILLING (Targeting Induced Local Lesions in Genomics).

(Discuss the process of TILLING and its usefulness in identification of candidate genes.)

TILLING is s new reverse genetic strategy that combines the high density of point mutations provided by traditional conventional mutagenesis with rapid mutational screening to discover induced lesions.

TILLING has the potential to become to become a standard reverse genetic strategy for plant functional genomics. Facile and efficient tilling depends on the availability of two resources. 1. A well mutanised population and 2. Genomic information.

Procedure:

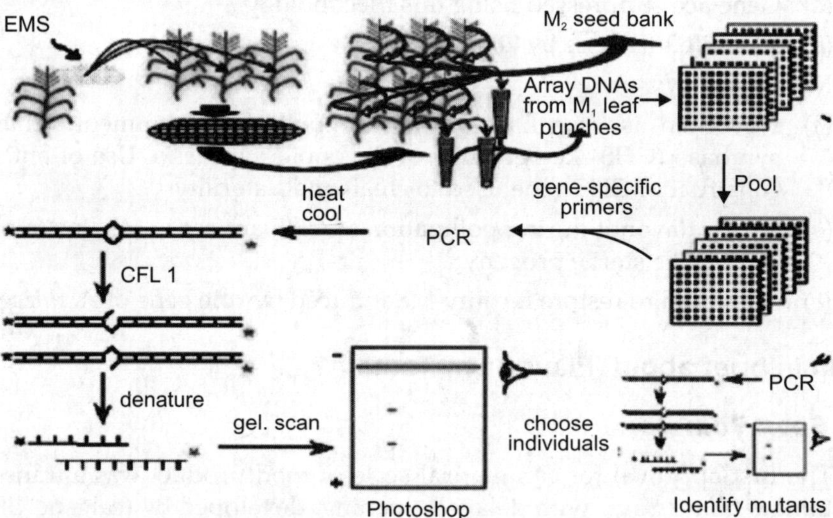

Advantages of TILLING:

- The likelihood of recovering deleterious mutations can be calculated in advance
- It minimizes efforts required to find mutations
- Appropriate for both small and large scale screening
- It is high through put. Most steps of tilling are suitable for automation
- It can be applied even if less information about selected targeted genes.
- As chemical mutagenesis is widely applicable and mutation duration for tilling dependent only on sufficient yield of PCR products.

Write short notes on Eco-TILLING?

Eco-TILLING: Use of high through put tilling technology for the detection of natural variation in DNA i.e. SNP discovery. It can be practical for SNP discovery and genotyping

- Screen many individuals at one locus
- Discover rare natural polymorphism
- Good in heterozygous populations.

What is meant by DNA sequencing? And what are the different methods of DNA sequencing. And explain the principle involved in Sanger-Coulson and Maxam- Gilbert method of DNA sequencing?

DNA sequencing: The basic objective of DNA sequencing is to determine the order of bases in a DNA molecule. There are two basic techniques of DNA sequencing.

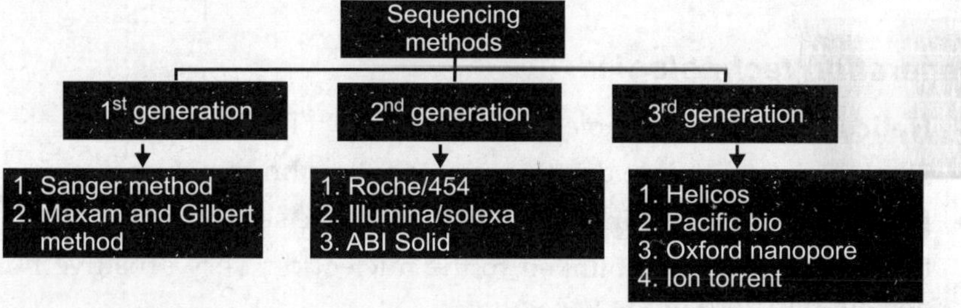

1. **Sanger-Coulson method:** This technique is also called as dideoxy-mediated chain termination method as it involves the use of chain terminating dideoxy nucleoside triphosphate (ddNTPs). Chain termination DNA sequencing is based on the principle that single stranded DNA molecules differ in length by just a single nucleotide can be separated from one another by polycrylamide gel electrophoresis.
2. **Maxam-Gilbert method:** This method is not commonly used and involves chemical degeneration of original DNA rather than synthesis. This method starts with a single stranded DNA that is labeled at one end (5′ or 3′) with radioactive ^{32}p. This method can be explained flow diagram as follows.

   Pyrosequencing is used for rapid determination of very short sequences. Pyrsequencing does not require electrophoresis or any other fragment separation procedure and so is more rapid than either chain termination or chemical degradation method. It can only generate a few tens of base paires per experiment, but it is becoming an important technique in situations where many short sequences must be generated as quickly as possible, for example SNP typing.

2nd Generation Sequencing Technologies:

Comparing sequencers:

Features	Roche (454)	Illumina	SOLiD
Chemistry	Pyrosequencing	Polymerase-based	Ligation-based
Amplification	Emulsion PCR	Bridge Amp	Emulsion PCR
Paired ends/sep	Yes/3kb	Yes/200 bp	Yes/3 kb
Mb/run	100 Mb	1300 Mb	3000 Mb
Time/run	7 h	4 days	5 days
Read length	250 bp	32-40 bp	35 bp
Cost per run (total)	\$8439	\$8950	\$17447
Cost per Mb	\$84.39	\$5.97	\$5.81

3rd generation technologies:

- **Helicos-** "true single molecule sequencing" (Tsms[tm]) system
  Single base, reversible dye terminator extension rreaction
- **Pacific Biosciences-** single molecule real time sequencing
  Dyes that are phospholinked to the nucleotide, very sensitive fluorescent detection in zero mode waveguides.
- **Oxford nanopores-** GridiON and MinION systems
  Direct reading of unlabeled DNA by threading it through a nanopore.

Explain in brief about the microRNAs and their regulatory role in plants.

MicroRNAs and their regulatory role in plants: MicroRNAs (miRNAs) are abundant new class of non coding 20-24 nucleotide small RNAs.

- Many plant miRNAs are evolutionarily conserved from species to species, some from angiosperms to mosses.
- miRNAs may originate from inverted duplications of target gene sequences in plants.
- Although miRNAs precursors display high variability, their mature sequences display extensive sequence complementarily to their mRNA sequences.

Regulatory role in plants:

- Plant post transcriptional gene regulation
- Involved in plant development

- Signal transduction
- Protein degradation
- Responses to environmental stresses and pathogen invasion
- Regulate their own biogenesis
- Regulate the expression of many important genes, a majority of these genes are transcriptional factors.

Explain in brief about role bioinformatics in plant breeding.

Bioinformatics in plant breeding: Bioinformatics may be defined as knowledge based theoretical discipline that attempts to make predictions about biological function using data from DNA sequence analysis. It is an application of information science to biology.

Types of bioinformatics databases: Information used in bioinformatics research may be grouped into two categories.

1. Primary databases: These databases consist of original biological data such as raw DNA sequences and protein structure information from crystallography.
2. Secondary databases: These databases contain original data that have been processed to suit certain specific applications.

General step in a bioinformatics: One purpose of searching a bioinformatics databases is to determine if the researchers unknown sequence, DNA or protein, matches any sequence in the database in terms of structure or function.

A search involves two key activities:

1. Sequence alignment scoring matches
2. Comparing sequences against a database Eg. typical homology search algorithms are used in this activity. Some of the most widely used software for this search is the Basic Local Alighnment Search Tool (BLAST) and FASTA.

Write short notes on the following.

1. DNA Microarray in plant breeding
2. Breeding by design
3. Map based cloning approach

DNA Microarray in plant breeding:

DNA microarray technologies have several different names Eg. DNA chips, genome chips, gene array and biochips technology. It allows researchers to adopt the whole picture approach in biological experimentation. The microarray technology is a tool for exploring the genome in a systematic and comprehensive fashion, to survey DNA and RNA variation. The basic principle of the technology is specificity and affinity of complementary base pairs. Studying the differential expression at the mRNA level is relatively straightforward.

Steps in a DNA microarray experiment: To conduct a DNA microarray experiment, six general steps are followed.

1. Probe: The experimenter first selects the genetic material of known identity (e.g. cDNA or small oligos) to be used as probes.
2. Fabrication of array: a fabrication format selected
3. Sample preparation: the sample (cDNA, RNA) to be used to integrate the spotted and immobilised probes is prepared and fluorescently labelled.
4. Assay: the assay is then conducted
5. Readout: the results are ready by, for example, electronic devices
6. Informatics: the data may be submitted to variety desired information according to the objectives of the researcher.

Applications of DNA microarrays in plant breeding: DNA microarrays may be used in two ways

1. For sequence identification and
2. To examine gene expression (level of abundance of gene expression).

Breeding by design:

The application of markers in breeding cannot only improve existing selection process but can aid in creating novel varieties bearing new characteristics of agronomic importance. Understanding of genetic basis of all agronomically important characters and the allelic variation at those loci would enable the breeder to design superior breeding lines *'in silico'*. This concept is referred to as 'Breeding by design' and uses two types of mapping methods to generate the knowledge required.

(*a*) Mapping traits using population which simplify complex characters, such as introgression libraries

(*b*) Mapping of traits and allelic variation by linkage disequilibrium (LD) mapping on germplasm.

The knowledge of the map positions of all loci of agronomic interest, allelic variation at those loci, and their contribution to the phenotype will enable the breeder to design superior genotypes consisting of a combination of favorable alleles at all loci.

Map based cloning approach: Positional cloning can be used to identify any gene, given an adequate map of the region of the chromosome in which it is located.

- Known as reverse genetic approach or positional or map based cloning
- First time in drosophila
- Other examples are gene from *Pseudomonas syringe* in tomato and xa21 in rice.

Steps:

1. Fine mapping of target trait.
2. Identification of tightly linked flanking DNA markers.
3. Physical mapping to know the approximate distance between marker and gene in base pairs.
4. Screening of large fragment genomic DNA library such as YAC, using flanking DNA markers as a probe.
5. Chromosome landing, walking, if needed.
6. Identification of large fragment DNA clones spanning the target gene region.
7. Screening of cDNA library.
8. Identification and characterisation of cDNA clone showing positive signal
9. Transformation of crop genotype using cDNA clone
10. cDNA that confers the expected trait phenotype in the transformed genotype is the target gene.

Write short notes on the following.

1. **Chromosomal walks**
2. **Chromosomal jumping**
3. **Gene library**

Chromosome walks: Positional cloning is accomplished by mapping the gene of interest, identifying an RFLP or other molecular marker near the gene and then walking or jumping along the chromosome until the gene is reached.

Major steps involved here are:

1. Prepare the restriction map of a clone
2. Sub clone the restricted fragment
3. Screen genomic library with sub clone as probe
4. Prepare restriction map for new clone
5. Sub clone restricted fragment
6. Rescreen library with new sub clone as probe
7. Repeat 1-3 steps for as many cycles as needed to reach the gene of interest

Chromosomal walks are initiated by the selection of molecular markers (RFLP) or known gene clone close to the gene of interest and use of this clone as a hybridisation probe to screen a genomic library for overlapping sequences.

Chromosome walking is very difficult in species with large genome and abundance of dispersed repetitive DNA.

Chromosome walking is easier in organisms such as Arabidopsis thaliana and *Ceanorabditis elegans* which have small genomes and little dispersed repetitive DNA.

Chromosomal jumping: When a distance from the closest molecular marker to the gene of interest is large, a technique called Chromosome jumping can be used to speed up an otherwise long walk.

- Each jump over a distance of 100 kb or more.
- Chromosome jumping as a short cut method for replacing long chromosomal walks.
- This procedure can also be used to jump over repetitive DNA sequences that block chromosomal walks.

Gene library: cDNA library and Genomic library

A gene library is a random collection of cloned fragments in a suitable vector that ideally includes all the genetic information of that species. It is also called as Shotgun collection. It is of two types i.e. cDNA library and Genomic library.

cDNA library: cDNA is DNA derived from mRNA. The mRNA in this case is processed transcript. It is used when it is known that gene of interest is expressed in a particular tissue or cell type.

Steps:

Tissue

↓

Extract mRNA

↓ Reverse transcriptase + dNTPs

Copy in to single stranded DNA (cDNA – RNA hybrid)

↓ RNAase H + DNA polymerase + dNTPs

Double stranded DNA

↓

Insert into plasmid vector

Transform bacteria with recombinant plasmid

Selection of transformed bacterial clone

Genomic library: It is also of great use because cDNA libraries lack introns and so not complete genome coverage is possible. In genome library it is necessary to contain at least one copy of every sequence of DNA in the genome.

Write short notes on the following?

1. Bt cotton
2. Bt brinjal
3. Features of rice genome project

Bt cotton:

- Bt cotton is a genetically modified variety of cotton producing an insecticide. It is produced by Monsanto. It is produced by Mahyco in India
- The cry gene of *Bacillus thuringiensis* (Bt) produces a protein delta endotoxin, which forms crystalline inclusions in the bacterial spores. These crystal proteins are responsible for the insecticidal activities of the bacterial spores against lepidopteran pests.
- India has become the number one global exporter of cotton and the second largest cotton producer in the world. India has also bred Bt-cotton varieties such *as Bikaneri Nerma* and hybrids such as (NHH-44).
- **Controversies:** In India, Bt Cotton has been enveloped in controversies due to its links with crop and seed monopolies and farmers' suicides. Bt cotton accounts for 93% of cotton grown in India. But these Bt cotton seeds are expensive and lose vigour after one generation, requiring farmers to buy new stock every year.

Bt brinjal:

- The Bt brinjal is a suite of transgenic brinjals (also known as an eggplant or aubergine) created by inserting a crystal protein gene (*Cry1Ac*) from the soil bacterium *Bacillus thuringiensis* into the genome of various brinjal cultivars against lepidopteron insects, in particular the Brinjal Fruit and Shoot Borer (*Leucinodes orbonalis*)(FSB).
- Mahyco, an Indian seed company based in Jalna, Maharashtra, has developed the Bt brinjal. The genetically modified brinjal event is termed Event EE 1. The Event EE 1 was introgressed by plant breeding into various local varieties by University of Agricultural Sciences, Dharwad and Tamil Nadu Agricultural University, Coimbatore. Some of the cultivars of brinjal include: Malpur local, Manjari gota, Kudachi local, Udupi local, 112 GO, and Pabkavi local.
- It was approved for commercialisation in India in 2009, but - after an apparent public outcry and rounds of debates in which representatives from Mahyco, the scientific community, and NGO's spoke on the topic - the Indian Environment Minister, Jairam Ramesh, facilitated a moratorium on its release until further, unspecified, tests were conducted.
- Many controversies surround the development and release of genetically modified foods, ranging from human safety and environmental impacts to ethical concerns such as corporate control of the food supply and intellectual property rights.
- Field trials conducted on research-managed farms carried out by Mayhco and the Indian Council of Agricultural Research suggested a 42% pesticide reduction and a doubling of the yield was possible. The economic gain for consumers, developers and farmers was estimated to

potentially be US$108 million per year with an additional $3–4 million saved due to health benefits associated with decreased pesticide use.

Features of rice genome project:

1. The rice genome is well mapped and well characterised, and it is the smallest of the major cereal crop genome at an estimated 400 to 430 Mb and consists of 32 to 50 thousand genes.
2. The International Rice Genome Sequencing Project (IRGSP) began in September 1997, at a workshop held in conjuction with the International symposium on Plant Molecular Biology in Singapore.
3. India, previously an unfunded member of the IRGSP, has a now Rice Genome Programme (represented by Akhilesh Tyagi of the University of Delhi and Nagendra Singh of the IARI) and will begin work on chromosome 11.
4. Sequencing of the rice genome is being performed mainly from genomic BAC or PAC libraries created from the *Nipponbare* (*Japonica*) variety, which was chosen as the common template throughout the IRGSP; China, working on the sequencing of chromosome 4, is the only member to use a different variety, *Indica Guang Lu Ai4*.
5. Rice genome composed of repetitive DNA is estimated to constitute at least 50% of the rice genome and as much as 70% of the maize genome.
6. The Rice Genome Research Programme uses a shot gun approach to sequence PAC or BAC clones.
7. Finger printing and physical mapping are used to create minimal tiling paths and to anchor BAC clones to physical positions along the length of chromosome.
8. After sequencing, softwares such as PHRED and PHRAP is used to order the subclone sequences and reassemble the entire BAC sequences.
9. Monsanto announced on April 4, 2000, that the company had completed a working draft of the rice genome, which would be made available to the IRGSP and helped to complete the sequencing in the year 2002.

Define Lethal Genes, Genetic Map, Physical Map, Dna Methylation, Dna Damage & Dna Damage?

Lethal Genes

One of the most important assumptions of inheritance of any trait is the equal survival of all gametes and zygotes produced as a result of segregation. Some genes affect the survival of those zygotes or individuals in which they are present in the appropriate genotype and are called as lethal genes. They may be grouped as: (i) lethal, (ii) semi lethal, (iii) sub vital, (iv) vital and (v) super vital genes.

Genetic Map

Genetic map (or linkage map) refers to chromosome maps calculated by using recombination frequencies. Alfred Henry Sturtevant constructed the first genetic map of Drosophila in 1913. Here, the distance between markers is determined by the frequency of recombination during meiosis, which is in turn determined by the relative distance between the loci. The genetic map provides the relative position of a gene according to the recombination frequency. It is expressed in centimorgans (cM).

Physical Map

Physical map refers to chromosome maps based on direct analysis of DNA and the distance between the gene loci is expressed by numbers of base pairs (bp), kilobases or megabases. Physical maps generally have higher resolution and are more accurate than genetic maps.

Dna Methylation

DNA methylation refers to the addition of methyl group ($-CH_3$) to the DNA bases namely, adenine and cytosine to yield 6-methyladenine and 5-methylcytosine, respectively. DNA methylation is mediated by enzymes called methylases. E.coli has three distinct methylation systems. i) The hsd system: This system methylates adenine. Occur in many bacteria. ii)The dam system: This system also methylates adenine and is involved in the control of DNA replication and in marking DNA strands for repair. iii). The dm system: It methylates cytosine., Its function is still unknown.

Dna Damage

The various damages to DNA may be grouped into the following two types: (i) Single Base Changes affecting a single base of a DNA molecule. It involves the conversion of one base into another. E.g., Deamination of 5 methyl cytosine results in thymine. (ii) Addition of a small group to a base affects its pairing behavior. Single base changes do not produce structural disturbance and do not affect replication and transcription of the affected molecules. (ii) Structural Distortions generally adversely affect the replication and transcription of the affected DNA molecule. They are represented by a single strand nick, removal of a base etc.

Dna Repair

One of the best understood repair mechanisms involves removal of pyrimidine dimers (usually covalently linked adjacent thymines in the same strand). Thymine dimers are easily induced in bacteria by ultraviolet (U V) light. These dimers are lethal if left unrepaired because they interfere with the normal replication of progeny DNA strands. There are at least three mechanisms known for repairing pyrimidine dimers.

Brief About Modern Concept of Gene ?

The fine structure of a gene is the linkage map of its various mutant alleles. In this map, the sites of mutational changes in different alleles of the same gene are determined on the basis of the frequency of recombination among these alleles. Thus, the fine structure map of a gene is essentially comparable to the linkage map of a chromosome [except for the number of genes involved (one in the case of gene fine structures and several in the case of chromosome maps)]. Understandably, the values of crossing over observed in the studies on fine structure of a gene are very low (e.g., 0.01 or even 0.0001%). Therefore, extremely large progeny populations have to be effectively screened for efficient identification of the cross over products. For this reason, fine genetic analysis is more convenient in prokaryotes than in eukaryotes.

What Is Jumping Genes ?

Most genes reside at a specific loçus or position on the chromosome. Some genes or closely linked sets of genes can mediate their own movement from one location to another location in a chromosome. These elements are called as "jumping genes," "mobile elements." "cassettes," "insertion sequences," and "transposons." The formal name for this family of mobile genes is transposable elements, and their movement is called transposition.

Write short notes on:

- Metabolomics
- Gene Therapy
- Chromosome Banding
- Chromosome Banding

Metabolomics

Metabolomics is one of the newest 'omics' sciences. The metabolome refers to the complete set of low molecular weight compounds in a sample. These compounds are the substrates and by-products of enzymatic reactions and have a direct effect on the phenotype of the cell. Thus, metabolomics aims at determining a sample's profile of these compounds at a specified time under specific environmental conditions. Metabolomics can be used to determine differences between the levels of thousands of molecules between a healthy and diseased plant. The technology can also be used to determine the nutritional difference between traditional and genetically modified crops, and in identifying plant defense metabolites.

Gene Therapy

Gene therapy is the direct transfer of genes into humans to treat disease. Gene therapy was first successfully implemented in 1990 and is now being used to treat genetic diseases, cancer, and infectious diseases. Gene therapy conducted to date has targeted only non-reproductive, somatic cells. Correcting a genetic defect in these cells (termed somatic gene therapy) may provide positive benefits to patients but will not affect the genes of future generations. Gene therapy that alters reproductive, or germ-line cells (termed germ-line gene therapy) is technically possible but raises a number of significant ethical issues, because it has the capacity to alter the gene pool of future generations. In 1990, gene therapy became reality. W. French Anderson and his colleagues at the U.S. National Institutes of Health (NIH) transferred a functional gene for adenosine deaminase to a young girl with severe combined immuno-deficiency disease, an autosomal recessive condition that produces impaired immune function.

Chromosome Banding

Chromosomes are visualized by the use of traditional dyes which stain them uniformly and provide uniform coloration to chromosomes. Caspersson et al in 1970, introduced newer staining procedures and used different dyes which produced dark and light cross bands of varying widths on chromosomes. Each chromosome in a haploid set has a unique banding pattern and can be distinguished from the other chromosomes of similar size and shape. Generally, there are four types of banding namely, Giemsa or G banding, centromeric or C banding, Quinacrine or Q banding and reverse or R banding.

- **G BANDING:** It involves Giemsa stain after trypsin treatment of metaphase chromosome to differentiate the chromosomal bands. When viewed through the microscope, the chromosomes show dark bands (G bands) and light bands (R bands). By this method approximately 400 dark and light bands could be resolved in a haploid set of chromosomes. This is the most widely used procedure.
- **C BANDING:** It requires heating in an alkali solution and staining with Giemsa. C bands are of constitutive heterochromatin located adjacent of the centromeres of all the chromosomes.
- **Q BANDING:** In Q banding, Quinacrine mustard is used to stain the chromosome. They are observed under fluorescent microscopy. The bright Q bands are equivalent to G bands.
- **R BANDING:** This is reverse banding and it uses Giemsa under elevated temperature to produce the reverse pattern to that seen in G banding.

Chromosome Painting

In 1980s, a technique called FISH (Fluorescent in situ hybridization) was developed to prepare brilliantly colored whole chromosomes. This technique requires complex mixtures of specially constructed DNA sequences called probes which are complementary to the different chromosomal regions, along with a set of fluorescent dyes. A special microscope with optical filters is also needed along with computer software to analyse the image and convert them into a dazzling multicolored display. Chromosome painting is useful for quickly detecting cells with extra or less chromosomes or with other structural abnormalities, especially in cancer cells.

CHAPTER - 10

Plant Genetic Resources and Their Regulatory System

What is meant by centre of origin of crop species? And explain the different centre of origins proposed by Vavilov?

Centre of origin:

Centre of origin are geographical areas where crop plants have originated while centre of diversity refers to a location where vast genetic variability for a crop and its wild species is found.

Thus centre of origin and centre of diversity may be same or may be different (Fehr, 1987).

Centre of origin (Vavilov, 1951) with important crops: Crop plants evolved from wild species in the areas showing great diversity and termed them as primary centers of origin. Later, crops moved to other areas primarily due to the activities of man and these areas generally lack the richness in variation found in the primary centers of origin, but show considerable diversity of forms; such areas are called secondary centers of origin. Vavilov originally proposed eight main centers of origin in 1926, which are China, Hindustan, Central Asia, Asia Minor, Mediterranean, Abyssinia, Central America and South America.

Eight main centres and three subsidiary centres of diversity.

1. The Chinese centre – Soyabean, peach, citrus sinensis.
2. The Indian centre – Rice, brinjal, mungbean.
3. The Central asian centre – Bread wheat, lentil, carrot, radish, chickpea.
4. The Near eastern centre – Einkorn wheat, durum wheat, melons, grapes, almond.
5. The Mediterranean centre – Durum wheat, lettuce, broad bean, cabbage.
6. The Abyssinian centre – Barley, lentil, pea.
7. The South Mexican and central American centre – Corn, common bean, upland cotton.
8. The South American centre – Sweat potato, potato, tomato, papaya, tobacco

Subsidiary centres:

9. The Indo Malayan centre – Banana, coconut, yam.
10. The Chilean centre – Potato.

11. The Brazilian centre – Paraguayan centre – peanut, rubber tree, pineapple, cocoa

China center of origin: It consists of the mountainous regions of central and western China and the neighbouring lowlands. The crops originated in this area are soybean, colacasia, buck wheat, brinjal, peaches/ apricots, plums etc., as primary center and for maize, rajma, cowpea, turnip and sesame as secondary center.

Hindustan center of origin: It includes Burma, Assam, Malaya Archipelago, Java, Borneo/ Sumatra and Philippines. The crops originated in this area are rice, pigeon pea, chickpea, cowpea, mung, Indian radish, noble canes, several species of cotton, black pepper, indigo, mango, orange, coconut, banana etc.,

Central Asia center of origin (Afghanistan center): It includes North west India, all of Afghanistan, the Soviet Republics of Tajikistan and Uzbekistan and Tian-Shan. The crops originated in this area are wheat, pea, club wheat, broad bean/ linseed, safflower, hemp, cotton, radish, carrot, onion, garlic, spinach, almond, grape etc., as primary center and for rye as secondary center.

Asia Minor center of origin (Persian center): It includes interior of Asia Minor, the whole of Transcaucasia, Iran and Highlands of Turkmenistan. The crops originated in this area are triticum, rye, alfalfa, Persian clover, carrot, cabbage, oat, lettuce, fig, pomegranate, apple, almond etc., as primary center and for rape, black mustard, leaf mustard, turnip and apricot as secondary center.

Mediterranean center of origin: It includes all the Mediterranean regions of the world. The crops originated in this area are durum wheats, emmer wheats, several species of Avena, barley, lentil, pea, broad bean, lupins, chickpea, clovers, asparagus, lavender, peppermint and sage etc., as primary center.

Abyssinian center of origin: It includes Ethiopia and hill country of Eritrea. The crops originated in this area are barley, durum wheat, jowar, bajra, gram, lentil, sem, pea, khesari, linseed, safflower, sesame, castor, coffee, onion and okra etc., as primary center and for broad bean as secondary center.

Central American center of origin (Mexican center): It includes the region of south Mexico and Central America. The crops originated in this area are maize, rajma, lima bean, melons, pumpkin, sweet potato, arrow root, chillies, cotton, papaya, guava and avacado etc., as primary center.

South American center of origin: It includes the high mountainous regions of Peru, Bolivia, Ecuador, and Columbia, parts of Chile and Brazil, and whole of Paraguay. The crops originated in this area are many species of potatos, maize, lima bean, peanut, pineapple, pumpkin, tomatos, guava, tobacco, quinine tree, cassava and rubber etc., as primary center.

Write short notes on foolowing.

1. Law of homologous series
2. Types of centres of diversity
3. Genetic erosion

Law of homologous series or Law of parallel variation:

- This was proposed by vavilov in 1951
- It states that a particular variation observed in a crop species is also expected to be available in its another related species
- Eg. i. New world and world cotton: lint characteristics, seed colour, stem colour, degree of hairiness
  (*ii*) And 28 contrasting character in diploid, tetraploid and hexaploid wheat
  (*iii*) Analogous to synteny in molecular terms
- This law serves as a guideline to the breeder and explores for collection, creation and utilization of plant genetic resources.
- Knowing the serves of forms within the limits of one species, we can predict the occurrence of parallel forms in other species and generation.

Types of centre of diversity:

1. Primary centre of diversity: These are regions of vast genetic diversity of crop plants.
   - They have wide genetic diversity
   - Have large number of dominant characters
   - Mostly have wild characters
   - Exhibit less crossing over
   - Natural selection operates

   These are original homes of the crop areas like mountains, hills, river valleys and forests etc.
2. Secondary centres of diversity: Valuable forms of crop plants found far away from primary centre of diversity are called secondary centre of diversity. These are generally cultivated areas.
   - Have less genetic diversity than primary centre of diversity
   - Large number of recessive genes
   - Mostly have desirable characters
   - Exhibit more crossing over
   - Both natural and artificial selection operates
3. Micro centres: Small areas within the centres of diversity exhibit tremendous genetic diversity of some crop plants.
   - The rate of natural evolution is faster
   - Important sites for study of crop evolution

Genetic erosion:

- H. V. Harlan was the first to express concern that there was a danger of losing genetic diversity (Harlan and Martini, 1936).
- The gradual loss of variability in the cultivated forms and in their wild relatives is referred to as genetic erosion.
- In genetic erosion, a few improved varieties become predominant and rapidly replace the local varieties leading to a rapid depletion of genetic variability.
- This, in turn limits the prospects of further improvement in the species since variability is the pre-requisite for any modification in the characteristics of a crop species.
- Germplasm collections aim at minimising the detrimental effects of genetic erosion by collecting and preserving the variability in crops and then-related species.
- Eg. Almost all the semi-dwarf wheat varieties have Rht 1 and Rht 2 genes for reduced height; which are derived from a single wheat variety, Norin 10.
- Thus the improved varieties of wheat are becomingly increasingly similar to each other due to commonness of one or more parents in their ancestry.
- This led to narrowing down of the genetic base of these varieties due to genetic erosion.

What is meant by germplasm collection? Describe in brief about the germplasm conservation.

Germplasm collection: The sum total of heredity material or genes present in a species is known as the germplasm of that species. Therefore, germplasm collection of a crop species consists of a large number of lines, varieties and related wild species of the crop. Such collections are also called as gene banks.

Germplasm Conservation: The germplasm has to be maintained in such a state that there is a minimum risk for its loss and that either it can be planted directly in the field or it can be prepared for planting with relative ease is called germplasm conservation. It can be conserved either *in situ* or *ex situ*.

In situ germplasm conservation: It is the conservation of germplasm in its natural habitat or in the area, where it grown naturally, which can be achieved by protecting this area from human interference.

- These protected areas are often called as Natural Park, biosphere reserve or gene sanctuary, which is best, located within the center of origin of crop species concerned, preferably covering the microcenter within the center of origin.
- Eg. Gene sanctuaries in Meghalaya for citrus, North-Eastern region for Musa and *Oryza* etc.

- The main advantage of gene sanctuary is that it not only conserves the existing genetic diversity present in the population, but it allows evolution to continue; and as a result, new alleles and new gene combinations would appear with time.
- The main limitation in gene sanctuary is the difficulty in establishment and its maintenance, especially in India, where it is facing the pressure of population increase and also they cannot conserve the variability found in crop plants, for which *ex situ* conservation is the only answer.

***Ex situ* conservation of germplasm:** Conservation of germplasm away from its natural habitat is called *ex situ* germplasm conservation, which can be achieved by seed gene banks, plant or field gene banks, shoot tip gene banks, cell and organ gene banks, and DNA gene banks.

Seed gene banks: In seed gene banks, germplasm is stored as seeds of various accessions and is quite easy, relatively safe and ordinarily needs minimum space.

- Seeds are classified, mainly based on the "basis of their storability, into two major groups like orthodox and recalcitrant seeds.
- Orthodox seeds: Seeds which can be dried to moisture content of 5 % or lower without lowering their viability.
- Recalcitrant seeds: The seeds for which viability drops drastically if their moisture content is reduced below 12 - 30%.
- The conditions for seed storage depends mainly on the duration of storage and are classified into base collections, active collections and working collections, which increases the efficiency of the use and the level of management of the collections.
- In base collection, the accessions of the crop are stored at about 18-20°C with 5±1 % moisture content and are distributed only for regeneration. These collections are for long storage and high quality orthodox seeds can maintain good viability upto 100 years.
- The accessions in an active collection are stored at 0 °C with seed moisture of 8%, and are stored for medium duration, i.e., 10 - 15 years. These active collections are used for evaluation, multiplication and distribution of the accessions, and are maintained by multiplying the seeds of their own accessions.
- The accessions being actively used in crop improvement programmes constitute working collections, which are stored for 3-5 years at less than 5-10°C and about 8-10 % moisture, and are maintained by the breeders using them.
- During storage in gene bank, it is necessary to control temperature and humidity to maximise the longevity of seeds.
- Harrington (1963) indicated that seed longevity is doubled for each 5°C fall in temperature or for each 2% drop in moisture content.
- This, he called 'rule of thumb'. This rule applies between 1°C and 50°C for temperature and between 5 and 14 percent for seed moisture content.

Field gene banks: A field or plant gene bank is an orchard or a field, in which accessions of fruit trees or vegetatively propagated crops are grown and maintained.

- The main limitations of field gene banks is that it require large areas, expensive to establish and maintain, prone to damage from disease and insect attacks, natural disasters and human errors in handling.

Shoot tip gene banks: The germplasm is conserved as slow growth cultures of shoot-tips and nodal segments in shoot tip gene banks, and it consists of subculturing the cultures every 6 months to 3 years.

- The main advantages of these gene banks are that the genotypes of the accessions can be conserved indefinitely free from diseases and pests, can be used for those crops which do not produce seeds or produce recalcitrant seeds, and regeneration i.e., sub culturing, requires a comparatively very short time.
- In addition, cuttings, bulbs and tubers can be maintained under controlled humidity and temperature conditions; however, this approach is practical for the short and medium term storage, and it should be used in conjunction with a field gene bank.

Cell and Organ gene banks: A germplasm collection based on cryopreserved (at -196 °C in liquid nitrogen) embryogenic cell cultures, shoot tips and or somatic/ zygotic embryos may be called cell and organ gene bank, i.e., cell or organ gene bank for potato in Germany.

DNA gene banks: The DNA segments from the genomes of germplasm accessions are maintained as cosmid clones, phage lysates or pure DNA is stored in the DNA gene banks.

- DNA segments can be evaluated and the desired ones may be used to produce transgenic plants.
- This approach is applicable to the conservation of genetic materials of already extinct species.

What is meant by quarantine? Write in detail about the necessity of quarantine and its procedure?

Quarantine means to keep materials in isolation to prevent the spread of diseases, etc. present in them to the other materials.

- According to the Destructive Insects and Pests Act, 1914, all plant produce imported in India must be free from diseases, insect pests and weeds.
- Plant produces may be grouped based on .consumption, cultivation and research and the quarantine of these groups or plant produces is carried out by separate agencies.

Plant produce imported for consumption: Directorate of Plant Protection, Quarantine and Storage is responsible for quarantine of plant produce imported for consumption. The directorate has quarantine and fumigation centers at 8 sea ports, 7 air ports and at the entry points of 7 land routes, which examine the plant produce imported for consumption and carry out their fumigation, if necessary.

Plant propagules imported for cultivation and research:

- The quarantine of plant propagules is done by the NBPGR, New Delhi for agricultural and horticultural species.
- Forest Research Institute, Dehradun for forest trees and
- Botanical Survey of India, Kolkata for remaining plant species.

Quarantine procedure: The procedure designed to prevent the entry of weeds, diseases and insect pests and followed by NBPGR, New Delhi is

- It is essential that the propagules must be clean, healthy and free from weeds and insect pests. The propagules should not be treated by the sender with fungicides or insecticides. If necessary, the sender may only fumigate the seeds / propagules before sending them, and indicate the same in the phytosanitary certificate.
- Each imported entry or sample must be accompanied by a 'phytosanitary certificate' from the scientist / institution sending the sample / entry, which certifies that the seeds / propagules being sent are free from weeds, diseases and pests.
- The entries accompanied by an authentic phytosanitary certificate are examined closely with the help of a magnifying glass / microscope and screened with X - rays. The healthy entries free from diseases, insect pests and weeds are identified and sent to the recipient scientists/institutions. Contaminated entries are detained by the bureau and attempts are made to free them from the contaminating weeds, insects and pathogens. Contaminated entries may be fumigated, if it is considered that such a treatment would rid them of the contaminating insect
- If needed or considered desirable, the contaminated entries may be grown in isolation in an effort to isolate some healthy plants. Seeds/propagules from such healthy plants may be collected and sent to the indentor. Entries suspected to be contaminated by pathogens, particularly viral pathogens, are grown or by doing the immunological assays for monitoring the disease symptoms and to detect the presence of pathogens, respectively.
- Samples / entries that are heavily contaminated are destroyed by the bureau.
- The process of quarantine (except growing in isolation) at the NBPGR, New Delhi takes at least 3 weeks.

What is meant by gene pool? Explain the Harlan and Dewet classification of germplasm resources?

Gene pool: Gene pool is nothing but sum total of genes present in a Mendalian/ Panmictic population or the whole library of different alleles of a species or sum total of genes in a species. And it also called germplasm, genetic stock and genetic resources.

- The total genetic information encoded in the sum total of the genes in a breeding population existing at a given time is known as gene pool.
- A gene pool, which preserves a constant composition throughout a series of generations, is denoted as existing in equilibrium.

Utility based classification of germplasm resources: Harlan and Dewet (1971), gave the concept of 'gene pools' for effective and judicious use of germplasm

1. Primary gene pool (GP1): The gene pool in which intermating is easy and leads to production of fertile hybrids. It includes genotypes of same species.
2. Secondary gene pool (GP2): The genetic material that leads to partial fertility on crossing with primary gene pool. It includes genotypes of related species.
3. Tertiary gene pool (GP3): The genetic material which leads to production of sterile hybrids on crossing with primary gene pool. Successful hybrid embryos can be produced through special technique of embryo culture, tissue culture, chromosome doubling or using bridging species.
4. Gene pool 4 (GP4): Sources of genes across general across general species and kingdom E.g. Bt gene from *Bacillus thuringiensis*. It is also called as Gene Ocean.

Write short notes on activities of IPGRI and NBPGR.

IPGRI:

- IPGRI strands for International Plant Genetic Resources Institute. And it is an autonomous international scientific organisation.
- Earlier it was known as IBPGR, which is established in 1974.
- The functioning of IPGRI is supervised by CGIAR.
- IPGRI was established by CGIAR in 1994 and it is situated in Rome, Italy at FAO of United Nations.

Functions of IPGRI:

- To conduct research and to promote and international network for PGR activities to ensure the collections, conservations, evaluation, documentation and utilisation of plant germplasm.
- It also helps in global collection and exchange of plant genetic resources. It promotes and coordinates the global germplasm activities.

NBPGR: National Bureau of Plant Genetic Resources

- NBPGR was established by ICAR in 1976 at New Delhi.
- In 1946, plant introduction started at IARI in the division of botany

- In 1961, separate division of plant introduction was established by H.B. Singh and huge collection of germplasm was made.
- In 1976, division of plant introduction was elevated to NBPGR.

Functions:

- It is a sole agency for germplasm exchange in India
- Promotion of national genetic resources activities i.e. collection, conservation, evaluation, documentation and utilisation (CCEDU) of crop plants and coordinates in all these activities.
- Organises short term training course on CCEDU.
- Organisation of exploration trips
- Provide guidelines about development of cold storage facilities for short term conservation of germplasm
- Takes decision about setting up of gene sanctuaries for endangered crop species
- It maintains core collection for quick utilisation of plant breeding activities.

Write short notes on the following.

1. CBD
2. Cartagena protocol
3. Biodiversity act - 2002

Convention on biological diversity (CBD):

The convention on biological diversity is also known as the biodiversity convention. It is an international treaty that was adapted in Riode Janeiro in June 1992, for signatute at the Earth summit and entered in to force on 29, Dec. 1993.

The biodiversity agreement has been signed by 189 countries. The convention has three main goals:

1. Conservation of biological diversity
2. Sustainable use of its components
3. Fair and equitable sharing of benefits arising from genetic resources.

The CBD reaffirms in its preamble that "states have sovereign rights over their own biological resources". Under CBD the access to CGIAR centres must be on the basis of prior informed consent (PIC) and mutually agreed terms (MAT). CBD's main decision making body is the Conference of Paris (CoP) which consists of delegates of all signatory countries.

Cartagena protocol: The Cartagena protocol on biosafety of the convention, also known as the biosafety protocol, was adapted in January, 2000.

- The biosafety protocol seeks to protect biological diversity from the potential risks posed by living modified organisms resulting from modern biotechnology.

- The biosafety protocol makes clear that products from new technologies must be based on the precautionary principles and allow developing nations to balance public health against economic benefits.
- It will be the example for lot of the countries to ban imports of a genetically modified organism if they feel there is not enough scientific evidence the product is safe and requires exporters to label shipments containing genetically altered commodities such as corn or cotton.

Biodiversity act – 2002:

In the year 1998, Ministry of Environment and Forestry (MoEF) of the government of india prepared a draft bill and it was passed by the Indian parliament in December 2002 and named as "Biological diversity act 2002". For protection of indias rich diversity and associated knowledge against their use by foreign individuals, institutions or companies without sharing the benefits arising out of such use and check bio-piracy.

Objectives of the act are to regulate collection, documentation, conservation and sustainable use of biodiversity.

The act also seeks to create a three tier structure:

1. National level – National biodiversity authority (NBA), Chennai.
2. State level – State biodiversity board (SBB)
3. Local level (panchayat/ gramsabha level) – Biodiversity management committe (BMC).

Criticisms of Biodiversity act 2002:

1. Act differentiate multinational corporations (MNCs) with domestic companies which is against the TRIPs of WTO.
2. Researchers used to register with SBB for utilizing resources.
3. Agrobiodiversity is kept outside the purview of act.

What is meant by UPOV? Write the criteria for protection of variety?

UPOV: Union International Pour la Protection Des Observations Vegetables (UPOV), International Union for Protection of Plant Varieties.

- UPOV came in to existence during convention of WIPO at Paris in 1961, (WIPO stands for World Intellectual Property Organisation) for granting protection to crop varieties to the respective breeders.
- Director general of WIPO is the secretary general of the UPOV.
- UPOV head quarter – Geneva along with WIPO.
- The 1961 convention came in to existence in 1968 and subsequently revised in 1972, 1978 and 1991.
- The 1978 act came in to effect in 1981. And 1991 act came in to force on April 24, 1998.
- UPOV: Field crops – 15 years (1978); 20 years (1991).

  Forest crops – 18 years (1978); 25 years (1991).

Provision	UPOV 1978	UPOV 1991
Protection coverage	Plant varieties of nationally defined species	Plant varieties of all genera and species
Requirements	Distinctness Uniformity Stability	Novelty Distinctness Uniformity Stability
Protection term	15 – 18 years	20 – 25 years
Breeders exemption	Yes	Not of EDV
Farmers privilege	Yes	Upto national law

Write short notes on the following:

1. Patent
2. Plant breeder's rights
3. Breeders exemption and farmers privilege

Patent:

A patent is the right granted by a government to an inventor to exclude others from imitating, manufacturing, using or selling the invention in question for commercial use during the specific period.

Patents are granted for an invention, innovation/improvement in an invention, the process of invention/product and a concept.

Patent requirements:

1. Novelty
2. Inventiveness
3. Industrial application and usefulness
4. Patent ability – the subject matter of the patent must be patent able under the existing law and its current interpretation.
5. Disclosure.

Limits of patent:

1. Limitation of period – generally 15-20 years
2. Limitation of space – a patent is valid only in the country of its award; it is not valid in other countries.

Plant breeders' rights: The Plant Breeders Rights programme was first established in 1987 under the Plant Variety Rights Act (PVRA), which was succeeded by the current Plant Breeders Rights Act 1994.

- Plant Breeders Rights (PBR) also known as Plant Variety Rights (PVR) are intellectual property rights granted to the breeder of a new variety of a plant.
- Basic requirements for protection of variety under PBR's are novelty, distinctness, uniformity and stability.
- The Plant Breeders Rights Act provides
  - Right to commercial seed production
  - Right to marketing
  - Right to export and import
  - Right to authorisation and
  - Right to prevent infringement for his variety.
- Period of protection varies with plant species. It is 15 years for field crops, 18 years for forest crops (UPOV, 1978).
- Plant Breeders Rights Act provides two exemptions i.e. breeders exemption and farmers exemption.
- PBRs started, in USA-1930, Europe-1964 and India-2001

  USA: 1930 – PPA – Plant patent act. (Potato not included)

  1970 – PVPA – Plant variety protection act – (cereals and potato not included)

  1985 – UPPA – Utility plant patenting act

  Europe 1964 – PVP – Plant variety protection act

  India 2001 – PPV & FRA – Protection of plant variety and farmers rights act

PPV&FR act 2001:

To fulfil the obligation of TRIPs under WTO, Indian government passed a legislation which is named as protection of plant varieties and farmers right act 2001.

- A requirement of TRIPs is that member nations provide for, beside other things. Plant variety protection by patents or by an effective sui-generis (self-development) system or by any combination of the two.
- PPV&FR act is sui-generis system which provides protection to the plant variety. It also includes the breeders exemption and farmers privilege.

Breeder's exemption: Under PBR regime, the use of material of a protected variety (the initial variety) for the development of new variety is exempted from protection.

- The PBR for these new varieties will be of the breeder who developed them and the holder of the PBR-title of the initial variety will have claim to it. This provision is called Breeders exemption

- Under UPOV 1978: all new varieties evolved using a protected variety was exempted from protection under this provision.
- But UPOV 1991: has limited scope of breeders exemption by bringing essentially derived varieties under the cover of PBR protection granted to the initial variety.

Farmers privilege: PBR system generally allow the farmers to use the material of a protected variety produced on their farm for planting new crop without any obligation to the PBR title holder. This exemption is usually referred to as farmers privilege.

- Under UPOV, 1978 – explicit provision for farmers privilege.
- Under UPOV, 1991 – this privilege has been made optional and each UPOV member state can either allow or disallow this privilege.
- Farmers privilege applied to the use of seed produced by a farmer for sowing 'his own' fields. It does not allow farmers to exchange seed of protected varieties produced on their farms.

Write short notes on the following?

1. EDV
2. Farmers rights
3. Global plan of action

Essentially derived variety (EDV):

- It is defined as a variety predominantly derived from initial variety, which retains the expression of essential characteristics from the genotype or combination of genotypes of initial variety.
- Thus a variety produced by, say mutation or transfer through backcross method/introgression by genetic transformation of a single gene will be considered as an "EDV". And will be protected under the PBR-title granted to the initial variety.
- The breeder of such variety will, therefore, be required to obtain permission from the PBR-title holder of the initial variety.

Farmers rights:

Farmers rights refers to the rights arising from the past, present and future contributions of farmers in conserving, improving and making available plant or animal genetic resources, particularly those in the centres of origin/diversity. Or Legal rights provided to farmers to save, use low, replant, exchange share or sell his farm produce including seed of a variety protected under plant variety protection act refers to Farmers rights.

- The purpose of these rights to "ensure full benefits to farmers and support the continuation to their contributions".

- Requirements: novelty, distinctness, uniformity and stability.
- Duration: 15 years for field crops and 18 years for forest crops.
- The FAO conference held in Rome from 11-29 Nov, 1989 endorsed the concept of farmers rights with a view to:
  (*i*) Ensuring global recognition of the need for conservation and the availability of sufficient funds for these purposes.
  (*ii*) Assisting farmers and farming communities throughout the world, especially those in areas of original diversity of plant genetic resources, in the protection and conservation of their PGR and of the natural biosphere.
  (*iii*) Allowing the full participation of farmers, their communities and countries in the benefits derived, at present, in the future, from the improved use of PGR.

Global plan of action (GPA):

Plan for the conservation and sustainable utilisation of plant genetic resources for food and agriculture (PGRFA) of FAO at global level is GPA.

- It was adopted by 150 countries in June 1996 at FAO, International technical conference on PGR.
- The GPA was also endorsed by CoP (Coference of paris) of CBD.
- The GPA, also referred to as the 'Leipzig plan' is the first global plan on plant genetic resources agreed at government level.
- GPA has support of CGIAR.
- The GPA aims to create a rational, coordinated global system for PGFRA.
- GPA activities includes:-
  1. *In situ* conservation
  2. *Ex situ* conservation
  3. Utilisation of PGR
  4. Institution and capacity building.

Write short notes on the following?

1. CoP
2. TRIPs
3. Geographical indicators.

Conference of parties (CoP):

- The CBD's main decision making body is the conference of parties which consists of delegates of all signatory countries.
- CoP has met four times so far, decisions made at CoP – 4 which have significant implications on economic and environment policies include:

(*i*) Conservation of biodiversity in different ecosystems like forests, arable agriculture, marine and coastal and inland waters etc.

(*ii*) Access and benefit sharing towards establishing legal policy and administrative measures which ensure that access to genetic resources lead to equitable sharing of benefits accruing from their use.

(*iii*) Indigenous and local communities to set up a working group to deal with issues relating to

(*a*) Indigenous and local community rights to biodiversity

(*b*) Traditional knowledge and practices, and

(*c*) Equitable sharing of benefits arising from use of such knowledge.

(*iv*) Trade and intellectual property rights

(*v*) Public education: CoP – 4 stressed that member countries should make concerted efforts to create awareness about provision of CBD.

(*vi*) Providing incentives for conservation and sustainable use.

(*vii*) Precautionary principle in decision making on issues related to CBD provisions.

Trade related aspects of intellectual property rights (trips) of the world trade organization:

- The TRIPS agreement establishes minimum standards for intellectual property rights protection.
- This protection takes the form of copyrights, patents, plant variety protection etc.
- It should be noted that reconciliation of IPR regime implementation through TRIPs and community rights has implications for the continued success of international research by CGIAR centres whose efforts in developing enhanced germplasm for variety developing for the developing countries has contributed significantly to increased food production in many countries.

Geographical indicators: GIs covers such qualities, reputation or other characteristics of a product that could be used to recognise that a product has originated from a particular territory, region or locality; these should be essentially attributable to the geographical origin of the product. Eg. scotch, whiskey, champagn.

- The GIs once enforced by legislation, exclude other from using a GI as a trade marks since such use in likely to mislead the consumer at the place of origin of product.
- India has enacted the GI act (1999) to claim GI for a variety of goods including 'basmati' rice.
- GI protection is recognised under the provision of TRIPs of GATT.

CHAPTER - 11

Seed: Classes and Certification

What is seed certification? Why it is essential to have seed certification programme?

Certification means attaching label of purity and authenticity to the seeds by the seed agencies.

The purpose of seed certification, as defined by the international crop improvement association is as under

(*a*) "To maintain, and to make available to the public sources, high quality seeds and propagating materials of superior varieties grown and contributed as to ensure gentic identity. Only these varieties that contain the superior germplasm are eligible for certification. The certified seed is high in varietal purity and of good seeding value."

(*b*) "Varieties eligible for certification have resulted either from natural selection or through systematic plant breeding. In either case, without a planned method for maintaining purity (genetic), there is a danger of losing varietal identity".

(*c*) "Varietal purity is the first consideration in seed certification, but other factores, such as weeds, diseases, viability, mechanical purity and grading are also important. A properly cleaned and graded seed is easier to plant and gives more uniform stands, and controls weed distribution and disease spread".

(*d*) "Seed certification is designed, therefore to maintain not only the gentic purity of superior crop varieties but also the reasonable standards of seed condition and quality".

Seed testing: Seed testing is evaluation of the planting value of seed. It includes following steps.

1. Seed sampling – Different types of sample like primary, composite, submitted and working sample.
2. Physical purity analysis – proportion of pure seed, other seed and inert matter is calculated. Weight of working sample for purity analysis varies from crop to crop. E.g. French bean – 700 gm, in tomato it is 7 gm.
3. Seed germination testing
4. Seed viability testing

   (*a*) Tophographical tetrazolium test (TZ): Given by lokan, 2,3,5 – triphenyl tetrazolium chloride, red stable, non-diffusable triphenyl formazon formed.

   (*b*) Embryo excision method 20°c, 14 days.

5. Ferric chloride test for mechanical damage.

   In legumes – damaged area turns black in this solution.

6. Seed vigour testing – seed vigour is the sum total of those properties that determine the potential activity and performance of the seed or seed lots during germination and seedling performance. It may be direct test like brick gravel test (the hiltner test), paper piercing test, conductivity test, accelerating aging test and cold test for corn etc. Or indirect like seedling growth rate, weight of seedling, speed of germination.

7. Seed health tests: sodium hydroxide seed soak method for kernel bunt for 0.2% solution. The proportion of seed infected NAOH kernel bunt exhibit a jet black shiny appearance contrasting to pale-yellow healthy portion of the seed. Also used for bunt of rice.

8. Other tests
   (*a*) Moisture tests
   (*b*) Varietal purity test
      (*i*) Grow out test – varietal purity and freedom from disease inoculums
      (*ii*) Molecular markers
      (*iii*) DNA fingerprinting if available.

Difference between nucleus, breeder, foundation and certified seed

Feature	Nucleus seed	Breeder seed	Foundation seed	Certified seed
Definition	It is the initial seed of improved variety	It is the progeny of nucleus or breeder seed	It is the progeny of breeder seed	It is progeny of foundation or certified seed
Produced by	Original plant breeder	Original or sponsored plant breeder	Produced by NSC	Produced by SSC and NSC
Produced at	Experimental forms of concerned research institute or Agricultural university	Experimental forms of concerned research institute or Agricultural university	Govt. Seed multiplication forms	The fields of progressive farmers
Genetic purity	100%	100%	100%	99.9%
Physical purity	100%	100%	98%	98%
Certification	Not required	Not required	Certification is done by SSCA	Certification is done by SSCA
Colour of tag	-	Golden yellow	White	Azur blue
Use	For production of breeder seed	For production of foundation seed	For production of certified seed	For distributed to the farmers for commercial crop production

Difference between certified and truthfully labelled seed

Certified seed	Truthfully labelled seed
The seed, which is being certified by seed certification agency, is called as certified seed.	Truthfully labelled seed is one which is being and produced and marketed by the producing company by maintaining the labelling standards.
Certified seed has to maintain both field and seed standards and if the seed lot meets both the field and seed standards then only the certification tag or label is issued.	The farmer or the user of the seed does not know the pedigree of the truthfully labelled seed and he has to relay on the seed producing company.
Colour of tag is azur blue.	Colour of tag is green.

Explain how seed certification is carried out.

Procedure for seed certification: Seed certification is voluntary and that too for the kind and variety notified by the government of India. It can be completed in six broad phases.

1. Receipt and scrutiny of the application.
2. Verification of seed source, class and other requirements.
3. Field inspection should be conducted to see that fields are up to the prescribed field standard.
4. Post harvest inspection, including processing and packing.
5. Seed sampling and testing to confirm that the seeds are up to the prescribed seed standards.
6. Grant of certificate, tagging and sealing.

Receipt and scrutiny of the application: All those persons who are interested in seed certification should submit an application in Form No 1 to the concerned seed certification officer with the prescribed fees of Rs 25/-. The fee is for one season for a single variety and for an area up to 25 acres (10 ha.) If the area is more than 25 acres or if more than one variety is planted separate applications should be made for each variety.

On receiving the applications the seed certification agency verifies for the following conditions:

(*i*) Eligibility of the variety: Only those varieties that are notified by the central govt. are eligible for certification.

(*ii*) Establishing the seed source: The seed producer should submit the tag, invoice, and a copy of Form No2.

(*iii*) There should not be any difficulty in reaching the field for carrying out timely field inspection.

(*iv*) Whether the required isolation and land requirement is followed or not.

(*v*) Whether the processing plant facility is available to the applicant.

(*vi*) Whether the applicant has paid the requisite registration fee or not.

If all the six conditions are fulfilled then the seed producer has to pay the field inspection fees.

Verification of seed source, class and other requirements: The seed should be from authentic source and from appropriate class and should be in accordance with Indian Minimum Seed Certification Standards.

Inspection of Seed Fields: The certified seed producers should grow and harvest the crop as per the guidelines issued by the seed certification agency. They must carefully and faithfully carry out the roguing and other operations as per the directions of the certification agency.

The certification staff conducts field inspections at appropriate stages of crop growth to ensure that minimum standards of isolation, preceding crop requirement, roguing and other special operations are maintained at all times. The inspection of seed crop is done at different stages of crop growth such as at the time of sowing (when new crop is introduced), vegetative stage or preflowering stage, flowering stage, post flowering or preharvest stages and at the time of harvest. The contaminants to be observed during field inspections are offtypes, pollen shedders, shedding tassels, inseperable other crop plants, objectionable weed plants and diseased plants. The field inspections are designated to ensure that the crop is up to the prescribed field standards. All the seed fields, which do not meet the required field standards, are eventually rejected.

Post Harvest Inspection: The personnel from the seed certification agency should inspect the fields during harvesting or post harvesting, so that there are no mechanical mixtures and the seed is not handled badly during threshing or afterwards. Then the seed is sent to seed processing plant with a threshing certificate. The personnel from the seed certification agency will be inspecting the seed processing plant to avoid mechanical mixtures and damage caused to the seed during processing.

Seed Sampling and Testing: The representative from seed certification agency draws a representative sample from the seed lot at the time of processing or after processing and sends the sample to official seed testing laboratory for evaluation. In the seed testing laboratory the samples will be evaluated for seed standards such as pure seed, inert matter, other crop seed, weed seeds, germination percentage and moisture percentage etc.

Grant of certificate, tagging and sealing: After receiving a satisfactory report from the seed testing laboratory, tagging and sealing of bags will be done under the supervision of seed certification agency. Under special circumstances, advance tags will also be issued to the extent of 75 per cent of the seed lot. Tags and seals should be in accordance with general seed certification requirements. Affixing of tags and seals on the containers completes the process of certification of seeds.

Validity period: The seed is initially valid for a period of nine months from the date of testing the samples. If the seed is not sold within the stipulated period, it can be revalidated for a period of six months if the seed lot meets the required seed standards. The seed can be revalidated as long as it meets the prescribed seed standards and for each revalidation the validity period will be extended for six months.

CHAPTER - 12

Economic Botany and Medicinal Plants

Botanical classification of different crops:

Rice (*Oryza sativa* L.) (2n=24) Class : Monocotyledon Sub class : Glumiflorae Series : Glumacea Family : Poaceae (Gramineae) Sub family : Pooideae Tribe : Oryzeaea Genus : *Oryza* Species : *sativa*	**Wheat** (*Triticum aestivum* L.) (2n=14, 28, 42) Class : Monocotyledon Sub class : Glumiflorae Series : Glumacea Family : Poaceae (Gramineae) Sub family : Pooideae Tribe : Hordeae Genus : *Triticum* Species : *aestivum*
Maize (*Zea mays* L.) (2n=20) Class : Monocotyledon Sub class : Glumiflorae Series : Glumacea Family : Poaceae (Gramineae) Sub family : Panocoideae Tribe : Maydeaea Genus : *Zea* Species : *mays*	**Jowar** (Sorghum bicolor L.) (2n=20) Class : Monocotyledon Sub class : Glumiflorae Series : Glumacea Family : Poaceae (Gramineae) Sub family : Panocoideae Tribe : Andropogonae Genus : *Sorghum* Species : *bicolor*
Pearl millet (Pennisetum americanum L.) (2n=14) Class : Monocotyledon Sub class : Glumiflorae Series : Glumacea Family : Poaceae (Gramineae) Sub family : Panicoideae Tribe : Paniceae Genus : *Pennisetum* Species : *americanum*	**Finger millet** (Eleusine coracana L.) (2n=36) Class : Monocotyledon Sub class : Glumiflorae Series : Glumacea Family : Poaceae (Gramineae) Sub family : Pooideaea Tribe : Chlorideaea Genus : *Eleusine* Species : *corocana*
Foxtail millet (Setaria italica L.) (2n=20) Class : Monocotyledon Sub class : Glumiflorae Series : Glumacea Family : Poaceae (Gramineae) Sub family : Panicoideae Tribe : Paniceaea Genus : *Setaria* Species : *italica*	**Red gram** (Cajanus cajan L.) (2n=22) Class : Dicotyledon Sub class : Polypetalae Order : Rosales Series : Calcyflorae Family : Fabaceae (Leguminoceaea) Sub family : Papiolionaceaea Tribe : Phaseoleae Genus : *Cajanus* Species : *cajan*

Bengal gram (Cicer arietinum L.) (2n=16) Class : Dicotyledon Sub class : Polypetalae Order : Rosales Series : Calcyflorae Family : Fabaceae (Leguminoceaea) Sub family : Papiolionaceaea Tribe : Phaseoleae Genus : *Cicer* Species : *arietinum*	**Black gram** (Vigna mungo L.) (2n=22) Class : Dicotyledon Sub class : Polypetalae Order : Rosales Series : Calcyflorae Family : Fabaceae (Leguminoceaea) Sub family : Papiolionaceaea Tribe : Phaseoleae Genus : *Vigna* Species : *mungo*
Green gram (Vigna radiata L.) (2n=22) Class : Dicotyledon Sub class : Polypetalae Order : Rosales Series : Calcyflorae Family : Fabaceae (Leguminoceaea) Sub family : Papiolionaceaea Tribe : Phaseoleae Genus : *Vigna* Species : *radiate*	**Garden bean** (*Dolichos lab lab* L.) (2n=22) Class : Dicotyledon Sub class : Polypetalae Series : Calcyflorae Order : Rosales Family : Fabaceae (Leguminoceaea) Sub family : Papiolionaceaea Tribe : Phaseoleae Genus : *Dolichos* Species : *lab lab*
Soya bean (Glycine max L.) (2n=40) Class : Dicotyledon Sub class : Polypetalae Order : Rosales Series : Calcyflorae Family : Fabaceae (Leguminoceaea) Sub family : Papiolionaceaea Tribe : Phaseoleae Genus : *Glycine* Species : *max*	**Ground nut** (*Arachis hypogaea* L.) (2n=40) Class : Dicotyledon Sub class : Polypetalae Series : Calcyflorae Order : Rosales Family : Leguminoceaea Sub family : Papiolionaceaea Genus : *Arachis* Species : *hypogaea*
Castor (*Ricinus commcinis* L.) (2n=20) Class : Dicotyledon Sub class : Monochlamydae Series : Curvembryae Order : Unisexualis Family : Euphorbiaceaea Genus : *Ricinus* Species : communis	**Sunflower** (*Helianthus annuus* L.) (2n=34) Class : Dicotyledon Sub class : Gamopetaleaea Series : Inferae Order : Asterales Family : Compositae/Asteraceae Genus : *Helianthus* Species : *annuus*

Safflower (*Carthamus tinctorius* L.) (2n=24) Class : Dicotyledon Sub class : Gamopetaleaea Series : Inferae Order : Asterales Family : Compositae/Asteraceae Genus : *Carthamus* Species : *tinctorius*	**Gingelly** (*Sesamum indicum* L.) (2n=26) Class : Dicotyledon Sub class : Gamopetaleaea Series : Bicarpellatae Order : Personales Family : Pedaliaceaea Genus : *Sesamum* Species : indicum
Niger (*Guizotia abyssinica* L.) (2n=30) Class : Dicotyledon Sub class : Gamopetaleaea Series : Inferae Order : Asterales Family : Compositae/Asteraceae Genus : *Guizotia* Species : *abyssinica*	**Cotton** (*Gossypium hirsutum* L.) (2n=26,52) Class : Dicotyledon Sub class : Polypetalae Series : Thalamiflorae Family : Malvaceaea Genus : *Gossypium* Species : *hirsutum*
Sun hemp (*Crotalaria juncea* L.) (2n=16) Class : Dicotyledons Sub class : Polypetalae Series : Calcyflorae Family : Leguminpceae Sub family : Papilionaceaea Genus : *Crotalaria* Species : *juncea*	**Kenaf** (*Hibiscus cannabinus* L.) (2n=36) Class : Dicotyledons Sub class : Polypetalae Series : Thalamiflorae Order : Malvales Family : Malvaceae Genus : *Hibiscus* Species : *cannabinus*
Roselle (*Hibiscus sabdariffa* L.) (2n=36/72) Class : Dicotyledons Sub class : Polypetalae Series : Thalamiflorae Order : Malvales Family : Malvaceae Genus : *Hibiscus* Species : *sabdariffa*	**Mustard** (*Brassica nigra* L.) (2n=16, 18, 20, 22, 36) Class : Dicotyledons Sub class : Polypetalae Series : Thalamiflorae Order : Perietales Family : Brassicaceae/Cruciferae Genus : *Brassica* Species : *nigra*
Coconut (*Cocosnucifera* L.) (2n=32) Class : Mocotyledons Sub class : Cocodea Series : Calycinae Order : Commelianales Family : Arecaceae (Palmae) Genus : *Cocos* Species : *nucifera*	**Chillies** (*Capsicum fruitiscence* L.) (2n=24) Class : Dicotyledons Sub class : Gamopetalae Series : Bicarpellatae Order : Polemoniales Family : Solanaceae Genus : *Capsicum* Species : *fruitiscence*

Economic importance of different crops:

Rice: Used as a food crop. Parched rice, rice flakes and puffed rice. Rice wine, which contains 10-15 percent alcohol, is usually made from glutinous rice. The husks are value less as an animal feed and little value as fertilizers. They are used as fuel for rice mills, as an aggregate for light weight concrete and for making hard board. Rice bran is valuable livestock and poultry feed. The rice straw is used for cattle feeding; rice bran oil is used as cooking.

Wheat: It is used as a food crop.

Maize: The grain is used as food for man, and livestock. The immature cobs are largely eaten after roasting or boiling. The grain is also used in making corn starch and industrial alcohol. Corn oil is used for soap making, lubrication and as a salad oil.

Jowar: Used as a food (millet) crop in the drier parts of tropical Africa, India and China. The threshed grain is ground into a whole meal flour and used for making thin porridge.

Pearl millet: Used as staple food in the drier parts of tropical Africa and in India. The green plants produce a useful fodder and it is sometimes grown for this purpose.

Finger millet: Staple food in parts of East and Central Africa and India. Particularly in Kamataka. It is used for making malting and brewing.

Foxtail millet: Used for human food and fodder. The grain is consumed as boiled or panched. It contains 10 percent protein, 8 percent fat and 72 percent carbohydrates.

Red gram: The green pods and green seeds are eaten as vegetable. The ripe dry seeds are boiled and eaten as pulse split, pulse, is used as dal. Dried husk and broken pieces of seed are used as cattle feed. Green leaf and tops of the plant is used as fodder and for making silage. Dried stake is used as fuel and as thatching material. It is source of protein (23%).

Black gram: Seed is used as a source of protein (23.4 percent). The cotyledon contains batter (aminoacid related substance). This makes the wet flour to swell up i.e. doughness. Because of this batter, it is suitable for making *idlis, dosa,* etc.

Green gram: Mungbean has highest digestibility among the grain legumes. It is free from causing flatulence, mostly used as sprouts, also as split *dal*. The rich source of protein (23.5 %)

Garden bean: Seed as a source of protein, fruit as a vegetable.

Safflower: Seeds have high protein (40%) and oil content (20%). Black seeds are rich in protein and low in oil yellow seeds are low on protein and rich in oil.

Sunflower: Sunflower oil is used as edible oil, yellow in colour semi drying oil. Oil is mixed with other drying oils and used for manufacturing of varnishes and paints. Fibre can be extracted from the stem. Fruit can be roasted and after removal of pericarp seed is eaten. Pericarp is used for cattle feed. Sunflower can be used as manure.

Gingelly: Sesamum oil, is rich in vitamin E and lacking in vitamin A content. The most useful property of this oil is its high stability and used in culinary purposes. Sesamum seeds are often used in sweet meats. They are roasted and ground into meal and made into cakes used in bakery industry.

Groundnut: Groundnut oil is extensively used in cookery as a salad oil and edibles fried in the oil have a pleasant flavour and keep for a comparatively long time. The oil is used in the manufacture of vanaspati. This oil is also used in shaving creams, pomades, cold creams, candles, glycerin and leather dressing's furniture creams. Oil cake is used as cattle feed and also applied as manure for paddy, vegetables, bananas and other fruit trees. Groundnut kernels are rich and cheap source of vegetable protein contain Vitamin A, B (Thiamin) and some members of the Ba group (riboflavin) and Nicotinic acid.

Castor: Castor oil is used in the manufacture of soaps and tooth pastes. Sodium ricinoleate and sulphuricinoleate are used in tooth pastes, due to their strong bactericidal action. Oil is also used as a solid contact lubricant in railway engine and locomotive bearings. Dehydrated castor oil is used in white surface finishes of refrigerators, manufacture of paints and enamels, high quality varnishes, linoleum and oil cloth manufacture.

Castor oil contains glycerides of the unsaturated hydroxy acid, 90 % of ricinoleic acid imports to the oil certain distinctive properties, high acetyl or hydroxyl value, high specific gravity, refractive index and viscosity, low saponification value, miscibility with ethyl alcohol and near insolubility in petroleum ether.

The finest quality of almost colourless castor oil is used in medicinal and aeroengine lubrication.

Cotton: Cotton is a major cash crop which gives three important these are products fibre, food and feed. Lint fibre is used for clothing, household and industrial articles. Articles of clothing include shirtings, gloves, hosiery etc. House hold articles include bed sheets, covers, towels etc. Industrial articles include bags, belting, awnings, tents tarpaulins, insulation, cellulose, plastics etc. Cotton seed oil is used in the manufacture of vanaspati; cotton seed cake after extraction of oil is good organic manure.

Kenaf: Fibre is widely used for rope and cordage. Large quantities are used in making fishing nets and strings for trying rafters. It is also used for coarse canvas, sacks and gunny bags, floor matting, rug backing etc. Leaves are used as green vegetable.

Mustard: The leaves of young plants of rape and mustard are used as green vegetables. The plants are used as green fodder for cattle. The seed contains 30-48 % oil. The chief use of oil in India is for edible purposes. The seed and oil are used as a condiment in the preparation of pickles and for flavouring curries and vegetables. Mustard oil along with rock salt is used as a dental solution in the diesases of the gum. It is also used as massage in many paralytic diseases of the nervous system.

Niger: The oil is mainly used for culinary and lighting purposes and also used for some pharmaceutical purposes as sesamum oil. The cake contains about 4.5 %. Nitrogen and it is used for manure and as a cattle feed.

Roselle: Roselle fibre is used for sacking, cordage, rope, fishing nets and generally for all purpose for which jute is used. Tender leaves and stalks are eaten as salad or used for seasoning curries. They are acidic in taste and may be used in soups, puddings cakes and pies and for colouring syrups and liquors.

Sun hemp: Sun hemp is essentially a cordage fibre and is used in the manufacture of ropes, twins, cords and marine cardage. It also used in the manufacture of sail cloth, canuas, matting and rope soles of shoes and sandals.

Chillies: Dry and green chilli is extensively used as a spice in all types of curries and dishes. It is also used for seasoning of egg, fish and meat preparations sauces, chutneys, pickles, frank furters, sausages etc.

Green chillies are rich in vitamin C and A so they serve as good vitamin supplement in addition to their being an appetizer. Simla mirch is rich in Vitamin C.

Capsicum preparations are used as counter-irritants in lumbago, neuralagia and rheumatic disorders.

Coconut: The wet meat or endosperm is used for culinary purposes in various forms. The use of tender coconut water is recommended in cases of gastro-enteritis and as a useful substitute for saline glucose coconut oil is used as a medium in cooking vegetables and as a hair oil. It is also used in soap manufacture, creams, cosmetics and production of palsticisers. Coconut cake is a valuable cattle and poultry feed. Coconut fibre is used for manufacturing of mats, ropes, twines, cusions, brushes, brooms, shock proof packing material and hard boards. Coconut leaves are used for thaching houses, fencing and for making brooms, baskets etc. The root is astringent and used as mouth wash and gargle in medicine.

Economic importance of crops:

Common name	Botanical name	Family	Description of useful economic part
CEREALS			
Barley	*Hordeum vulgare* L.	Gramineae	Caryopis, Grain is used for preparing breakfast foods. It is used for the manufacture of beer and it is also of medicinal use.
Rye	*Secale cereale* L.	Gramineae	Caryopsis, Grain used as a stock feed. Manufacture of whisky and alcohol. The rye grains have about 13 % protein and aglutinous like wheat. The rye flour is used for making bread and much used in the black bread in Germany, Poland and Russia. Most of the rye cultivation is for hay and the grains for stock feed. The straw is used as packing material, thatching for paper making and animal bedding. The grains are also used in breweries making whisky and for alcohol. Rye can also be grown as a green manure crop.
Oats	*Avena byzantina* L.	Gramineae	Caryopsis, Oat meal is prepared by grinding the seed. This is used for preparing cakes, biscuits, breakfast food. Rates are fed to Race horses.
SUGARS AND STARCHES			
Beet Root	*Beta vulgaris L.*	Chenopodiaceae	The napi form root and tender leaves. The tender leaves are used as a cattle feed and a manure. Beet root is used for extracting sugar. Molasses are used for preparation of alcohol, filter cake as manure and feed stock for animals.
Tapioca	*Manihot esculenta* L.	Euphorbiaceae	Tubers, Sago made from tapioca are used in the preparation of various food dishes, *khir, kanji, payasam, kesari, uppuma, pappad* etc. Starch is used as food stuff, textile and paper industry and manufacture of plywood and veneer, adhesives, glucose and dextrin etc. Minor industrial applications include, manufacture of explosives dyes, drugs, chemicals, carpets, for the production of alcohol and for the coagulation of rubber latex. Tapioca flour is used in preparation of bread, biscuits, leaves are used as a vegetable and cattle feed. The pith of stem is used for section cutting of plant material for anatomical studies in laboratories.
Sago	*Metroxylon sagus*	Palmae	Pith removed from stem. Pith is grounded, mixed with water strained through sieve. The pellet that comes out are the sago of commerce. These are used mostly as a medicine for preventing dysentry. Sago is used for the preparation of puddings and other dishes. It is also used as an invalid food.

Arrow Root	*Maranta arundinaceae* L.	Marantaceae	Tuberous roots, Roots are peeled, washed, crushed and the pulp that comes out is passed through perforated cylinders. A stream of water carries this starch into tanks where it settles. This starch is used as a valuable food for children. Commercial good quality arrow root starch, due to its high digestibility it is valued as a food stuff particularly for infants and is used in the preparation of biscuits, cakes and puddings. The starch possesses demulscent properties and is sometimes used in the treatment of diseases of the intestine. It is also used as a base of face powders and in the prepration of certain specialised glues. The fibrous material left over from the rhizomes can be used as cattle feed or manure.
Yam	*Dioscorea alata* L.	Dioscoreaceae	Root, Root is used as a staple food in some countries like S. America and West Indies. Roots are baked and boiled grounded to flour. This flour is used for preparing many foods (preparations).
Beet root	*Beta vulgaris* L.	Chenopodiaceae	The napi from root and tender leaves. The tender leaves are used as a cattle feed and a manure. Beet root is used for extracting susgar. Molasses are used for preparation of alcohol, filter cake as manure and feed stock for animals.
Sweet Potato	*Ipomoea batatas* L.	Convolvulaceae	Tuberous root, Vegetable root crop. Manufacture of starch, flour and glucose syrup. Tubers yield high proportion of alcohol. Plants good food for live stock besides its importance as food crops. Rich in vitamin A, B & C. Sweet potato is used as food, after boiling, baking or frying. The starch is suitable for sizing paper and textiles and for use of laundries. Sweet potato flour prepared by dehydrating and grinding tubers is used as a supplement to cereal flours in the preparation of bakery products, puddings and milk jelly. Sweet potato is a good raw material for the fermentation production of industrial alcohol, lactic acid, acetone, vinegar etc. The vines are used as a cattle feed.
Potato	*Solanum tuberosum*	Solanaceae	Stem tubers, Tubers are used for table purposes (vegetable), used in the manufacture of starch, dextrose glucose and alcohol. Tuber used in the manufacture of dehydrated products. Chips, wafers etc. Plants as live stock feed. Potato flour is used in the baking industry in the preparation of bread, pastries, cakes, biscuits etc. potato peels are used in the production of synthetic cork.
Saffron	*Crocus sativus*	Irdaceae	Dried stigma and tops of styles constitute saffron as spice. An Yellow dye is extracted.
Cinnamon	*Cinmomum zeylanicum*	Lauraceae	Dried and split bark usually called quill is used as spice.

Cumin	*Cyminum cyminum*	Umbelliferae	Dried fruits used as spice
Fenugreek	*Trigonellafoenum graecum*	Papilonaceae	Seeds and young plants used as spice, flavour is due to coumarin.
Coriander	*Coriandrum sativum*	Umbelliferae	Dried fruits and young plants used as spice.
Poppy	*Papver somniferum*	*Papaveraceae*	Seeds used as spice. The dried latex from unripe fruit forms opium.
Turmeric	*Curcuma longa*	Zingiberaceae	Cured thizomes used as spice. An orange dye is extrated from rihizome. The colouring matter is curcumin powder used as medicine.
Cardamon	*Elettaria cardamomum*	Zingiberaceae	Fine curred fruits with seeds are aromatic and used as spice. Dried fruits are used as a spice, as masticatory and in medicine. The seeds have a pleasant aroma and characteristic slightly pungent taste. A volatile oil extracted from the seeds is used in perfumery for flavouring liguors and bitters.
Anise	Pimpinella anisum	Umbelliferae	Seeds possess sweet aromatic taste. Used as spice. Fruit a stimulant, carminative, diuretic and milk expectorant.
Caraway	*Carum carvi*	Umbelliferae	Fruits used as spice. A milk carminative.
Nutmeg and Mace	Myristica fragrans	Myristicaceae	Nutmeg is stimulant tringent, used as spice. Mace is the 'aril' used as spice. (Nut Megs are the dried seeds and Mace is the dried aril surrounding the seed).
Asafoetida	*Ferula foetida*	Umbelliferae	The dried latex eudes from the rhizome of tap root used as spice. Highly medicinal.
Cloves	*Eugema caryophyllata*	Mytaceae	Dried unopen flower buds constitute condiment. Clove oil is medicinal and used as clearing agent in cytological work. The taste is due to Euggenol.
Garlic	*Allium satsvum*	Liliaceae	Bulbs used as condiment. The active principle is an antibiotic *allicin* which is pungent.
Onion	*Allium cepa*	Liliaceae	Bulbs used as condiment taste due to a glycoside (ally/propyl-disulphide) in various ways in all curries, fried, boiled, baked, used in soup making in pickles etc.
Ginger	*Zingiber officinale*	Zingibereceae	Rhizomes harvested before used set and are used as condiment, dried and cured rhizome is called sonti.

Pepper	*Piper nigrum*	Piperaceae	Dried unripe fruits used as condiment. Pungency is due to an alkoloid piperin, responsible for the biting taste of black pepper.
BEVERAGES			
Coffee	*Coffea arabica* L.	Rubiaceae	Perisperm of the seed. Seeds are dried, fried, powdered, decoction prepared from the powder mixed with milk and sugar is taken as a beverage. Coffee is mainly used as a beverage in the form of aqueous extract prepared from roasted and powdered beans. The extract has a bitter taste and characterstic aroma and a stimulating effect. It is used in medicine as diuretic and beverage to convalescent people. The disturbances in digestion by use of coffee are due to volatine toxic substance called cafecotixin. Coffee beans are used in the manufacture of a plastic called caffelite. The wood of coffee tree is hard and can be used as a box wood and the branches of the coffee plant are used as walking sticks and hammer handles.
Tea	*Thea sinensis* L. / *Camellia senensis kuntz*	Theaceae	Cured tender leaves are used in preparing the decoction. This decoction along with milk and sugar is used as a beverage. The stimulative effects of tea are due to the presence of coffeine the vegetable tannin and tiny amounts of several volatile oils. Coffeine taken in small doses increases mental and muscular power and has no ofter depressing effect on the system. The vegetable tannin gives the pungency and colour to the tea.

Cocoa	*Theobroma cacao* L.	*Sterculeaceae*	Seeds, seeds are fried, powedered and this powder is used for making chocolates and also used as beverage. Cacao shell is the most considerable by product of the cocao and chocolate industry. It is used as a filler for thermo setting resins in the plastic industry. It is also used as a fertiliser and even as fuel. Cacao butter is used in the production of chocolate, confectionery cosmetics and some medicinal preparations. Chocolate flavoured coatings prepared with lower melting vegetable fats or hydrogenated vegetable fats other than cacao butter marketed for use in the ice cream industry.
FUMITORIES AND MASTICATORIES **Fumitories:** Narcotic substances used for smoking purposes. **Masticatories:** Narcotic substances used for chewing purposes.			
Fumitories:			
Tobacco	*Nicotiana tabacun*	Solanaceae	Tobacco leaves are used for making cigars, cigarettes and sniff. This contains an alkoloid 'Nicotine' used as an insectide. Oil is extracted from seeds, a semi drying oil used in soap industry it is also used in special paints, varnishes and enamil paints. Seed cake is used as organic manure.
Betelvine	*Piper betel*	Palonaceae	Leaves used as masticatory, contains riboflavin and betel phenol.
Arecanut or Betelnut	*Areca catechu* L.	Arecaceae	Leaf whole leaf is used for chewing; it improves salival activity and digestion. Leaves contain riboflavin of betal phenol.
PLANTATION CROPS			
Rubber	*Hevea brasiliensis*	Euphorbiaceae	Latix obtained from the lower part of the stem. Latex after cooking is used for the manufacture of rubber and rubber products like balloons, erasers, rubber bands, tubes, tyres etc
FIBRES			
Flax	*Linum usitatissimum*	*Linaceae*	Fibres extracted from primary phloem mainly cellulose
Hemp (True hemp)	*Cannabis sativa*	*Cannobinaceae*	Fibre obtained from primary and secondary pheloem cellulose in nature.

Coir	*Cocos nucifera* L.	Palmae	Mesocarp of the Drupe/Endosperm. Coir is used for making ropes, marine cables, coasrse and rough fibres, brushes, door mats floor coverings, wall coverings, beds and upholsetery. The endosperm is used for extracting oil and the oil is edible and also as a hair oil.
Kapok	*Cieba pentandra* L.	Bombacaceae	Pods containing of floss used for stuffing pillows, mattresses, beddings etc. Kapok finds use in the production of life saving equipment, and in the construction of thermally insulated and sound proof covers and walls. On account of its buoyancy, freedom for water logging and weight bearing capacity, it is the material par excellence for the manufacture of life buoys and belts, waist coats and other naval life saving appliances. The wood is light and soft and is suitable for making canoes and toys. It is used for matches. The root bark yields a fibre.
Manila hemp	*Musa textilis*	Musaceae	Outer portion of leafstalks used in the preparation of Gunny bags door mats, nets etc. The fibre is used mainly for twines, ropes and cable. It is much valued for hawkers, ship's cables and riggings. It is also used in the manufacture of tough paper.
Agave	*Agave sisalana*	Amaryllidaceae	The leaves of agave yield a valuable fibre. The fibre is mostly used for making ropes, cordage and twine. The shorter and istles are used for making mops and brushes. It is also used for the manufacture of coarse fabrics.
OIL SEEDS			
Linseed	*Linum usitaltissimum* L.	Linaceae	The seed contains 40 % oil, a drying oil used in varnishes, paints and linoleum. Linseed oil mixed with lime water, used as application in bums. This oil with landanum and spirits of turpentine is considered to be one of the best remedies for colic in ponies. Flowers are used as nervine and cardiac tonic. Lin seed cake is used as good organic manure. It is also used a cattle feed.
Mustard	*Brassica campestris* L.	Cruci-fereae	Seeds are used as a condiment and spice seed yield oil which is edible. It is also used in preparation of pickles. Mustard oil cake is used feeding cattle and also as a manure. Musturd oil along with rock salt is used as dental solution in the diseases of the gum. It also forms an ingredient in many Ayurvedic medicated oils used as liniment or massage in many paralytic diseases of the nervous system. The leaves of young plants are used as green vegetable.

Question and Answers

Give the botanical classification of Ground nut and describe its various uses?

For botanical classification see the above table.

Bunch type: The main axis is long and exert, and the laterals are oblique to the main axis in the bunch type.

Spreading type: The main axis is very short and erect, the laterals spreading horizontally along the ground, usually lie within an inch of the ground except at the tips which are somewhat ascending.

Uses:

1. As cattle feed: Ground nut haulms are used as cattle feed in the green state after drying or after converting in to silage
2. As human food: Ground nut kernels are rich source of vegetable protein. (Vitamin A, B and some members of B_2 group (Riboflavin and Nicotinic acid))
3. Ground nut milk and curd.
4. Ground nut oil: It belongs to non-drying group of vegetable oils. Oil is used in the manufacture of vanaspathi (Vegetable ghee)
   - Used as salad oil i.e. for cookery.
   - Oil is used limited extent in soap making
   - Also used in leather dressings, furniture creams, candles, glycerine and tallow substitutes. Tallow is an important solid fat of animal origin, longely used in textile industry.
   - The oil can also be used as an illuminitant, lubricant for the manufacture of olein, stearin and their salts.
   - Oil is substitute for olive oil in the preparation of official liniments, salicylate and mercury.
   - Ground nut emulsion is used as insecticide and the emulsion also found to increase the toxicity of insecticides like rotenone, nicotine, nicotine sulphate etc.
5. Oil cake or Poonac: As cattle feed.
6. Ground nut shell is used in manufacture of activated carbon, acetic acid, alcohol, acetone etc.
7. Ground peanut shells used for polishing tin plate, for cleaning carbon from air-craft engines and plastics.

Discuss various industrial and special uses of maize?

Uses of maize:

1. Forage and feed: Maize plant is used as cattle feed both as fresh green fodder, silage and as stover.
   - Maize plant does not have any problem of hydro cyanic acid and prussic acid production; hence crop can be harvested and fed to cattle at any stage of its growth.
2. Food: In the form of chapathis, porridges, corn flakes and pop corn and roasted green ears, corn meal, corn hominy and grils.
3. Industry: The starch industry using the wet willing process produce starch, maize oil and gluten for feed. Starch intern further processed to produce a series of products like dextrose, corn syrup etc.
   - Maize heats can be used as a cattle feed when ground and fortified with molasses and urea. Industrial processing of maize heat also yields products like xylan, furfurol etc.
   - Corn is also used in the distilling and fermentation industries in the manufacture of ethyl and butyl alcohol, acetone and whisky.
   - Maize oil is highly poly unsaturated oil and is highly valuable as cooking media and also used in soap making and paints.
   - Zein, a special kind of protein in maize, is spyn into a soft, durable artificial fibre. Zein is also employed as a foaming agent in fire extinguishers, an ingredient of linolien and oil cloth, a sizing for textiles, a binding material for composition – cork manufacture and as an adhesive to bind papers, glass, wood and veneer.
   - Corn sugars used in manufacture of chemicals, leather preparation, dyes and explosives.
   - Maize cobs when cooked under pressure with acids produce furfural, a compound used in the production of adiponitrile (nylon), in the refining of diesel, vegetable and lubricating oils, in the refining of butadiene and in the manufacture of plastics.
   - Maize leaves used as cigarette wrapper.
   - Pulverized maize cobs are used extensively as mild abrasive for removing carbon from aeroplane motors.

Mention various Gossypium sps and highlet their distribution and economic importance?

1. Diploid old world species:

   Gossypium arborium: Distributed throughout the rain fed areas from Africa, through Arabia and India to China, Japan and East indies.

   Gossypium herbaceum: This species occurs in Africa, Middle east countries, Central asia and Western India.

Races of species based on distribution	
Gossypium arborium (6)	***Gossypium herbaceum*** (5)
bengalense	persium (South central asia)
burmanicum	kuljinum (Chinese)
cernuum	acerifolium (North africa)
indicum	wightianum (Western India)
sinense	africanum (South africa)
sudanense	

2. Diploid new world species:

   Gossypium thurberi: North america,

   Gossypium raimondi: South America

3. Tetraploid new world species:

   Gossypium hirsutum: North America especially central America

   Gossypium barbadense: South America, particularly Columbia, Ecuador and Peru.

Uses of cotton:

- Cotton is major cash crop which gives three important products fibre, food and feed.
- Cotton seed oil is used in manfacture of vanaspathi.
- Cotton seed cake is good organic manure
- Cotton flour in bread and biscuit making.
- Cotton plant
  - Root bark – ergot like drug
  - Stacks – fuel, paper pulp, bast fibres and pyroligneous acid
  - Leaves - fodder.
- Cotton seed – meats, hulls and linters.
  - Meats – Cake & meal and crude oil

    Cake & meal – flour (bread, cake and crackers), feed for cattle and fertilisers.

    Crude oil – retined oil (salad and cooking oil), soap, glycerine, fattyacids, rubber, plastics, tinishes, insecticides, fungicides, metallic soaps and water proofing.
  - Hulls - feed for cattle, fertiliser (mulch and soil conditioner), bran (live stock feed), fibre (pulp) and furfural (synthetic rubber, petroleum refining and plastics).
  - Linters – chemical and non chemical.

Chemical – viscose (films and rayons), cellulose esters and ethers, cellulose nitrate and cellulose acetate (Yarn (clothing), plastics and films (x-ray and photographic))

Non chemical – absorbent cotton medical supplies, yarns (lamp and candle nicks, twine, rugs and maps) and for automotive upholstery, pads, cushions, comforts and mattresses.

Describe the important and crop plants belonging to the families' leguminaceae and solanaceae?

Leguminaceae: Three tribes

1. Viciae
   - Lentil (*Lens esculenta*)
   - Pea (*Pisum sativum*)
   - Chick pea (*Cicer ariatinum*)
   - Khesari (*Lathyrus sativum*)
   - Horse bean (*Vicia faba*)
2. Hydysareae
   - Ground nut (*Arachis hypogeae*)
3. Phaseoleae
   - Soya bean (*Glycine max*)
   - Lab lab bean (*Lablab purpurium*)
   - Common bean (*Phaseolus vulgaris*)
   - Cow pea (*Vigna unguilata*)
   - Pigeon pea (*Cajanus cajan*)
   - Horse gram (*Dolichos uniflorus*)

Another tribe, Galegae (Indigoferae) which includes Guar or Cluster bean used for extraction of gum.

Solanaceaea:

- *Atropa belladona* (Belladona)
- *Capsicum frutescence* (Chilly)
- *Cestrum nocturnum* (Night queen)
- *Cestrum diurnum* (Day king)
- *Datura metel* (Thorn apple)
- *Lycopersicuon esculentum* (Tomato)
- *Nicotiana tabacum* (Tobacco)
- *Petunia alba* (Petunia)
- *Physalis minima* (Sun berry)
- *Physalis peruviana* (Goose berry)

- *Solanum melongena* (Brinjal)
- *Solanum nigram* (Kamanchi)
- *Solanum surattense* (Vaakudu)
- *Solanum tuberosum* (Potato)
- *Withania somnifera* (Aswagandha)

What is the centre of origin of sugar cane? List various species of saccharum and comment on their utility in sugar cane breeding?

Sugar cane, the primary source of sugar is believed to be originated in the south pacific (probably new guinia) and spread throughout most of South East Asia at an early date.

Species:

- *Saccharum officinarum* (Noble or thick canes)
- S. *barberi* (Indian canes)
- S. *sinense* (Chinese or Japanese canes)
- S. *spontaneum* (Wild cane)
- S. *robustum et brandes* (Wild cane)

Utility: Reason for crossing noble canes with wild type was to produce hybrids that combine the best features of both the parents i.e. thick, soft, long barreled cane and high sucrose content of S. *officinarum* as well as deep penetrating root system, vigour, drought and disease resistance of wild species S. *spontaneum* and S. *robustum.*

Explain in brief about the commercial uses of castor?

- Soaps and tooth pastes: The sodium soap of castor oil, sold as technical sodium ricinoleate has special use in manufacture of soaps.
- Lubricants: In railway engine and locomotive bearings
- Surface coatings: Dehydrated castor oil (DCO) forms a solid film on exposure to air and DCO used in white surface finishes of refrigerators, manufacture of paints and enamels, high quality varnishes, linoleum and oil cloth manufacture.
- Wetting agents and detergents: Sulphonation with H_2SO_4 gives materials whose sodium salts have valuable wetting and detergent properties.
- Production of sebacic acid and secondary octyl alcohol: Sebacic acid used in manufacture of synthetic rubbers, synthetic resins and synthetic fibres.

  Secondary octyl alcohol is used in production of plasticisers, as an antifoam agent, in lacquers and as solvent for some resins.
- Production of undecylenic acid and heptaldehyde: Heptaldehyde used in various perfumes i.e. violet, ionone, lilac etc, and undecylenic acid used in perfumery.

- Hydrogenation: Hydrogenated castor oil is useful in lithium based lubricating greases, in polishes and plastics and as a shellac blend.
- Fuel: Motor and diesel fuels.
- Chemical contraceptives: Ricinoleic acid is used as ingradient in chemical contraceptives; it functions by so lowering the surface tension that the sperms are disrupted.
- Insecticides: amides from castor oil yields on pyrolysis undocyclamides used in fly sprays and insecticides.
- Castor seed cake contains at least three toxic substances, Ricin - an extremely poisonous substance in large amount, Ricinone – a slightly toxic alkaloid, an allergen – a protein polysaccharide. Ricin acts a blood coagulant.

Describe the various types and sources of commercial fibres and their economic importance in the Indian context?

Commercially the term 'fibre' refers to practically all small, thin fragments of any substance.

Classification of fibres on the basis of nature and substance:

1. **Bast fibres:** Fibres associated with the phloem, pericycle and cortex are referred to as stem, bast or soft fibres. These are found in dicotyledonous plants.
   Source: Flax, Jute, Hemp, Kenaf, Roselle and Ranie.
2. **Structural fibres:** Are also known as hard or leaf fibres and strands of small, short lignified cells ensheathing both xylem and phloem and are primarily found scattered in the leaves of monocotyledonous plants.

   Source: Manila hemp, Sisal and New zea land hemp.
3. **Surface Fibres:** Born on the surface of stems, leaves, fruits and seeds.

   Source: Cotton and Kapok fibres

Classification of fibres on the basis of their use:

1. **Textile fibres:** Used for manufacture of fabrics, netting and cadrage.

   Source: Cotton, Flax and Hemp
2. **Brush fibres:** Making brushes and brooms.

   Source: Istle and sisal (hard fibres), piassava (surface fibres of palm leaves and stems) and broom corn (the dense bristly inflorescence of *Sorghum vulgare* var *technicum*)
3. **Painting and rough weaving fibres:**

   Source: Panama hats are made by weaving and braiding strips of the palm like leaf of *Carludovica palmate.* And Bamboo stripes are used in the manufacture of fishing rods, furniture and baskets etc.

4. **Filling fibres:** Used in upholstery and for stuffing cushions, mattresses and life preservers etc.

   Source: Kapok, Cotton, Jute and Spanish mass.

5. **Natural fibres:** Best known bark cloth is 'tapa cloth' obtained from bark of paper mulberry, *Broussonetia papyrifera.*

6. **Paper making fibres:** For making paper and paper products, wood fibres, textile fibres and various grasses and sedges (such as Esparto and Cyperus) are much used.

Briefly describe what useful nutrients are mainly or typically found in the following plant parts? A. Cereals B. Fruits C. Leafy vegetables D. Grain legumes

Cereals: Are sources of carbohydrates

Fruits: The highly nutritious and delicious fruits rich in minerals and vitamins.

Leafy vegetables: Roots, stems, leaves, flowers and fruits of number of plants are eaten raw or cooked as vegetables. Some of them are rich sources of carbohydrates (Starch etc.). While others are rich in minerals and vitamins.

Legumes: Sources of protein. They supplement carbohydrates in human diet.

Nuts: Rich in fat or protein

Give botanical classification, economic uses and important varieties of pearl millet?

For botanical classification, refer the above table.

Uses:

- Seed for *chapatti, roti, sangati, hittu* and *muddle.*
- Flour for thatching and fuel
- Inferior for fodder not as like sorghum, paddy or ragi.
- But Napier grass (P. *purpurium*) is close relation of the *bajra,* is one of the favourable fodder grasses.
- Used as stable food in the dry parts of tropical Africa and in India.
- Green plants produce a useful fodder and it is sometimes grown for this purpose.

Improved varieties: HB1, HB2, HB3, HB4 and MAU 451

Composites: Vijay, Nagarjuna and Balaji

Discuss origin and various uses of cultivated Soyabean?

Origin: China

Uses: Seeds are consumed locally after splitting as *dhal.*

- Soyabean oil meal: It is sound residue after extraction of oil. And it is used as live stock feed and source of industrial protein used in manufacture of adhesives.
- Soyabean flour: Used as fertilizer.
- Soyabean oil: Used in margarine industry and also used in soap industry, lecithin, paints, varnishes and related proteins.
- Soya milk is made in to curd, butter milk, cheese and other milk products.

Describe briefly the binomial system of classification and its significance? And enlist the ISBN rules for naming of the plants?

The earliest scientific names of plants were 'polynomials' in which several Latin words were used giving the description of the plant. Subsequently Gaspard Bauhin (1623) used only two words for the first time and introduced the binomial system. But the credit of using binomial system consistently for the first in his 'Species Plantarum' goes to Carolus Von Linnaeus (1753).

In binomial system every plant name will have two names and invariably they should be in Latin. The first word will be the name of the genus and the second will be the name of the species. The generic name will be in noun form and always begins with capital letter. The specific name will be in the adjective form and starts with small letter.

E.g. *Solanum tuberosum,* is the name of potato plant in which *Solanum* is the genus and *tuberosum* is the species.

International Code of Botanical Nomenclature (ISBN): Naming of all plants should be done in accordance with the principles laid down by ICBN. These rules are formed at the periodical meetings of International Botanical Congress held once in five or ten years. The first meeting was held at Paris in 1867 which resulted in 'Paris code'. According to 'ICBN' the following rules must be followed while naming the plants.

1. Every plant should have only one correct scientific name.
2. The scientific name must be in Latin.
3. Binomial nomenclature system must be used for naming the plants.
4. The scientific name should be underlined or published in *italics*.
5. The author's name may be given in abbreviated form at the end of scientific name
   E.g. *Pisum sativum* L. ('L' stands for Linnaeus)
6. If the generic name and species name are exactly the same. It is known as tautonymy. According to 'ISBN' tautonyms are not valid in plant nomenclature. E.g. *Malus malus*.

Discuss the features of plant classification by Bentham and Hooker?

Bentham and Hooker's classification was based on the system proposed by de Candolle. They identified and described 97,205 species and grouped them in to 202 natural orders (= present families). Of these, 165 belong to *dicotyledonae*, 3 belong to *gymnospermae* and 34 belong to *monocotyledonae*. This system is popularly used in all commonwealth countries even today.

Bentham and Hooker divided the flowering plants into three classes namely *dicotyledonae, gymnospermae* and *monocotyledonae*. Each class was divided into different sub classes. The sub classes were divided into series. Each series comprises of different cohorts (= present orders) and each cohort consists of different natural orders (= families).

I. Class – Dicotyledonae: The *dicotyledonae* are characterised by tap root, reticulate venation, tetramerous or pentamerous flowers and two cotyledons in a seed. On the basis of the number of whorls in the perianth and the condition of petals, the dicotyledons are divided into three sub classes namely polypetalae, *gamopetalae* and *monochlamydae*.

1. **Sub class – Polypetalae:** The perianth is in two whorls and the petals are free. It is devided in to three series on the basis of the nature of thalamus.
   (*i*) **Series – Thalamiflorae**: Thalamus is elongated, conical or convex. It consists of six cohorts.
   (*ii*) **Series – Disciflorae:** Thalamus is disc shaped. It consists of four cohorts.
   (*iii*) **Series – Calyciflorae:** Thalamus is cup shaped and it consists of five cohorts.
2. **Sub class – Gamopetalae:** The perianth is in two whorls. The petals are fused and the stamens are epipetalous. Based on the nature of the ovary and merosity of the flower it is devided into three series.
   (*i*) **Series – Inferae:** Ovary is inferior and it consists of three cohorts.
   (*ii*) **Series – Heteromerae:** Superior ovary and more than two carpels. It has three cohorts.
   (*iii*) **Series – Bicarpellatae:** Superior ovary with two carpels. It has four cohorts.
3. **Sub class – Monoclamydae:** It has been created on the basis of single character, the perianth. This is not divisible in to calyx and corolla. There are eight series in this sub class. The series is not divided into cohorts and the families are directly included under each series.

II. Class – Gymnospermae: The Gymnospermae were divided into three families – *Cycadaceae, Coniferae* and *Gnetaceae*. Placing gymnosperms which are distinct by having naked seeds in between the fruit bearing groups dicots and monocots is considered inappropriate.

III. Class – Monocotyledonae: The Monocotyledonae are characterised by adventitious root system, parallel venation, trimerous flowers and a single cotyledon in every seed. Monocotyledons were divided into seven series. In monocotyledon families are included directly under each series without the creation of cohorts or orders. E.g. Family *Liliaceae*.

Important medicinal plants with uses:

Name of thedrug	Botanical name	Family	Place of origin	Part used	Active constituents	Uses
Atropine, hyoscyanine	*Atropa belladona*	*Solanaceae*	Central and southern europe	leaves	Hyoscyanine and scopalamine	Stimulant to sympathetic nervous system
Quinine	*Cinchona succiruba*	*Rubiaceae*	South america	Bark roots and trunk	Quinine, quinidine, cinchonine and cinchonidine	Treatment of malaria fever
Digitoxin	*Digitalis purpurea*	*Scrophulariaceae*		Dried leaves	Digitoxin, sitoxin and sitalin	Curing dropsy (unwanted accumulation of liquid in body cavities)
Ephedrine	*Ephedra equisetina*	*Ephedraceae*	China	Dried stem		Treatment of bronchial coughs asthma, hay fever and cold
Morphine, codeine, thebaine and papaverine	*Papaver somniferum*	*Papaveraceae*	Persia	seeds		Pain reliever
Reserpine	*Rauvolfia serpentine*	*Apocynaceae*		Roots, leaves and stems		Curing of epilepsy, high blood pressure, insanity, intestinal and cardiac diseases and snake bites etc.

Strychnine and brucine	*Strychnos nuxvomica*	*Loganiaceae*		Seeds		Treatment of nervous disorders and paralysis
Aconite	*Aconitum napellus*	*Ranunculaceae*	Europe and Western Africa	Dried roots	Aconitine, Aconite and Bezoylaconite	
Serna	*Cassia angustifolia*	*Leguminaceaea*	Somalia and Arabia	Dried leaves	Aloe-emodine, chyophanic acid	Laxative used in habitual constipation
American worm seed	*Chenopodium antrosiodus var. Anthelmintium*	*Chenopodiaceae*	W. indies, Central and south America	Fruits	Ascaridol	Anthelmintic – for hook worm infections
Cascara	*Ramnus purshiana*	*Rhamnaceae*	North America	Dried tree bark	Emodine, iso-emodine, aloe-emodine, chryso panic acid.	Tonic and laxative
Asvagandha	*Withnia samnifera*	*Solanaceae*	India	Dried roots	Nicotine, samniferin, samniferinine and withananine	Diunetic, used in rheumatism and applied to ulcer, carbunches and painful swellings

CHAPTER - 13

Statistical Methods and Field Plot Techniques

Define statistics? And write about degrees of statistics?

The word statistics may be defined as aggregates of facts affected to a marked extent by multiplicity of causes numerically expressed, enumerated or estimated according to reasonable standards of accuracy, collected in a systematic manner for predetermined purpose and placed in relation to each other. Statistics may also be defined as the science of collection, organisation, presentation, analysis and interpretation of numerical data.

Degrees of Statistics: Before dealing with statistical methods, it is necessary to be familiar with the degrees of statistics. These are three types of statistics, viz., first degree, second degree and third degree (Mather and Jinks, 1971), which are used in biometrical genetics. These are briefly described below.

1. **First degree statistics:** It includes mean, which is used for the measurement of all types of parameters. In biometrical genetics, first degree statistics is used for the study of generation means, heterosis and inbreeding depression, combining ability effects, metroglyph analysis, stability analysis etc.
2. **Second degree statistics:** It includes estimates of variances and covariances. The precision of variances can be increased in two ways viz., 1. By increasing the sample size and 2. By the use of computers for the analysis of such estimates. In biometrical genetics, second degree statistics is used for the estimation of correlations, path coefficients, discriminant function, D^2 statistics, heritability, coheritability, genetic advance and components of variance in diallel, partial diallel, line × tester cross, triple test cross, triallel cross, quadriallel cross and biparental cross.
3. **Third degree statistics:** It includes complex interactions like curtosis and skewness, which are used in fitting frequency curves, surfaces and flatness. In biometrical genetics, high order statistics is used to study the mean skewness of F_2, F_3, biparental progenies etc. Thus major part of biometrical analysis is based on first and second degree statistics.

Explain in detail about the measures of central tendency?

Measures of Central tendency: One of the most important aspects of describing a distribution is the central value around which the observations are distributed. Any mathematical measure which is intended to represent the center or central value of a set of observations is known as "measure of central tendency" (or) The single value, which represents the group of values, is termed as a "measure of central tendency" or a measure of location or an average. It includes arithmetic mean, median, mode, geometric mean and harmonic mean etc.

Arithmetic mean: It is defined as sum of all observations in a sample divided by their number of observations and it is calculated as follows

$$A.M = \Sigma x/N$$

Where, Σ – summation, x – an observation and N- number of observations in a sample

In case of frequency distribution, the mean is worked out by multiplying the observation by its frequency, adding the products and dividing by number of observations i.e.

$$A.M = \Sigma(fx)/N$$

Where 'f' is the frequency distribution

In genetics and plant breeding, arithmetic mean is commonly used for the study of generation means, heterosis, inbreeding depression, combining ability effects, metroglyph analysis and stability analysis etc.

Median: It is defined as the middle or the mostly central value of the variate when the observations are arranged in ascending orders of their magnitude. It is positional value but not calculated value.

For the ungrouped data, median is the value of the middle most observations, if there is odd number of observations. If the number of observations is even, then the median is the mean of two central observations. However, in both the above cases the observations are to be arranged in ascending or descending order.

$$\text{Median from the grouped data} = L + \{[(N/2)\text{-}C.F]/f\} \times h$$

Where, L= lower limit of the median class in which $(n/2)^{th}$ item falls, N= total frequency, f= frequency of the median class, C.F= proceeding cumulative frequency, and h= size of the class interval of the median class.

E.g.

Case-i: When the number of observations (n) is odd:

The number of runs scored by 11 players of a cricket team of a school is

5, 19, 42, 11, 50, 30, 21, 0, 52, 36, 27

To compute the median for the given data, we proceed as follows:

In case of ungrouped data, if the number of observations is odd then median is the middle value after the values have been arranged in ascending or descending order of magnitude.

Let us arrange the values in ascending order:

0, 5, 11, 19, 21, 27, 30, 36, 42, 50, 52

∴ Median = [n+1] /2th value = 6th value

Now the 6th value in the data is 27.

Median = 27 runs

Case-ii: When the number of observations (n) is even:

Find the median of the following heights of plants in cms:

6, 10, 4, 3, 9, 11, 22, 18

In case of even number of observations, there are two middle terms and median is obtained by taking the arithmetic mean of the middle terms.

Let us arrange the given items in ascending order

3, 4, 6, 9, 10, 11, 18, 22

In this data the number of items n = 8, which is even.

Median = average n/2 th [(n/2)+1]th terms.

= average of 9 and 10

Median = 9.5 cms

Mode: Mode is the value which occurs most frequently in a set of observations or mode is the

value of the variable which is predominant in the series.

E.g. Find the mode value for the following data:

27, 28, 30, 33, 31, 35, 34, 33, 40, 41, 55, 46, 31, 33, 36, 33, 41, 33.

As seen from the above data, the item 33 occurred maximum number of times i.e. 5 times. Hence 33 is considered to be the modal value of the given data.

Geometric mean: The central number in a geometric progression (E.g. 9 in 3, 9, 27), also calculable as the nth root of a product of n numbers.

Harmonic mean: The mean of a set of positive variables. Calculated by dividing the number of observations by the reciprocal of each number in the series.

The harmonic mean is one of the three Pythagorean means. For all positive data sets containing at least one pair of non equal values, the harmonic mean is always the least of the three means, while the arithmetic mean is always the greatest of the three and the geometric mean is always in between, i.e. A.M > G.M > H.M.

Average	Definition	Uses
Arithmetic mean	Sum of all values/No. of values	To calculate average yield, SD, Correlation and regression coefficients
Harmonic mean	Mean of set of positive variables	Deals with rate, price and speed
Geometric mean	Central number in a geometric progression	Deals with relative changes E.g. Bacterial growth, cell division
Median	Middle most item of all values	For ended class data's, intelligence, ability and efficiency
Mode	Most frequently occurred item	Typical soil type, cropping pattern in a locality and shoe and shirt size in business

Explain in detail about the measures of dispersion?

Dispersion: The averages help the breeder/experimenter in condensing a mass of data in to a single value which represent the entire population but fail to reveal the entire picture of distribution. Therefore, the measure of central tendency alone is not sufficient for the distribution of frequency distribution. In addition to there is another measure of scatteredness of observation. "The degree of the scatteredness or variance of the individual observations for their averages is called dispersion".

The measure of dispersion includes Range, Standard deviation, Variance, Standard over and Coefficient of variation. These are also called as simple measures of variability.

Range: Difference between the lowest and highest values presenting the observation in a sample and it is commonly used as a measure of variability in breeding population.

Standard deviation: It is the squares root of arithmetic mean of square of all deviations measured from the mean or it is the square root of the variance.

$$s = \sqrt{\frac{\Sigma\left(x-\bar{x}\right)^2}{n-1}}$$

- It gives more weight to estimate to extreme item and less weight there are nearer to mean.

Variance: Sum of square of the deviations of all observations of a sample from its mean and divided by degrees of freedom.

$$S = \sum (x-x)^2 / n-1$$

- Average of square deviations from the mean or it is the square of standard deviation.
- It is a measure of variation among families/treatments or replications etc, obtained by dividing the sum of squares (S.S) by corresponding degrees of freedom (n-1) to get mean sum of squares (MSS) i.e. variances.

MSS or Variances = S.S/n-1

- It provides a unified measure of pooled variation of all the treatments/ factors and is symbolically represented by σ^2. It has squared root of that of mean E.g. (number of tillers)^2 or cm^2

Standard error:

It is the measure of the mean difference between sample estimate of mean and the population parameter i.e. it is measure of uncontrolled variation present in a sample.

- In fact, Standard error is the standard deviation of means.
- Like SD, SE_m again is a unified measure of pooled variation (around X). However, SE can also be calculated for difference between two family means, Then it is called as SE of difference (SE_d)

$$SE_d = 2EMS/r$$

- Both SD and SE have the same unit as their means (X).

Coefficient of variation (C.V):

When variation has to be compared for different characters, each represented by different units (E.g. flowering time in days, plant height 'x' cm, tillering in numbers and so-forth), variance, SD and SE are not adequate. However, by converting units of all characters on the same scale, the job can be done neatly. The CV does the same.

It is the ratio of standard deviation of a sample to its mean. A sample in which C.V is higher would have greater variation (or) C.V. is high, the sample is less consistent (or) more variable.

$$CV\ (\%) = SD/\ mean \times 100$$

- CV is a relative measure of variability in percentage.
- CV has the same unit (%) for all characters; hence it can very well indicate which character is more variable than other. It is employed for following purposes.
  - (*a*) It is used to compare different characters under selection experiments.
  - (*b*) It can be deployed in mutation experiments to compare the effects of doses of mutagens for the same characters.
  - (*c*) It is most useful in deciding the veracity of a locations data under multilocational trails. Large CV associated with any location would indicate defective experimentation. But here SD is calculated from error variance to develop CV.
  - (*d*) In D^2 statistic, relative contribution of different characters to the total divergence can also be determined by CV of cluster means for each character.

Write short notes on the following.

1. Degrees of freedom
2. Level of significance
3. Test of significane

Allied statistical parameters:

Related to the measures of variation, there are some allied parameters which must be properly understood before embarking upon statistical treatment of data. These allied statistical parameters are Degrees of Freedom, Level of significance and Test of significance etc.

Degrees of freedom: The number of independent comparisons or randomisation is the degrees of freedom.

- This is very frequently used in various kinds of analysis of variance.
- The degrees of freedom is represented by n-1, where n being the number of families.

Level of significance (Confidence limit):

To know the real genetic difference among treatments (families/ varieties) is the primary objective of statistical analysis. However, these differences are often inflated due to the influence of uncontrollable chance errors. How much allowance can be granted to such chance error is purely arbitrary conventionally 1% or 5% error is allowed. This implies that if the experiment is repeated 100 times under the same (identical) conditions, the same results should be obtained 99 or 95 time, respectively. That means we may allow error in performance to occur once (1%) or five times (5%) in100 times

Any variation beyond this level is real and hence significant (as 1% or 5% level of significance).Thus 99% or 95% is the confidence limit. In other words, we can assert with confidence that in 99% or 95% cases, variation among treatment is real (due to genetic factor) and is not due to chance error.

Tests of significance

Statistical procedures which are used to decide whether the differences under study are significant or non-significant are known as tests of significance.

- Commonly used tests of significance are z-test, t-test and f-test
- These tests are used to decide whether the difference under study is significant or might be attributed to chance or fluctuation of sample.
- These tests are based on a hypothesis called null hypothesis.

Null hypothesis: This hypothesis was given by Fisher which is denoted by H_o. A hypothesis of no difference is termed as null hypothesis. If the sample result fails to support the null hypothesis then H_o is selected and another hypothesis (H_1) that is when the difference under study are significant the null hypothesis is rejected and H_1 is used.

Applications of tests of significance

- In plant breeding, significance of test between the selected plant sample and base population indicates the effectiveness of selection .
- It helps in deciding the significance of correlation or regressive relationship between two or more characters thus useful in determining the yield contributing characters in important crop species.
- It is also used to compare the variation present in two different samples of a population / two different population there by helping in choosing between two samples Eg. to determine the degrees of difference in bio diversity in two different ecosystems

Write short notes on the following?

1. F- Test 2. t- Test 3. Critical difference and 4. Unreplicated tests

F Test:

- It is one of the tests of significance used for testing the significance of difference among several treatments or means.
- Thus it is the ratio of treatment (family/genotype) variance to the error (environmental) variance.

  F (calculated) = Var. due to family or treatment/Var. due to error

  = Treatment MSS/Error MSS

- A calculated F value, say for family/treatment variation greater than Table F reveals significant (real) differences among the means of treatments.
- It differs from z and t test, where there is comparison between sample mean and population mean.
- For more than two groups, Bartlett's test for homogeneity of variance is useful.
- Example of test of significance based on F-distribution

  (*a*) Testing the significance of equality of variance

  (*b*) Linearity of regression.

t-Test:

Many a times, a breeder is faced with two kinds of situations of testing the difference of means. In the first kind, he wants to test whether the mean-difference between two related populations/ treatments/ genotypes are significant or not, for instance, between a parent and data on corresponding sample size recorded.

In the second kind, breeder wishes to tests the significance of difference between two unrelated means. For instance, in the above case, if the data on varying sample sizes are recorded in parent grown in one year and in the mutant grown in subsequent year.

- In first case, student – t test and in 2nd case fishers-t- test are used.

Student's t-test: To test the means difference of related samples. Student's t-test is employed.

$$\text{Student's } t = \overline{X}_d / SE(\overline{X}_d) \text{ (at n-1 d.f)}$$

Where $\overline{X}_d$ is the mean difference between two sets of related samples, SE ($\overline{X}_d$) is the standard error of mean difference and n is the number of related samples

Fisher's t-test: To test the difference between two unrelated means.

Fisher's t = x_1-x_2.................. at (d.f = n_1+n_2-2)

Critical difference (CD) or Least significant difference (LSD):

- Once the F-test reveals significance of difference, it becomes desirable to know the best or the worst treatment /family /variety on terms of mean performance.
- The significance of difference between two means can be tested by the statistic-critical difference (CD) or LSD, calculated as

$$\text{CD or LSD} = SE_d \times t_{5\%} \text{ or }_{1\%} \text{ (at error degrees of freedom).}$$

  Where, SE_d is the standard error of difference and't' is the table value of t at 5% or 1% level of significance corresponding to F-test.
- The CD has the same unit as the mean or SE_d of the trait in point.
- CD is a value beyond which difference between two means are significant.
- Based on this, families /varieties /treatments in an experiment can be arranged in significant groups.

Unreplicated tests:

Unreplicated test refers to planting single rows of genotypes with check cultivers strategically located for easy comparison.

A breeder may use this design to evaluate a large number of genotypes quickly. In order to eliminate inferior ones prior to more comprehensive field trials.

Advantages

- They save space.
- They are less expensive to conduct.
- Large number of genotypes can be quickly evaluated.

Disadvantages:-

- Such tests are susceptible to the effects of field heterogeneity there is no other plot of the genotype in the test for conformity performance poor soil may mask the genotype potential of a superior genotype by causing it to perform poorly.
- Experimental error estimation is problematic.

Write brief note on multivariate statistics in plant breeding?

Multivariate statistics in plant breeding:

Multivariate analysis is the branch of statistics concerned with analysing multiple measurements that have been made on one or several samples of individuals. Because these variables are interdependent among themselves they are the best considered together. Unfortunately handling data with multicolinerity can be unwidely and hence some meaningful summation is needed

- The multivariate technique in use may be divided in to two groups

  Interdependence models – Principal components analysis factor analysis

  Dependence models – multivariate analysis of variance, classification functions discriminate functions analysis, cluster analysis, multiple correlation and canonical correlation.

- W.W Cooley and P. R. Jones further classified multivariate procedures in to four categories according to the number of populations and number of variables as follows

  1. One set of variables, one population: Principal component analysis, factor analysis
  2. One set of variables ,two or more populations: Multivariate analysis of variance, discriminate functions classification functions
  3. Two or more sets of variables, one population: Polynomials multiple correlations, canonical correlation and multiple partial correlation
  4. Two or more sets of variables, two or more sets of population: Multivariate covariance.

  Multivariate analysis is done on computers because of their complexicity.

Define correlation? And write the silent features and role in plant breeding?

Correlation: Correlation refers to the degree and direction of association between two or more variables

- It measures the mutual relationship between two or more variables
- Correlation coefficient is represented by "r"
- When 'r' is positive, then the change of two variables in the same direction. When 'r' is negative, then the movement of two variables is in opposite direction
- It is independent of the unit of measurement
- It is value lies between -1 and +1
- It is confined to the study of linear relationship and does not involve cause and effect relationship.

Role of correlation in plant improvement

1. A positive correlation between desirable characters is favourable to the plant breeder, because it helps in simultaneous improvement to both the characters. Negative correlations, on the other hand will hind the simultaneous expression of both the characters. In such situations, some economic compromise has to be made.
2. The genetic improvement in a dependent trait can be achieved by applying strong selection to a character which is genetically correlation with the different character. This is called correlated response.

3. Sometimes, a character can have low heritability under such situations, another character having high heritability and high correlation with the character is chosen to make selection more effective.

Limitations:

- Sometimes, correlation coefficients give misleading results.
  E.g. A correlation obtained between two variables may be simply due to a third factor not used in the study or one may get high association between variables, when no one exists in practice. This may happen due to small sample size.
- Correlation, in some cases is zero between two variables, but one cannot presume only based on correlation, that no relationship exists between these two variables because correlation indicates only a linear relationship and thus it should be further checked.

Write in detail about the types of correlation coefficients?

Types of correlation coefficients:

Correlation coefficients are of three types, viz., 1. Simple or Total 2. Partial and 3. Multiple correlations.

1. **Simple correlation:** The association between any two variables is termed as simple correlation or total correlation or zero order correlation coefficient. Main features of simple correlation are
   - Involves two variables
   - Denoted as r_{12}
   - Ignores effects of other independent variables.
   - Estimated from variances and covariances
   - Coefficient of determination cannot be obtained directly
   - Its value is always lower than multiple correlation
   - Simple correlation is of three types, i.e. phenotypic, genotypic and environmental correlations

Phenotypic correlation:

The association between two variables which can be directly observed is termed as phenotypic correlation. It includes both genotypic and environmental effects.

Genotypic correlation:

The inherent or heritable association between two variables is known as genotypic correlation. This type of correlation may be either due to pleiotrophic action of genes or due to linkage or both.

If the association between two traits remains the same in the parental population as well as segregation populations, it means the association is due to pleiotropy. If it changes in the segregating population, it is most likely due to linkage between two genes. The first situation enhances the genetic improvement and this type of correlation is more stable and it is more important for a plant breeder to bring about genetic improvement in one character by selecting the other characters.

Simple correlation coefficient can be estimated from both unreplicated and replicated data.

From unreplicated data: $r = [\Sigma XY - (\Sigma X. \Sigma Y)/N] / [(\Sigma X^2 - (\Sigma X)^2)/N. (\Sigma Y^2 - (\Sigma Y)^2)/N]^{1/2}$

From replicated data: $r = MSPt / [MSt_x.MSt_y]^{1/2}$

2. **Partial correlation:** When the correlation between two variables is worked out by eliminating the effect of third variable it is called as partial or net correlation. It is a study of relationship between one dependent variable and one independent variable by keeping the other independent variable constant. Main features of partial correlation coefficients are
   - It involves three or four variables
   - It is denoted as $r_{12.3}$ or $r_{12.34}$
   - It is estimated from simple correlations
   - Its value is always lower than multiple correlations
   - It does not ignore effects of other variables
   - Coefficient of determination cannot be obtained from partial correlation
   - Partial correlation coeffient is calculated from estimates of simple correlation coefficients. The first order partial correlation coefficient is estimated from formula

     $$r_{12.3} = r_{12} - r_{13}\, r_{23} / [(1-r^2_{13})\,(1-r^2_{23})]^{1/2}$$

     The second order partial correlations are calculated from the estimates of first order partial correlation coefficients using formula:

     $$r_{12.34} = r_{12.3} - r_{14.3}\, r_{24.3} / [(1-r^2_{12.4})(1-r^2_{24.3})]^{1/2}$$

3. **Multiple correlations:** The estimate of joint influence of two or more independent variables on a dependent variable is called multiple correlation. So in case of multiple correlations, the effect of all the independent variables is studied on a dependent variable. Main features of multiple correlation are
   - It involves several variables
   - It is denoted by $R_{1.23}$, where R is the coefficient of multiple correlations.
   - It is of one type only
   - It is estimated from simple correlations
   - Its value is always higher than simple and partial correlations
   - Coefficient of determination can be directly obtained from multiple correlation i.e. R^2.

- It is a non negative estimate. It can never be negative; hence its value lies between 0 and 1.
- Multiple correlation coefficient is calculated from the estimates of simple correlation coefficients using formula:

$$R_{1.23} = (r^2_{12}\, r^2_{13} - 2\, r_{12}\, r_{13}\, r_{23} / 1\text{-}r^2_{23})\ 1/2$$

Define Regression coefficient? And write the salient features and role in crop improvement?

Regression analysis:

Regression coefficient is the statistical measure of the average functional relationship between two or more variables.

- In regression, one variable is regarded as dependent and other is independent.
- It is clearly indicates the cause and effect relationship between variables
- It studies both linear and non-linear relationship between variables
- It is denoted by 'b'
- It is expressed in terms of original unit of data.
- Between two variables (say x and y) two values of regression coefficient can be obtained.
- Both regression coefficients must have the same sign.
- If b_{yx} is positive, then b_{xy} will also be positive and vice versa
- $b_{yx} \neq b_{xy}$ i.e. both regression coefficients will never be equal.
- The geometric mean between two regression coefficients is equal to the coefficient of correlation

$$r = (b_{xy}.b_{yx})^{1/2}$$

- Prediction is possible in regression analysis. The independent variable is the prediction variables and the dependent variable is the predicted variable.

Role in crop improvement:

- It helps in finding out a cause and effect relationship between two or more plant variables.
- It is useful in determining the important yield contributing characters.
- It helps in the selection of elite genotypes through in direct selection.
- It is useful in predicting the performance of selected plants in the next generation.
- It helps in measuring the stability and adaptability of a plant species to particular environments.

Write short notes on the following?

1. χ^2 test
2. Probability

χ^2 test:

Karl Pearson developed the chi-square test of statistical significance which is commonly used in Mendelian and Population genetics. χ^2 test is a test of statistical significance which is used to test the significance difference between observed and expected frequencies or ratios

$\chi^2 = \sum(O-E)^2/E$

$\sum$= summation

O=observed frequencies

E=expected frequencies

Applications:

- Testing of segregation ratios
- Detection of linkage
- Study of gene frequencies

Limitations:-

- Chi-square can usually be applied only to numerical frequencies themselves. It cannot be ratios or percentages derived from the frequencies.
- Chi-square cannot properly be applied to distributions in which the frequency of any class is less than 5.

Probability:

Probability is the likelihood of occurrence of an event.

Eg. The likelihood of victory in a test match or obtaining good marks in an examination etc.

- Probability of mutually exclusive events is (p) = 1/n.

  Where, p = probability of a single mutually exclusive event, and n = the total number of events.

  In this, probability of an event is the average frequency of that event. Conversely, the average frequency of an event is its probability.

  Eg. A new born can be either a boy or a girl, while the occurrence of boy excludes that of girl and vice versa.

- Probability of occurrence of two or more independent events together is the product of their individual probabilities.

  Eg. If two coins, A and B, were tossed together once, the combination of head in both coins is ¼.

The number of all possible combinations doubles when the number of independent events is increased by one.

- Therefore, probabilities are determined by applying the binomial expression. However, binomial expression is applied to only those events in which the number of mutually exclusive events is two.

Define experimental design? And explain its principles?

Experimental Design:

The term experimental design refers to set of rules for allocating treatments (varieties) to experimental units (plants) for an unbiased comparison of treatment means and valid test of significance.

Principles of experimental design

Principles:

1. Replication
2. Randomization
3. Local control

Replication: The number of times a treatment is repeated in the experiment constitutes replication.

Functions of replication are:-

- To provide an estimate of error
- To reduce the magnitude of error
- To broaden the scope of inferences.

Two main advantages of replications are.

1. Increase in replication increase the precision by reducing the error to a great extent. The increase in precision results in two ways. In case of field experiments, standard error is calculated as $(2V_E)^{1/2}/r$. Where, V_E- error variance and r-number of replications. Thus it is clear that the standard error of mean decreases as the number of replication increases.
   - Increase in replication increases the degrees of freedom of error which decreases the value of 't' due to decrease in confidence interval. The shortening of confidence interval is a good proof of increased precision.

     $$r = t_\alpha s^2/d^2$$

     Where, r is the number of replications, t_α is the table value of t at α level of significance, s^2 is the estimate of error mean square and d is the actual difference

   - At least 12 degrees of freedom is needed for estimation of error.
2. Error of experiment arises from the difference between the plots of the same treatment. Thus without replications estimate of errors is not possible and without estimate of error comparison are not possible.

Randomization:

It refers to random allocation of treatment to different plots. The functions of randomization are:

- To estimate bias in estimating the mean performance and difference of treatment means.
- To make error independent of treatment effects this ensures valid test of significance.

Local control: The principle of making use of greater homogeneity in groups of experimental units for reducing the experimental error is known as local control.

Functions:

- To reduce error.
- To make comparison among treatment, more uniform and meaningful.
- To sample wide range of condition.

What is meant by experimental error? And write the source and approaches for reduction of experimental error?

Experimental error and source of experimental error and approaches to reduce experimental error:

The variation among plots treated similarly is called experimental error. It is the result of difference in soil fertility, conduct of experiment. The basic objective of experimental design is to assess and reduce experimental error.

Three major sources of error are:

1. Soil heterogeneity
2. Lack of uniform conduct of experiment
3. Interplant competition and border effect.

Approaches to reduce error:

1. Assessment of soil fertility of different patch of experimental site.
2. Uniformity trial and fertility map
3. Conduct of experiment should be uniform like taking data at a unit in single day, appropriate preparation of land, uniform sowing practice etc.
4. Border effect refers to effect due to aggressive competitor, effect of neighbour plant, effect of fertilizer at border. So, data should be taken excluding observation of border rows.

Measures adopted to reduce error:

- For a particular plot, plot shape with minimum coefficient of variation (CV) value is chosen
- Design should have minimum 12 degrees of freedom for error.

- For large number of treatments, incomplete block design should be used
- Differential treatment should be avoided
- Uniform conduct of experiment
  (*a*) Field uniformity prepared
  (*b*) Plot should be accurately measured
  (*c*) Data recording completed at same time
  (d) Measuring scale, equipment should be accurate
- Use of supplementary data, error is reduced by use of additional data through "Analysis of covariance".

What is meant by Analysis of Variance? Describe its utility in identifying potential genotypes in yield trials?

The statistical procedure which separates the total variation into different components is known as analysis of variance.

It has following two main advantages

1. It is useful in estimating the components of variance
2. It provides basis for the test of significance

The analysis of variance is carried out with replicated data. In plant breeding, such analysis divides the total variation into two main plants, viz., variation between varieties and variation within varieties, i.e. environmental variation. Thus it helps in partitioning of phenotypic variation into genotypic and environmental components.

Genotypic variance (V_G) = ($Ms_t - Ms_e$)/r

Environmental variance (V_E) = Ms_e

Phenotypic variance (V_P) = $V_G + V_E$

Where, Ms_t = Mean square for the treatments, Ms_e = Mean error variance and r = number of replications.

ANOVA is useful in two main ways:

1. ANOVA permits estimation of phenotypic, genotypic and environmental coefficients of variability for various polygenic traits.
2. It also permits the estimation of broad sense heritability and expected genetic advance under selection.

Types of experimental designs:

The complete block designs are suitable for when small number of entries are to be tested. In these designs each block includes at least one complete set of varieties. The incomplete block design, on the other hand, is characterised by blocks, of each includes only a part of the total number of varieties. Such designs are applicable for testing a relatively large number of entries.

In plant breeding, mainly three types of designs are used:

1. Screening designs: Augmented design
2. Evaluational designs: CRD, RBD and LSD
3. Genetic designs: NCD I, NCD II and NCD III designs

Explain the different types of complete block designs with situations where to be used, layout, Anova table, advantages and disadvantages?

Complete block designs includes CRD, RBD and LSD.

Completely randomized design (CRD): Completely randomized design is the simplest basic single experimental layout with the assumption that the entire experimental material is limited and supposed to be homogeneous.

Situations:

- It is used when experimental material is limited and homogeneous (uniform).
- This design is specially used for glass house or pot experiments.
- Where some of the units are likely to be destroyed or fail to respond.
- It is generally used in chemical and biological experiments where the material is homogeneous.
- When rapid conclusions are needed.
- In small experiments, where the increased accuracy from an alternative design does not compensate the loss of error degrees of freedom.
- It is not frequently used under field conditions because variation for soil fertility among the experimental plots is usually quite high that results in large estimates of error. Any treatment can appear anywhere i.e. there is no expectation in randomization. But all plots are homogeneous i.e. uniformly except for factor of comparing.

Layout: (Layout of CRD using five treatments and four replications)

D^1	D^2	E^3	E^4	C^5
D^{10}	A^9	A^8	C^7	C^6
B^{11}	B^{12}	A^{13}	B^{14}	B^{15}
E^{20}	B^{19}	A^{18}	C^{17}	E^{16}

Anova table:

S.V	DF	SS	MSS
Between treatment	t-1	tSS	tMS
Within treatment	n-t	eSS	eMS
Total	n-1		

Advantages:

- It is simplest in adaption and in analysis.
- Unequal number of replications for different treatments can be accommodated without any complicacy in layout and analysis.
- The missing data for any replication of some treatments does not complicate analysis.
- It provides the maximum number of degrees of freedom for error and hence a relatively low magnitude of error.

Limitations:

- It is not suitable for large number of varieties as it would require large experimental area which is likely to be variable in soil fertility.
- If experimental area is not uniform it gives relatively large magnitude of error.

Randomized Block Design (RBD): The experimental design which controls the fertility variation in one direction only is known as Randomized Block Design.

Situations:

1. It is highly suitable if the number of entries to be tested is not very large and fertility gradient of the experimental area is known.
2. RBD is most widely employed field design by plant breeders and is useful under following three situations
   (*a*) For evaluating new introductions and for their suitability (test of adaption) for early release.
   (*b*) For gathering genetic information through biometrical techniques from the course of breeding programme.
   (*c*) For local (at experimental site) testing of strains evolved through breeding (Testing of adaption) and also for multilocation testing of varieties in coordinated field trials. (Testing of adaptability).

Layout: (Layout of RBD using 7 entries and 3 replications)

C	F	A	D	G	B	E
B	D	E	F	A	G	C
G	E	B	C	F	D	A

Anova table:

S.V	**D.F**	**SS**	**MSS**
Replications	r-1	rSS	rMS
Varieties (g)	g-1	gSS	gMS
Error (r×g)	(g-1)×(r-1)	eSS	eMS
Total	rg-1	tSS	

Advantages:

- Flexibility is an important advantage of RBD. This design allows any number of treatments and replications. The only restriction is that the number of replications is proportional to the number of blocks. Generally r = t but if any extra replication is desired to some treatments, they may be applied to two or more units in each block. If 't' is the number of treatments, then the minimum number of replications ensuring at least 12 degrees of freedom for error i.e. r = 1+ 12/t-1
- It helps to reduce error by eliminating the effect of soil heterogeneity.
- Avoids practical difficulties encountered in uniform application of cultural practices.
- The scope of application of inferences can be widened by placing blocks at different places.

Limitations:

- It is not applicable for large number of treatments because with increased block size the chances of variation among plots within blocks are increased.
- The unequal number of replications and missing values for some replications create problems in analysis.
- It is less efficient if source of variation moves in more than one direction.
- It is also less efficient than CRD if the experimental material is homogenous where it unnecessarily reduces degrees of freedom for error.

Latin Square Design (LSD):

The experimental design which simultaneously controls the fertility variation in two directions is called Latin Square Design.

Latin square designs are adopted for eliminating the variation of two factors which are generally called rows and columns.

Layout: Layout of LSD using 5 rows and 5 columns.

Rows	Columns				
	1	2	3	4	5
1	A	B	C	D	E
2	B	C	D	E	A
3	C	D	E	A	B
4	D	E	A	B	C
5	E	A	B	C	D

Anova table:

S.V	D.F	SS	MSS
Rows	r-1	rSS	rMS
Columns	c-1	cSS	cMS
Treatments	t-1	tSS	tMS
Error	(t-1)(t-2)	eSS	eMS
Total			

Advantages:

- It helps to eliminate simultaneously the effect of two known gradient of variation from error.
- It is useful for small group of treatment. The maximum number of treatments for which we can use Latin square design is 12.
- When data from some of the plots are missing. Yates missing plot technique can be used for analysis of data.

Limitations:

- Lack of flexibility in the choice of varieties and replication of not or large number of varieties.

Explain in detail about the incomplete block designs?

Incomplete block designs (Lattice designs):

Situations:

In a plant breeding programme, the number of new strains to be evaluated for the first time is usually quite large. When the increased number of varieties cannot be accommodated in a single block without adversely affecting the homogeneity among plots within blocks the use of complete block designs become less efficient. The only alternative is to constitute smaller blocks of homogenous plots with the condition that only a plant of all the varieties is included in each block.

There are two important designs:-

1. BIBD- Balanced incomplete block design
2. PBIBD- Partially balanced incomplete design

BIBD: A balanced lattice design is most efficient design for the particular combination of block size and number of genotypes but in most practical circumstances we are restricted to fewer replicates than required for balance, in such cases, the incomplete lattice with two, three or four replicates is still the ideal design.

- In this design treatment comparisons have same precision because these are present in same blocks equal in number of time.

- It is an arrangement of 'V' treatments in 'B' blocks each of size 'K' such that each treatment will appear at most once in each block and every treatment occurs exactly 'r' blocks and every pair of treatment occurs together in exactly 'λ' blocks and they satisfying relationship

$$Vr = bK \text{ and } \lambda (v\text{-}1) = r(K\text{-}1) \text{ and } b \geq v$$

Layout:

I	II	III	IV
A	B	C	D
B	C	D	A
C	D	A	B

$v = 4$, $b = 4$, $k = 3$, $r = 3$ and $\lambda = 2$

PBIBD: In this design the number of varieties to be tested is perfect square and the number of plots per block is 'q' square root of number of varieties but there is more flexibility with respect to the number of replications. Unlike the balanced lattice the number of replications is not a function of number of varieties. But here all the pairs of varieties do not occur together in the same block equal number of times as a result of which precision of all comparisons is not same. The varieties which occur together in each block are compared with better precision than those which are present in different blocks.

The PBIBD with two replications is simple lattice, three replications is triple (cubic) lattice and four replications is Quadruple lattice

Layout:

A	B	C	D
E	F	G	H
I	J	K	L
M	N	O	P
A	E	I	M
B	F	J	N
C	G	K	O
D	H	L	P

Write short notes on compact family block design.

Compact family blocks design:

In situations the treatments can be combined together on the basis of certain common characteristics to make comparisons between groups and within groups more meaningful. The experimental designs for such situations are also called group balanced block designs. Here randomization is done at two stages. At first the groups of treatments are randomized to blocks within each replication and the treatments within each group are then randomized separately.

Analysis of such designs is done at two stages i.e. first for comparison of groups is exactly the same as routine randomized complete block design which helps to detect the differences between groups of treatments. Analysis of treatments into each group is performed separately considering each group as an independent experiment.

Explain in detail about Augmented design?

Augmented design: The concept of augmented design was developed by Federer in 1956. This is an experimental design which is usesd to test a large number of germplasm lines in a limited area.

Situations:

- When seeds of test genotypes are small in quantity so that replications are not possible. For instance some newly introduced stocks, crossed seeds.
- When the large number of germplasm collections is to be evaluated, limited facility of uniform land becomes a bottle neck. Such a huge collection apparently cannot be accommodated in RBD as soil heterogeneity becomes unimaginable. In absence of error term in the ANOVA without replication test of significance cannot be applied.

  Both these problems can be surmounted by employing augmented designs as propounded by Federer (1956)

Analysis of this design provides the

- An estimate of experimental error for comparing new selections with checks.
- A way to adjust the yield of new selections for overall differences among blocks to bring these at comparable levels.

Augmented Design –Three models

1. Augmented completely randomized design (Augmented Design-I)
2. Augmented randomized complete block design (Augmented Design-II)
3. Augmented Latin square design (Augmented design-III)

Only the first two different designs satisfy and only widely used by plant breeders.

Augmented design-I:

When number of seeds is limitation, the augmented completely randomized design is most suited. In this the whole experimental area is divided into N plots (where N is the number test genotypes (v) + number of checks (c) which are standard varieties or hybrids of known performance) repeated 'b' times.

A representative field design with v = 8 (v_1 to v_8), c = 4 (c_1 to c_4) b = 3, e = 12 and n = 20

Field layout:-

C_3	V_6	C_4	C_3	V_1	C_4	V_7	C_4	V_4	C_2
C_1	V_3	C_2	C_1	C_3	V_5	C_1	V_8	C_2	V_2

Anova table:

S.V	D.F	SS	MSS
Among entries(e)	e-1	eSS	eMS
Among checks(c)	c-1	cSS	cMS
Among varieties(v)	v-1	vSS	vMS
c *Vs* v	1	cvSS	cvMS
Error	c(b-1)	eSS	eMS
Total	N-1		

Augmented design-2:

This is a slightly modified version of augmented design-I where instead of repeating all the checks 'b' times throughout the field, the whole field is divided distinctly into 'b' blocks. Then randomization is done such that all the checks and a part of test genotypes fall only once in each block. Thus, the total number of plots (N) remains the same as in design1 (if same number of v and c are used) but the number of plots in each block is variable. The advantage of such stratification (block formation) of area is to reduce the EMS by apportioning the block effect which is cofounded with error in design I. Actually, what RBD does by replicating the whole entries (e), the augmented design II accomplishes the same by replicating only the checks (c), thus saving space.

Field plan:

V=8 (V_1 to V_8), c = 4 (c_1 to c_4), b=3, e=12 and n = 20, Now the whole experimental field into b=3 distinct blocks.

Block 1, consisting (say) of 7 plots (c_1 to c_4 + V_1, V_4 and V_8).

Block 2, consisting (say) of 6 plots (c_1 to c_4 + V_2 and V3) and

Block 3, consisting (say) of 7 plots (c_1 to c_4 + V_5, V_6 and V_7).

So as to construct relatively homogeneous stratification of are in each block. Thus number of test varieties falling in block are n=3, 2 and 3, respectively.

Randomize all the entries into each block as depicted below

C_2	C_4	V_8	C_1	V_1	C_3	V_4	V_3	C_3	C_2
V_2	C_1	C_4	V_1	C_3	C_1	V_5	C_2	V_6	C_4

Anova table:

S.V	D.F	SS	MSS
Blocks	b-1	bSS	bMS
Entries	e-1	eSS	eMS
Checks	c-1	cSS	cMS
Varieties	v-1	vSS	vMS
Checks *vs* varieties	1	cvSS	cvMS
Error	(c-1) (b-1)	eSS	eMS
Total	N-1	tSS	

CHAPTER - 14

Comparison

Write the difference between the following?

1. Prokaryotic cell and Eukaryotic cell?
2. Plant cell and Animal cell

S.No	Feature	Prokaryotic cell	Eukaryotic cell
1.	Meaning	Prokaryotes are primitive organisms (Pro – primitive and karyon - nucleus)	Eukaryotes are higher organisms (Eu – good or true and karyon - nucleus)
2.	Number of cells	Generally uni cellular	Generally multi cellular
3.	Diameter of cell	Average diameter of prokaryotic cell ranges from 5 – 10 μm	Average diameter of eukaryotic cell ranges from 10 – 100 μm
4.	Envelope system	Posses only one envelope system.	Posses two envelop system.
5.	Cell organelles	Don't posses well defined cytoplasmic organelles. Pigments are distributed throughout the cytoplasm.	Posses well defined cytoplasmic organelles like endosplasmic reticulum, golgi bodies, chloroplast, and mitochondria. Pigments are present in plastids
6.	Nucleus condition	Lack nucleus and chromosomes. Nucleolus and nuclear membrane are absent. Spindle fibres are absent Histone proteins are absent	Well developed nucleus and chromosomes. Nucleolus and nuclear membrane are present. Spindle fibres are present Histone proteins are present
7.	DNA	DNA is circular and lies free in the cytoplasm	DNA is linear and lies within the nucleus
8.	Cell division	Cell division is by amitosis (binary fission)	Cell division is by mitosis and meiosis

9.	Ribosomes	Posses ribosomes of 70 S type	Posses ribosomes of 80 S type
10.	Cell wall	Cell wall is made up of polysaccharides Eg. Muramic acid	Cell wall is made up of cellulose, hemicellulose and pectins
11.	Respiration by	Mesosomes support respiration	Mitochondria support respiration
12.	Examples	Eg. Bacteria, blue green algae, E. coli, PPLOs (Pleuropheumonia like organisms)	Eg. Plant and animal cells

S. No	Feature	Plant cell	Animal cell
1.	Cell	Usually larger in size. Cannot change its shape.	Comparatively smaller in size. Cannot change its shape.
2.	Cell wall	Plant cell has a rigid wall on the out side	The cell wall is absent
3.	Cell organelles	Plastids are found. Posse's chlorophyll containing plastids called chloroplasts. A mature plant cell contains a large central vacuole. Mitochondria are comparatively fewer. Cristae are tubular in plant mitochondria. Golgi apparatus consists of a number of distinct / unconnected units called dictyosomes. Cytoskeleton does not contain intermediate fibres. Lysosomes are rare and their activity is performed by specialized vacuoles. Glyoxysomes may be present.	Plastids are usually absent. Chromatophores are present. Chloroplasts are absent. Vacuoles are numerous and very small Mitochondria are generally more numerous. Cristae are plate like in animal mitochondria. Golgi apparatus is either localized or consists of a well connected single complex. Cytoskeleton contains intermediate fibres. Typical lysosomes occur in animal cell. Glyoxysomes are absent.
4.	Nucleus	Nucleus lies on one side in the peripheral cytoplasm. Centrioles are usually absent in lower plants. Spindle fibres formed during nuclear division are anastral	Nucleus usually lies in the centre. Centrioles are found in animal cell. Spindle fibres formed during nuclear division are amphiastral

5.	Placing in hypotonic solution	Plant cells do not burst if placed in hypotonic solution due to presence of cell wall	Animal cells burst if placed in hypotonic solution unless and until it posses contractile vacuole
6.	Other cracteristics	Crystals of inorganic substances may occur inside the cells. Reserve food is generally starch and fat. A tissue fluid does not bathe the individual cells. Adjacent cells may be connected through plasmadesmata.	Crystals usually do not occur in animal cells. Reserve food is usually glycogen and fat. A tissue fluid contain in a NaCl bathes the cells. Adjacent cells are connected through a number of junctions.

Write the differences between the following

i. Linkage and crossing over

ii. Epistasis and dominance?

S.No	Linkage	Crossing over
1	It keeps the genes together	It leads to separation of linked genes
2	It involves individual chromosomes	It involves exchange of segments between non-sister chromatids of homologous chromosomes
3	The number of linkage groups can never be more than haploid chromosome number	The frequency of crossing over can never exceed 50 %
4	It reduces variability	It increases variability by forming new gene combinations
5	It produces higher frequency of parental types than recombinant types in test cross progeny	It provides equal frequency of parental and recombinant types in test cross progeny
S.No	**Epistasis**	**Dominance**
1	Refers to interaction of two or more genes	Refers to the interaction of two alleles of the same gene
2	Involves both homo and heterozygotes. Hence it is fixable in homozygotes	Always involves heterozygotes. Hence it is not fixable
3	It is of several types such as recessive, dominant and duplicate etc.	It is of three types i.e. incomplete, complete and overdominance.
4	Modifies the normal dihybrid phenotypic ratio	Partial dominance alters the normal segregation ratio of 3:1 into 1:2:1
5	Also known as inter genic or interlocus gene interaction	Also known as intragenic or intralocus gene interaction
6	Recessive gene also exhibit masking effect	Recessive gene will express only in hemizygous condition

Write the Difference between Chromosomal and Extra Chromosomal Inheritance?

S. No	Character	Chromosomal inheritance	Extra chromosomal inheritance
1	Location of hereditary Factors	Nucleus	Cytoplasm
2	Associated with	Chromosomes	Chloroplasts and Mitochondria
3	Pattern of Inheritance	Can be explained by Mendelism	Cannot be explained by mendelism
4	Individual hereditary factors are known as	Genes	Plasmagenes
5	Hereditary factors are collectively known as	Genome	Plasmon
6	Characters of F_1 Progeny	May show dominance or may be intermediate between the parents	Exhibits only the characteristic of the female parent
7	Reciprocal differences	Not observed	Observed
8	Segregation of factors and recombination	Present	Absent
9	Attributes of progeny	Under the control of their own genes	Under the control of cytoplasm of female parent
10	Action of mutagen	Non-specific	Very specific
11	Frequency of occurrence	Most common	Rare
12	Gene mapping	Easy	Difficult

Write the difference between the following?

1. Purines and Pyramidines
2. DNA and RNA
3. Different forms of DNA

S.No	Purines	Pyramidines
1	These are double ring (nine member) compounds.	These are single ring (six member) compounds.
2	They are of two types, viz., adenine and guanine.	They are of three types, viz., cytosine, thymine and uracil.
3	They occupy more space in DNA structure.	They occupy less space in DNA structure.
4	Deoxyribose is linked at position 9 of purine.	Deoxyribose is linked at position 3 of pyrimidine.

S.No	Feature	DNA	RNA
1	Strands	Usually two, rarely one.	Usually one, rarely two.
2	Sugar	Deoxyribose.	Ribose.
3	Bases	Adenine, guanine, cytosine and thymine.	Adenine, guanine, cytosine and uracil
4	Pairing	AT and GC	AU and GC
5	Location	Mostly in chromosomes, some in mitochondria and chloroplasts.	In chromosomes and Ribosomes
6	Replication	Self replicating	Formed from DNA. Self replication only in some viruses
7	Size	Contains up to 4.3 million nucleotides	Contains up to 12,000 Nucleotides
8	Function	Genetic role	Protein synthesis, genetic in some viruses
9	Origin	Replication of pre-existing DNA In case of infection by RNA viruses, reverse transcription of genetic RNA	Genetic RNA either through transcription of DNA or Through replication of RNA by RNA dependent RNA polymerase Non-genetic RNA from transcription of DNA
10	Native form	Double stranded DNA usually in B-form	Double stranded RNA usually in A-form
11	Types	There are several forms of DNA like A, B, C and D.	Three types, viz., mRNA, tRNA and rRNA

Main difference between A, B, C and Z form of DNA

Character	A-DNA	B-DNA	C-DNA	Z-DNA
Coiling	Right-handed	Right-handed	Right-handed	Left-handed
Pitch	28 A°	34 A°	31 A°	60 A°
Base pairs per turn	11	10	9.33	12
Diameter	26 A°	20 A°	19 A°	18 A°
Vertical rise per base pair	2.56 A°	3.38 A°	3.32 A°	3.71 A°
Sugar-phosphate back bone	Regular	Regular	Regular	Zig-zag
Major groove	Narrow and deep	Wide and deep	-	Flat
Minor groove	Wide and shallow	Narrow and deep	-	Narrow and deep

Write the Differences between Replication and Transcription of DNA?

S. No	Feature	Replication	Transcription
1	Rate	500-1000 nt/min	300 nt/min
2	Enzymes involved	DNA polymerase and RNA polymerase	Only RNA polymerase
3	Occurrence	In both strands	In one strand
4	Specific start and end sites	Don't have specific start and end sites	Specific start and end sites i.e. promoter
5	Nature	Continuous in leading and discontinuous in lagging strand	Always continuous

Write the Differences between Prokaryotes and Eukaryotes in Relevance to the Following?

1. DNA replication
2. Transcription
3. Translation

Differences in replication between prokaryotes and eukaryotes

S.No	Prokaryotes	Eukaryotes
1	RNA primers and okazaki fragments are longer	RNA primers and okazaki fragments are shorter
2	DNA synthesis is continuous through cell cycle	Only in 'S' phase
3	Single origin per chromosome	Multiple origins control the replication of DNA molecule

Transcription differences between prokaryotes and eukaryotes

S.No	Feature	Prokaryotes	Eukaryotes
1	RNA polymerase	Holoenzyme consists of five polypeptides ($\alpha 2\beta\beta'\sigma$)	Three types, RNA polymerase I, II and III
2	Transcription factors	Not known	Many different transcription factors involved in transcription and its regulation
3	Promoter	Simpler and relatively smaller sequence	Longer sequence: has variable module

4	Enhancer	A part of promoter	Present variable distance from promotor
5	Transcription initiation	Holoenzyme bind to promoter and initiate transcription	Transcription factor bind to promoter, then RNA polymerase associates with them and initiate transcription
6	Transcription complex a. Composition b. Separation of components	 Core polymerase + σ factor σ factor dissociates after transcription initiation	 RNA polymerase + Transcription factor Transcription factor dissociates when transcription is initiated
7	End product	Polycistronic RNA	Monocistronic
8	Regulation Type Genes under same regulation	 Mainly negative control, inducible or repressible Located in a single operon forming a single transcription unit and all are driven by single promotor	 Mainly positive control and inducible Located in divergent sites. Regulation results since all of them have same response element

Differences in protein synthesis between prokaryotes and eukaryotes

S. No	Feature	Prokaryotes	Eukaryotes
1	mRNA	• Poly cistronic • No processing after transcription • Very short half life (2-5 min) • No poly A tail and 5′ cap	• Monocistronic • Processing after transcription • Longer half life (6 hrs) • Poly A tail and 5′ cap
2	rRNA	• Uses 70S ribosomes with 30S and 50S sub units • Ribosomes free in cytoplasm • rRNA consists of 16S (30S) and 23S, 5S (50S)	• Uses 80S ribosomes with 40S and 60S sub units • Ribosomes attached to endoplasmic reticulum as well as free in cytoplasm • rRNA consists of 18S (40S) and 28S, 5.8S and 5S (60S)
3	tRNA	Starting amino acid is N-formyl methionine	Starting amino acid is methionine

4	Translation	• Simultaneous with transcription • Translation occur when mRNA attached to chromosomes • Translation initiation involves base pairing between 16S rRNA and shine-dalgarno sequence in mRNA • mRNA first binds to 30S subunit of ribosome and then accepts the formyl-methionine (tRNAf met)	• Not simultaneous with transcription • mRNA has to be processed and transported from nucleus to cytoplasm • Metheonyl tRNA first binds to 40S subunit to mRNA molecule to form initiation complex • Metheonyl tRNA first binds to 40S subunit to mRNA moleculc to form initiation complex
5	Coding dictionary	Isoleucine coded by AUU not AUA	Isoleucine coded by AUA

Write the basic difference between prokaryotic and eukaryotic gene control system?

S. No	Prokaryotes	Eukaryotes
1	As the life span of prokaryotes is very short so it cannot have multiple stages/levels of gene regulation	Eukaryotes have multiple stages/levels of gene regulation
2	Prokaryotes exhibit lower levels of genome organisation	Eukaryotes exhibit higher level of genome organisation
3	Cellular and organ level diversity. It is a product of genetic and epigenetic control	Spatial and temporal regulation of genes
4	Abiotic stresses like nutrients are lacking for survival of prokaryotes i.e. nutrients are limiting factors for prokaryotes helps in induction and repressing of expression of genes	Nutrients and temperature are driving forces behind the gene reguation
5	Operon – group of structurally/ functionally related genes that are under common genetic control and transcribed as a single unit (Polycistronic unit)	Operons are not present in eukaryotes but operon like pathways is present, but the transcript is not a polycistronic unit/ monocistronic unit.
6	In prokaryotes, group of genes have same promoter	In eukaryotes, these clusters of genes will have separate promoter sites of themselves.

Write the differences between monosomics and nullisomics?

Feature	Monosomics	Nullisomics
Definition	Individual lacking one chromosome set from diploid set (2n-1)	Individual lacking one pair of chromosomes from a diploid set (2n-2)
Method of production	(*a*) From haploids (Sears, 1939) (*b*) From backcrosses of interspecific hybrids Eg. Tobacco and Wheat (*c*) From partially asynaptic plants (*d*) Irradiation treatment (*e*) Spontaneous production	(*a*) Selfing of monosomics (*b*) Monotelosomics or Monoisosomics
Identification	(*a*) By using morphological traits (*b*) Presence of univalent at meiosis or chromosome count during mitotic metaphase eg. Cotton and tobacco	By using morphological traits eg. Shape of spike in wheat
Meiotic behaviour	(*a*) Monosomics usually form bivalents in addition to a solitary univalents at diakinesis and metaphase I. Eg. $23^{II} + 1^{I}$ in tobacco, $25^{II} + 1^{I}$ in cotton and $20^{II} + 1^{I}$ in wheat (*b*) Rarely, a trivalent may be formed due to association of a univalent with a non-homologue. (*c*) Univalent shift: Monosomic progeny plants are monosomic for a different chromosome as compared to the monosomic parent of the cross.	Presence of bivalents and no univalents at metaphase I of meiosis E.g. 20^{II} bivalents in wheat
Breeding behaviour	On selfing, progeny of monosomics consists of disomics, monosomics and nullisomics are in the ratio of 24:73:3	Nullisomics cannot maintained by selfing because nullisomics are not always fully fertile
Uses	(*a*) Locating genes on chromosomes (*b*) Suitable for locating hemizygous ineffective genes through analysis of F_3	(*a*) Compensating nullisomic tetrasomic lines (*b*) Locating genes on chromosomes (*c*) Hemizygous ineffective genes cannot be located using nullisomics only
Example	Complete set of monosomics available in wheat, oat, cotton, tobacco and maize.	Complete set of nullisomics available only in wheat and oat.

Write the difference between self pollination, cross pollination and asexual propagation?

Character	Self pollination	Cross pollination	Asexual propagation
Genetic constitution of natural population	Homozygous	Heterozygous	Heterozygous
Inbreeding depression	Absent	Present; severity variable	Present; severe
Genetic constitution of varieties	Usually homozygous	Heterozygous	Heterozygous
Types of varieties	Purelines, sometimes hybrids	Hybrids, synthetics and composites	Clones
Pollination control			
1. Selfing	Occurs naturally	Requires bagging/ special techniques	Requires bagging/ special techniques
2. Crossing	Difficult	Easy, greatly facilitated by male sterility	Easy
Creation of genetic variation	Easy	Easy	Easy, only in species sexual reproduction occurs
Gene transfer	Easy	Easy	Easy, only in species sexual reproduction occurs
Genetic stability of varieties	Stable (pure lines)	Hybrids: unstable	Clones: stable
Examples	Rice, wheat, barley. Oat and chickpea etc	Maize, sunflower, bajra and rye etc.	Potato, sugar cane, coffee, tea and banana etc.

Write the differences between the following?

1. Gametophytic and sporophytic system of self incompatibilities
2. Self incompatibility and male sterility

S. No	Gametophytic system	Sporophytic system
1.	It is controlled by genetic constitution of pollen	It is controlled by genotype of pollen producing plant
2.	Pollen tube growth is usually inhibited in style or ovary	Pollen tube growth is inhibited on the stigma
3.	Plant species of this system have binucleate pollen	Plant species of this system have trinucleate pollen
4.	Reciprocal differences are not observed	Reciprocal differences are observed
5.	Recovery of only male parent is possible	Recovery of both male and female parents was possible.
6.	Does not permit the production of homozygotes	Permits the production of homozygotes
7.	Crosses may be sterile, partially fertile or fully fertile.	Crosses would be either fully fertile or fully sterile
8.	Biochemical substance associated with this is formed during pollen formation	Biochemical substance associated with this is formed before pollen formation
9.	This system can be overcme by polyploidy	This cannot overcome by polyploidy
10.	Eg. Red clover, white clover, rye, potato and tomato etc.	Eg. Radish, Cabbage, Cauliflower and Sunflower etc.

S. No	Self incompatibility	Male sterility
1.	Inability of a viable pollen to fertilize the eggs of same flower or same plant	It is a condition where the pollen grains are non functional
2.	Pollen production is sufficient and are functional	Pollen production is not sufficient. The anthers are shriveled and abortive.
3.	It leads to heterozygosity	It also leads to heterozygosity
4.	It can be used for hybrid seed production in Brassica	It is also used in hybrid seed production (Rice, cotton, bajra and maize)
5.	It creates problem in development of inbred lines	Inbreds can be produced by before transferring male sterility
6.	It is of two types: Hetero and Homomorphic	It is of three types: CMS, GMS and CGMS.
7.	During seed production, seed harvested from two cross compatible but self compatibles are hybrids seeds	Seed harvested from male sterile lines are only hybrid seeds.
8.	Artificial induction is difficult.	Artificial induction is easy.

Write the difference between directional, stabilizing and disruptive selection.

Directional selection	Stabilizing selection	Disruptive selection
It is the selection that favours the extreme types at the cost of intermediates	It is the selection that favours the intermediate forms at the expense of extremes	It favours those members of a population which in respect of some metric traits are well above or below the average of the population at the expense of more typical members (i.e. intermediate) of population.
In this 1. Mean of selected population is difference from that of population mean 2. Variance of selected population reduced rapidly	1. Mean of selected population is essentially same as that of original population mean. 2. Variance is reduced. Change in variance is the function of selection intensity.	It enlarges the range of variation because genetic heterogeneity (diversity) increases rapidly
It utilises only additive genetic variance	Predominance of heterozygotes i.e. dominance genetic variance	It utilises both additive and non additive variances
Will not observe any polymorphism	It causes population polymorphism at genic level through overdominance	It acts on larger gene complexes and causes a polymorphism on higher super-gene level and it causes the chromosomal polymorphism.
It will not maintain continuity of the population	Maintain the continuity of the distribution of the population	It disrupts the continuity of the population
Inferior fitness compared to disruptive selection	It is prerogative of natural selection i.e. it favours the intermediate forms so it satisfy the maximum fitness	Disruptive selection lines have superior fitness than directional selected lines.

Explain in brief about the differences between pedigree, bulk and single seed descent methods of breeding?

Features	Pedigree method	Bulk method	Single seed descent method
Definition	Method of breeding in which individual plants are selected in F_2 and the subsequent generations and a pedigree record is maintained.	Method of breeding in which F_2 and the subsequent generations are grown in bulk usually without artificial selection, in the end, pureline varieties are developed through individual plant selection.	Method of breeding, in which one seed from each F_2 plant of a cross and the subsequent generations, is bulked to rise the next generation.
Principle	Early generation selection and only artificial selection operates between and within progenies. As a result of which most of the progenies with undesirable characters are eliminated at early stages	Selection in advanced generations i.e. waiting for opportunity for selection and only natural selection operates. Natural selection does shift the population by reducing proportion of lines with reduced fecundity. Thus it capitalizes both between lines as well as within line genetic variation, relatively more emphasis on within line selection.	Maintenance of variability in homozygous form and there is no natural and artificial selection. Random selection of single seed will starts from F_2, until all the lines become homozygous.
Individual plant selection	Starts in early generation (F_2) generation.	Starts in advanced (F_6/ F_7) generations.	Starts in F_7 generation
Progeny row evaluation	Starts in F_3 generation	Starts in F_7 / F_8 generation	Starts in F_7 / F_8 generation
Maintenance of records	Pedigree record may be used as a basis for selection in advanced generations.	No pedigree records	No pedigree records
Other features	• Takes less time (12 years) for development of variety • Pedigree records provide the information about inheritance of characters	Takes more time (14 – 15 years) for development of variety compared to pedigree method.	Most suitable for development of Recombinant inbred lines (RILs) development as it maintains the F_2 spectrum of variability up to F_6 in homozygous form.

Write the difference between inbreeding coefficient and inbreeding depression?

S. No	Inbreeding coefficient	Inbreeding depression
1	It is the measure of relationship between mating individuals	Deterioration in performance of the subsequent generation by continuous inbreeding
2	Can be defined as the correlation between uniting gametes	Inbreeding depression varies crop to crop
3	Represented by F	It depends on nature of pollination and mode of reproduction
4	Also defined as the probability that the two genes at any one locus in an individual are identical by descent.	Higher in cross pollinated crops and lower or nill in self pollinated crops
5	It is also a measure of the proportion by which heterozygosity is reduced by inbreeding	Result due to increased recessive homozygosity

Write the difference between hybrid, variety and population?

Distinctive feature	Variety	Hybrid	Population
Origin	By pure line selection either in land races or hybrid materials developed by crossing between two parents	By hybridisation between two inbreds, F_1 seed is immediate value to faremer	By intermating a number of judiciously chosen components involving inbreds, single crosse, double crosses or open pollinated varieties followed by selection for productivity and uniformity.
Breeding behaviour	True breeding, no segregation, mean is conserved in filial generation	Maximum segregation in F_2 and later generations, free pollination causes lowering of yield levels from F_2 onwards.	Low to high amount of segregation, population mean tends to remain constant due to random mating
Occurrence	In self pollinated crops	Predominantly in cross pollinated crops but occur in selfers also	In cross pollinated crops, but depending upon the degree of allogamy and availability of male sterility, it can occur in selfers too.
Phenotypic variability	Nearly perfect	Nearly perfect	Low to moderately variable

Genetic homogeneity	Nearly homogeneous, variation due to environment only	Homogenous, variation only environmental	Highly heterogeneous due to variable genotypic frequency
Homozygosity	Nearly homozygous	heterozygous	Variable mixture of homo and heterogeneous components
Gene frequency for different loci	p = 1 & q = 0 or p = 1 & q =1	p = q= 0.5	p = q= 0.5
Inbreeding coefficient	F ≈1	F = 0	F = 0 ≈ 0.5
Genetic correlation between parent and offspring	$r_g \approx 1$	$r_g = 0$	$r_g \approx 0.5$
Growers need	Seed lost for 6-8 years	Seed renewal required every year	Seeds may lost for 3-4 years

Write the difference between polygenic and oligogenic traits.

S. No	Difference	Polygenic	Oligogenic
1.	Variation	Continuous	Discontinuous
2.	Individual effect	Small and undetectable	Large and detectable
3.	Number of genes	Several	One or few
4.	Grouping	Not possible	Possible
5.	Environmental role	High	Little
6.	Analysis	Based on means, variance and covariance	Ratio and frequencies
7.	Metric measurement	Possible	Not possible
8.	Stability	Low	High

Write the difference between GCA and SCA?

S. No	General combining ability	Specific combining ability
1	It is average performance of a strain in a series of crosses.	It refers to the performance of specific cross in relation to gca.
2	gca is due to additive genetic variance and additive x additive epistasis.	sca is due to dominance genetic variance and all the three types of epistasis.
3	It is estimated from half sib families	It is estimated from full sib families
4	It helps in the selection of suitable parents for hybridisation.	It helps in the identification of superior cross combinations

5	It has relationship with narrow sense heritability.	
6	GCA variance arises from difference in the gca effects of parents involved and corresponds to the variance of breeding value i.e., additive genetic variance $\sigma^2 g = \frac{1}{2} VA$	It arises from difference in SCA effects of crosses and corresponding o mainly dominance variance. $\sigma^2 s = VD$

Write the difference between vertical and horizontal resistance?

Term vertical and horizontal resistances were given by 'Vander plank'.

S. No	Vertical resistance	Horizontal resistance
1	Specific resistance of host to the particular race of pathogen is called V.R.	The resistance of a host to all the races of pathogen is called H.R.
2	It is also known as monogenic/ oligogenic/ qualitative/non uniform/ major gene/race specific / seedling/ differential/ complete/ hypersensitive resistance.	It is also known as minor gene/ poly gene/non-specific/ general/ adult/ field/ partial/ uniform/ incomplete resistance.
3	It has high heritability	It has low heritability
4	It is less durable and it can easily breakdown by new race of pathogen	It cannot be easily overcome by new races of pathogen.
5	It is applicable for gene for gene hypothesis	Not applicable for gene for gene hypothesis.
6	It affects the disease onset by intercepting amount of initial inoculums.	It has rate reducing effect after establishment of pathogen in the host.
7	Here r = 0 for avirulent (immune reaction) type and for virulent type r = 1 (susceptible reaction). Here, 'r' means rate of reproduction of pathogen.	Here 'r' never zero $r > 0$ but < 1

Write the difference between gene pyramiding and gene deployment?

Gene pyramiding	Gene deployment
Gene Pyramiding refers to the incorporation of two or more diverse genes for resistance into a single cultivar that provide resistance to a common spectrum of races.	It refers to strategic use of major gene, by deploying these over space and time to avoid phenomenon of boom and bust.
It provides durable resistance, used in oat against crown rust.	It helps in preventing disease and maintaining diversity of major genes in different geographical region.

Demerits:- • Lot of efforts needed • Extensive testing method • Chances of development of super race • Agronomic performance of cultivar restricted.	Drawback is difficult to find out different genes with similar effectiveness.
Eg. Successful pyramid of four genes Xa4, Xa5, Xa13 and Xa21 conferring resistance to four different races of bacterial leaf blight pathogen has been achieved in rice (Huang *et. al* 1997) by using molecular markers.	Frey *et. al* (1973) suggested sub division of puccinia path in central North America three areas to control crown rust of oat. The placement of VR genes in pathway path of movement has to be strategy VR gene should be deployed in varieties where epidemic ends. Another approach-deployment over time, such a cycling of VR genes provide in contrast to spatial discontinuity.

Write the difference between PCR and Gene cloning:

S .No	Parameter	PCR	Gene cloning
1	Manipulation	*In vitro*	Step *in vitro* and *in vivo* last
2	Selectivity of specific Segments from complex DNA	1st step	Last step
3	Biological reagents require	DNA segment 2 primers DNTP technique polymers	Restriction enzymes ligases, Vector DNA, host cells
4	Automation	Yes	No
5	Labour intensive	No	Yes
6	Error probability	Less	More
7	Applications	More	Less
8	Cost	Less	More
9	Users skill	Not required	Required
10	Time required for Conducting an experiment	4 hrs	2-4 days

Write the difference between Linkage analysis and Association mapping.

S. No	Linkage analysis	Association mapping
1	Family mapping	Population mapping
2	Structured population	Unstructured population
3	Based on two diverse mapping populations	Based on diverse collections
4	Biparental mapping	Multiparental mapping
5	Only two alleles i.e. from both parents	Multiple allelic variation
6	Few cycles of recombination i.e. one which leads to shuffling of genome	Several cycles of recombination i.e. extensive shuffling of genome because of more meiotic events
7	Less resolution	More resolution
8	Cannot go much nearer to the gene	Can go more nearer to genome
9	High power and moderate marker density	Low power and high marker density
10	Genome scan	Candidate gene approach

Write the difference between correlation and regression analysis?

S. No	Correlation	Regression
1.	Proposed by Karl pearson	Proposed by Galton
2.	It measures the mutual relationship between two or more variables	It measures functional relationship between two or more variables
3.	Does not involve cause and effect relationship	It clearly implies cause and effect relationship
4.	It is confined to the study of linearity between variables	It studies both linear and non- linear relationship
5.	It does not predict any performance of variables	It can predict the performance of a dependent variables based on the independent variables
6.	It is independent of the unit of measurement	It is recorded in the unit of measurement of variable
7.	Correlation coefficient between x and y and between y and x are the same i.e., $r_{xy} = r_{yx}$	Regression coefficient between x and y and between y or are not the same i.e., $b_{yx} \neq b_{xy}$
8.	Range of r between -1 and +1	'b' can take any value i.e., <1 or >1
9.	Correlation is unaffected by both change of origin and change of scale	Unaffected by change of origin whereas it is affected by change of scale
10.	Correlation studies be misleading with no practical relevance	Regression does not give any misleading may results.

Abbrevations

1. **AFLP:** Amplified Fragment Length Polymorphism
2. **AMMI:** Additive Main effects and Multiplicative Interaction
3. **ANOVA:** Analysis of Variance
4. **AMOVA:** Analysis of Molecular Variance
5. **AUDPC:** Area Under the Disease Progress Curve
6. **BAC:** Bacterial Artificial Chromosomes
7. **BILs:** Backcross Inbred Lines
8. **Bp:** Base pair(s)
9. **cDNA:** Complementary DNA
10. **CHA:** Chemical Hybridizing Agent
11. **cM:** Centimorgan
12. **CMS:** Cytoplasmic Male Sterility
13. **CSSLs:** Chromosome Segment Substitution Lines
14. **dATP:** Deoxyadenosine triphosphate
15. **dCTP:** Deoxycytidine triphosphate
16. **dGTP:** Deoxyguanine triphosphate
17. **DH:** Doubled haploid
18. **Dnase:** Deoxyribonuclease
19. **dNTP:** Deoxyribonucleotide
20. **dUTP:** Deoxuridine triphosphate
21. **EDTA:** Ethylene Diamine Tetra Acetic acid
22. **EGMS:** Environment-induced Genic Male Sterility
23. **ELISA:** Enzyme-Linked Immunosorbent Assay
24. **EMS:** Ethyl Methane Sulfonate
25. **EST:** Expressed Sequence Tag
26. **FNP:** Functional Nucleotide Polymorphism
27. **GCA:** General Combining Ability
28. **GEAC:** Genetic Engineering Approval Committee
29. **GIS:** Geographic Information System
30. **GM:** Genetically Modified

31. **HEPA:** Higrowth hormone Efficiency Particulate Air
32. **HFCS:** Higrowth hormone Fructose Corn Syrup
33. **HPLC:** High-Performance Liquid Chromatography
34. **IRAP:** Inter-Retrotransposon Amplified Polymorphism
35. **ISSR:** Inter-Simple Sequence Repeat
36. **Kb:** Kilobases
37. **LD50:** Lethal dose/fifty LD50
38. **MAS:** Marker-Assisted Selection
39. **Mb:** Mega basepairs
40. **mRNA:** messenger RNA
41. **mt DNA:** Mitochondrial DNA
42. **NILs:** Near-isogenic lines
43. **PAGE:** Polyacrylamide Gel Electrophoresis
44. **PCR:** Polymerase Chain Reaction
45. **PEG:** Polyethylene Glycol
46. **PFGE:** Pulsed-Field Gel Electrophoresis
47. **PMCs:** Pollen Mother Cells
48. **QTL:** Quantitative Trait Loci
49. **RAPD:** Random Amplified Polymorphic DNA
50. **RCB:** Randomized-Complete Block
51. **REMAP:** Retrotransposon-Microsatellite Amplified Polymorphism
52. **RFLP:** Restriction Fragment Length Polymorphism
53. **RILs:** Recombinant Inbred Lines
54. **RNase:** Ribonuclease
55. **RT-PCR:** Real-time Polymerase-Chain Reaction
56. **SAMPL:** Selective Amplification of Microsatellite Polymorphic Loci
57. **SAUDPC:** Standardized Area Under the Disease Progress Curve
58. **SCA:** Specific Combining Ability
59. **SCAR:** Sequence-Characterized Amplified Region
60. **SDS:** Sodium Dodecyl Sulfate
61. **SDS-PAGE:** Sodium dodecyl sulfate Polyacrylamide Gel Electrophoresis
62. **SNP:** Single Nucleotide Polymorphism
63. **SSD:** Single Seed Descent
64. **SSLP:** Simple Sequence Length Polymorphism
65. **SSR:** Simple Sequence Repeat
66. **STMS:** Sequence Tagged Microsatellite Site

67. **STS:** Sequence Tagged Site
68. **Ti plasmid:** Tumer inducing
69. **Tm:** Melting temperature of DNA
70. **TRIM:** Terminal-repeat Retrotransposon in Miniature
71. **VIGS:** Virus-Induced Gene Silencing
72. **WGA:** Whole Genome Amplification
73. **YAC:** Yeast Artificial Chromosome

Abbreviations related to national and international research institutes:

74. **BARC:** Baba Atomic Research Center – Mumbai
75. **CARI:** Central Agricultural Research Institute – Port Blair (Andman & Nicobar)
76. **CAZRI:** Central Arid Zone Research Institute – Jodhpur
77. **CCMB:** Center for Cellular and Molecular Biology – Hyderabad
78. **CDFD:** Center for DNA Finger Printing and Diagnostics – Hyderabad
79. **CDRI:** Central Drug Research Institute – Lucknow
80. **CFTRI:** Central Food Technology Research Institute, Mysore
81. **CGIAR:** Consultative Group on International Agricultural Research
82. **CIAT:** International Centre for Tropical Agriculture – Columbia
83. **CICR:** Central Institute of Cotton Research – Nagpur
84. **CIMAP:** Central Institute of Medicine and Aromatic plants - Lucknow and Hyderabad
85. **CIMMYT:** International Wheat and Maize Improvement Centre, Mexico (Centro International de Mejoraminento de Maiz Y Trigo)
86. **CIP:** International Potato Centre – Peru
87. **CPMB:** Center for Plant Molecular Biology – 7′ centers
88. **CPRI:** Central Potato Research Institute – Shimla
89. **CRIDA:** Central Research Institute for Dryland Agriculture – Hyderabad
90. **CRIJAF:** Central Research Institute for Jute and Allied Fibres – Barrackpore
91. **CRRI:** Central Rice Research Institute – Cuttack
92. **CSSRI:** Central Soil Salinity Research Institute – Karnal
93. **CTCRI:** Central Tuber Crops Research Institute – Trivendrum
94. **DMR:** Directorate of Maize Research – New Delhi
95. **DOR:** Directorate of oilseed Research – Hyderabad
96. **DRR:** Directorate of Rice Research – Hyderabad
97. **DWR:** Directorate of Wheat Research – Karnal
98. **FAO:** Food and Agricultural Organization – Geneva

99. **IARI:** Indian Agricultural Research Institute, New Delhi
100. **ICAR:** Indian Council of Agricultural Research – New Delhi
101. **ICARDA:** International Centre for Agricultural Research in Dryland Areas, Italy
102. **ICRISAT:** International Crop Research Institute for Semiarid Tropics, Hyderabad
103. **ICRO:** International Cell Research organization
104. **IGFRI:** Indian Grassland and Fodder Research Institute – Jhansi
105. **IIB:** International Institute of Biotechnology – Conterbury kent in UK
106. **IIHR:** Indian Institute of Horticulture Research – Banglore
107. **IIPR:** Indian Institute of Pulses Research – Kanpur
108. **IISC:** Indian Institute of sciences – Bangalore
109. **IISR:** Indian Institute of Spice Research – Calicut
110. **IISR:** Indian Institute of Sugarcane Research – Lucknow
111. **IIT:** Indian Institute of Technology – Kanpur, New Delhi
112. **IITA:** International Institute for Tropical Agriculture – Nigeria
113. **IIVR:** Indian Institute of Vegetable Research – Varanasi
114. **IMTECH:** Institute of Microbial Technology – Chandigarh
115. **IRRI:** International Rice Research Institute – Philippines
116. **IVRI:** Indian Veterinary Research Institute, Izatnagar
117. **JNU:** Jawaharlal Nehru University, New Delhi
118. **MRC:** Malaria Research Center – New Delhi
119. **NCL:** National Chemical Laboratory – Pune
120. **NDRI:** National Dairy Research Institute - Karnal - Haryana
121. **NIM/NII:** National Institute of Immunology – New Delhi
122. **RRL:** Regional Research Laboratory – Jammu
123. **SBI:** Sugarcane Breeding Institute – Coimbatore
124. **UNEP:** United Nations Environment Programme
125. **VPKAS:** Vivekanand Parvatiya Krishi Anusandhan Shala – Almora

Terminology

A line: The seed-bearing parent line used to produce hybrid seed that is male sterile

Aberration: Variation of chromosome structure caused by induced or spontaneous mutations; in

Accessory chromosome: A chromosome that is present in addition to the normal chromosome complement i.e. B chromosome

Acclimatization: Adaptation of a introduced variety to the new environment

Acentric: Chromosome, chromosome segments, or chromatids that show no centromere

Acetolactate synthase (ALS): An enzyme essential for amino acid production

Addition line: A cell line or line of individuals carrying chromosomes or chromosome arms in addition the normal standard chromosome set

Additive genes: Gene interaction without dominance (if allele), or without epistasis (if nonallele)

Additive resistance: Resistance governed by more than one gene, each of which can be expressed independently, but which is reinforced by the expression of each of the additional genes

Additive variance (V_A): The proportion of the genetic variance due to additive effects

Adenine: A nitrogenous base, one member of the base pair A-T (adenine-thymine).

Adenosine diphosphate (ADP): High-energy phosphoric ester (nucleotide) of the nucleoside adenosine that functions as the principal energy-carrying compound in the living cell

Adenosine triphosphate (ATP): High-energy phosphoric ester (nucleotide) of the nucleoside adenosine that functions as the principal energy-carrying compound in the living cell

Adjacent segregation: A reciprocal translocation heterozygote in which during meiosis the segregation of a translocated and a normal chromosome happens together, giving unbalanced gametes with duplications and deficiencies leading to nonviable zygotes; adjacent segregation is of two kinds depending on whether nonhomologous (adjacent-1) or homologous (adjacent-2) centromeres segregate together; adjacent-1 segregation is the usual type of adjacent segregation and adjacent-2 segregation is rare

Adjacent-1 segregation: Segregation of nonhomologous centromeres during meiosis in a reciprocal translocation heterozygote such that unbalanced gametes with duplications and deficiencies are produced, as opposed to alternate segregation and adjacent-2 segregation

Adjacent-2 segregation: Segregation of homologous centromeres during meiosis in a reciprocal translocation heterozygote such that unbalanced gametes with duplications and deficiencies are produced, as opposed to alternate segregation and adjacent-1 segregation

A-DNA: The dehydrated form of right-handed helical DNA obtained under nonphysiological conditions

Adult resistance: Resistance not expressed at the seedling stage; it increases with plant maturity (mature-plant resistance or age resistance); generally attributed to horizontal resistance

Adventitious embryony: A condition in a seed in which the embryo arises from somatic rather than reproductive tissue; the development of a diploid embryo from nucellary or integumentary tissue (sporophyte tissue)

Agricultural Biotechnology: A range of tools, including traditional breeding techniques, that alter living organisms, or parts of organisms, to make or modify products; improve plants or animals; or develop microorganisms for specific agricultural uses. Modern biotechnology today includes the tools of genetic engineering.

***Agrobacterium rhizogenes*:** A species of Gram -negative, rod-shaped soil bacteria, often harboring large plasmids, called Ri plasmids; it can cause a tumorous growth known as hairy

***Agrobacterium tumefaciens*:** A bacterium that causes crown gall disease in some plants; it infects a wound, and injects a short stretch of DNA into some of the cells around the wound; the DNA comes from a large plasmid the Ti (= tumor induction) plasmid

***Agrobacterium*-mediated transformation:** *Agrobacterium* is the generic name of a soil bacterium that frequently causes crown gall in many plant species; besides *A. rhizogenes, A. tumefaciens* is one species that is most used in DNA transfer by manipulating the Ti (tumor inducing) plasmid that is harbored by these bacteria

Alien addition line: A line (strain) of plants with one or more extra chromosomes of an alien species

Alien chromosome transfer: Cytogenetic methods that facilitate the transfer of individual chromosomes from one species to another

Alien chromosome: A chromosome from a more or less related species transferred to a crop plant

Alien gene transfer: The transfer of genes between species or genera by different means

Alien germplasm: Genes introduced from a wild relative or nonadapted species

Alien species: An organism that has invaded or been introduced by man and is growing in a new

Alien substitution line: A line of plants in which one or more alien chromosomes from a certain donor species replace one or more chromosomes of a recipient species

Allele frequency: A measure of the commonness of an allele in a population of alleles

Allele mining: An approach to access new and useful genetic variation in crop plant collections; it focuses on the detection of allelic variation in important genes and/or traits within a germplasm collection

Allele: One of the different forms of a gene or DNA sequence that can exist at a single locus.

Allergen: A substance, usually a protein, that can cause an allergy or allergic reaction in the body.

Allodiploid: Cells or individuals in which one or more chromosome pairs are exchanged for one or more pairs from another species

Allogamy: Cross-fertilization

Allohaploid: A haploid cell or individual derived from an allopolyploid and composed of two or more different chromosome sets

Alloheteroploid: Heteroploid individuals or cells whose chromosomes derive from various genomes

Alloplasmic: An individual having the common nucleus, but an alien cytoplasm (E.g. alloplasmic rye containing a wheat cytoplasm); usually leads to meiotic disturbances and sterility

Allopolyploid: Plants with more than two sets of chromosomes that originate from two or more parents; the sets contain at least some nonhomologous chromosomes

Allosome: A chromosome deviating in size, form, or behaviour from the other chromosomes autosomes), such as the sex chromosome or B chromosome

Allosteric transition: A change from one conformation of a protein to another conformation **Allosubstitution:** The replacement of a chromosome or chromosome arm by an alien chromosome or chromosome arm

Allosteric: An enzyme whose activity is altered when its structure is distorted by an organic compound at a nonsubstrate site

Allosyndesis: Chromosome pairing of completely or partially homologous (homoeologous) chromosomes

Allotetraploid (*syn* amphidiploid): A plant that is diploid for two genomes, each from a different species

Alternate segregation: At meiosis in a reciprocal translocation heterozygote, the segregation of both normal chromosomes to one pole and both translocated chromosomes to the other pole, giving genetically balanced gametes, or segregation of centromeres during meiosis in a reciprocal translocation heterozygote such that genetically balanced gametes are produced

Alternation of generation: The alternation of two or more generations, reproducing themselves in different ways (i.e., alternation of gametophyte and sporophyte stages in the life cycle of a plant)

Alternative disjunction: The distribution of interchange chromosomes at anaphase I of meiosis is determined by their centromere orientation; in the case of alternative disjunction, chromosomes located alternatively in the pairing configuration are distributed to the same spindle pole

Alternative splicing: Formation of diverse mRNAs through differential splicing of the same RNA precursor; it may result in proteins with different composition of amino acids, or it may involve just the length of 3′ UTR; a reason for alternative/differential splicing is base modification during RNA editing causing a change in slice sites

Amber codon: Amber suppressor mutation that changes anticodon of amino acid-carrying tRNA

Ameiosis: The failure of meiosis and its replacement by nuclear division without reduction of the chromosome number

Ameiotic parthenogenesis: Parthenogenesis in which meiosis has been entirely suppressed amino acid, usually resulting in an elongated protein

Amino acid: Any of a class of 20 molecules that are combined to form proteins in living things. The sequence of amino acids in a protein and hence protein function are determined by the genetic code

Aminoacyl-tRNA: A tRNA molecule covalently bound to an amino acid *via* an acyl bond between the carboxyl group of the amino acid and the 3′-OH of the tRNA

Amitosis: Nuclear division by a process other than mitosis

Amplification: An increase in the number of copies of a specific DNA fragment; can be *in vivo* or *in vitro*. See cloning, polymerase chain reaction

Amplified fragment length polymorphisms (AFLPs): Polymorphic DNA fragments are amplified through PCR procedure; their differences are used for genotype identification and linkage studies

Analysis of variance (ANOVA): A statistical method that allows the partitioning of the total variation observed in an experiment among several statistically independent possible causes of the variation; among such causes are treatment effects, grouping effects, and experimental errors

Anaphase separation: The disjunction of the chromatids of each chromosome during mitosis and anaphase II of meiosis or the separation of chromosomes in anaphase I of meiosis

Androdioecious: Describes a species having male and hermaphroditic flowers on separate individuals

Androecium: A collection of stamens that form the male reproductive organs of flowering plants

Androgenesis: Development of a haploid embryo from a male nucleus

Aneuhaploid: When the chromosome number deviates from the haploid standard chromosome number of the species or individual

Aneuploid reduction: Reduction of the genetic variability by decreasing the chromosome number

Aneuploid: A cell or organism whose nuclei possess a chromosome number that is greater or smaller by a certain number than the normal chromosome number of that species; an aneuploid results from nondisjunction of one or more pairs of homologous chromosomes

Annealing: Formation of double-stranded nucleic acid from single stranded forms.

Anther culture: Culturing of anthers containing pollen or of single pollen grains; the method is used for the production of haploid plants, for the production of doubled haploids (which are homozygous) after spontaneous or induced rediploidisation

Anther: The terminal portion of a stamen of a flowering plant; the pollen sacs containing pollen are borne on the anther

Antibiosis: The phenomenon whereby a natural organic substance secreted by one organism has an injurious effect on normal growth and development of another organism when the two organisms are brought together

Anticodon: A triplet sequence of nucleotides in tRNA that during protein synthesis binds by base pairing to a complementary sequence, the codon, in mRNA attached to a ribosome

Antimutator gene: Mutant genes that decrease the mutation rates

Antiparallel orientation: The normal arrangement of the two strands of a DNA molecule, and of other nucleic-acid duplexes (DNA–RNA, RNA–RNA), in which the two strands are oriented in opposite directions so that the 5′-phosphate end of one strand is aligned with the 3′-hydroxyl end of the complementary strand

Antipodal cells: Three haploid nuclei that are formed during megasporogenesis in plants; all are located opposite the micropylar end of an ovule

Antisense DNA: Noncoding DNA of one of the double-stranded DNA, as opposed to sense strand DNA, which is the coding DNA (i.e., which is transcribed as mRNA)

Antisense gene: A gene construct placed in inverted orientation relative to a promoter

Antisense RNA: A complementary RNA sequence that binds to a naturally occurring (sense) mRNA molecule; in this way it thus blocks its translation

Apomixis: Asexual reproduction in plants without fertilisation or meiosis

Apoptosis: Programmed cell death (PCD); a process in which cellular DNA is degraded and the nucleus condensed; then cell is then devoured by neighboring cells or phagocytes.

Apospory: The development of a diploid embryo sac in some plants by the somatic division of a

Artificial chromosome: A chromosome experimentally created and constituted, in addition to genetically coding DNA sequences, by ligating origin of replication, autonomous replicating sequences, and telomeric and centromeric sequences

Artificial selection: Plant selection by human or agronomic means; it is the practice of choosing individuals from a population for reproduction, usually because these individuals possess one or more desirable traits

Asexual reproduction: A propagation without formation of zygotes by sexual organs and genetic recombination; in plants, there are two types of asexual reproduction: vegetative propagation (by stolons, rhizomes, tubers, tillers, bulbs, bulbils, or corms) and apomixis (by vegetative proliferation or agamospermy)

Association mapping: A population-based survey of molecular marker analysis in order to identify trait-marker relationships based on linkage disequilibrium; the association between a pair of linked markers is also called linkage disequilibrium (LD) or, less frequently, gametic disequilibrium

Association of Official Seed Certifying Agencies (AOSCA): Established in 1919 in the United States; it sets minimum standards for genetic purity and identity and recommends minimum standards for seed quality; its goal is to standardise certification regulations and procedures internationally so companies compete less than one set of standards

Assortative mating: Occurs if the plants mating resemble each other, with regard to some traits

Asynapsis: Chromosomes of meiosis I in which pairing either fails or is incomplete

Autoallopolyploid: Cells or individuals whose genomes show characteristics of both auto- and alloploidy

Autogamy: Obligatory self-fertilisation

Autogenous control: The action of a gene product that either inhibits (negative autogenous control) or activates (positive autogenous control) expression of its gene coding

Autonomous apomixis: Agamic seed formation that does not depend on pollination

Autoploid: A cell or individual with genomes characteristic of the species itself

Autopolyploid: A polyploid organism that originates by the multiplication of a single genome of

Autopolyploidization: The occurrence of doublings of chromosome number by failure of chromosomes to divide equationally in a mitosis following chromosome replication

Autoradiography: A technique that uses X-ray film to visualize radioactively labeled molecules or fragments of molecules; used in analyzing length and number of DNA fragments after they are separated by gel electrophoresis.

Autoreduplication: Biological systems that generate the template for their own reproduction and duplicate themselves

Autosome: Any chromosome in the cell nucleus other than a sex chromosome

Autosyndesis: The pairing of complete or partial homologues of chromosomes

Autotetraploid: An autopolyploid with four similar genomes; if a given gene exists in two allelic forms *A* and *a,* then five genotypic classes can be formed: *AAAA* (quadruplex), *AAAa* (triplex), *AAaa* (duplex), *Aaaa* (simplex), and *aaaa* (nulliplex)

B chromosome: Any chromosome of a heterogeneous group of chromosomes present in several plant species, which differ in their morphology, numerical variation, meiotic pairing, and mitotic behavior from normal A chromosomes; they are also called supernumerary chromosomes, accessory chromosomes, or extra chromosomes; a B chromosome derives from the A chromosome complement by aberrant division processes and subsequent modifications

B line: The fertile counterpart or maintainer line of an A line; does not have fertility restorer genes; used as the pollen parent to maintain the A line; used in hybrid seed production

Bacillus thuringiensis (Bt): A soil bacterium that produces toxins that are deadly to some pests. The ability to produce Bt toxins has been engineered into some crops. See Bt crops. Biopharming: The production of pharmaceuticals such as edible vaccines and antibodies in plants or domestic animals.

Back mutation: A reverse mutation in which a mutant gene reverts to the original standard form

Backcross breeding: A system of breeding whereby recurrent backcrosses are made to one of the parents of a hybrid, accompanied by selection for a specific characters

Backcross: A cross of an F_1 hybrid or heterozygote with an individual of genotype identical to that on one or the other of the two parental individuals

Backcross-assisted selection (BCAS): A method that allows the selection of plants carrying a favorable recessive allele at each generation, limiting the need for a progeny test, which is common in traditional backcrossing; in cases where the traditional means of selection are limited by environmental conditions (Eg. the presence of an abiotic or a biotic stress such as drought) this selection strategy is superior to conventional ones; particularly in genetic transformation approaches, where the transgenes can be used as markers, BCAS may show a considerable advantage

Backward selection: Selection of parent plants based on results from a progeny test

Bacterial artificial chromosome (BAC): Pieces of plant DNA that have been cloned inside living bacteria; they can be used as probes to detect complementary

DNA sequences within large pieces of DNA *via* hybridisation techniques, or for marker-assisted selection by faster selection of segregant-bearing genes for a particular trait and to develop future crop varieties faster

Balanced design: An experimental design in which all treatment combinations have the same number of observations; there are various different kinds of balance (variance balance, efficiency balance, etc.), none of which are equivalent to the combinatorialist's pairwise balance, but become equivalent to it if some extra conditions are satisfied (Eg. in binary equireplicate uniform block designs)

Balanced diallelic: The genotype involving a multiple allelic locus in an autotetraploid where two different alleles are represented an equal number of times (Eg. *u1u1u2u2*)

Balanced incomplete block design (BIB): A design in which one constant value for the residual variance of the difference between candidates for all pairs of candidates is indicated

Balanced lattice: A special group of balanced incomplete block; allows incomplete blocks to be combined into one or more separate complete replicates

Balanced lethals: Recessive lethals at different loci, so that each homologous chromosome carries at least one lethal, and associated with inversions, so that no recombination occurs between the homologous chromosomes

Balanced polymorphism: A genetic polymorphism that is stable, and is maintained in a population by natural selection, because the heterozygotes for particular alleles have a higher adaptive fitness than either homozygote; it is referred to as overdominance, as opposed to underdominance, where the heterozygotes have a lower fitness, giving rise to unstable equilibrium

Balanced tertiary trisomic (BTT): A specific interchange trisomic spontaneously selected or experimentally designed in a way that it is heterozygous (*Aaa*); its trisomic progeny after selfing is genetically similar to the parent; the dominant allele is present on the translocated chromosome linked to the break point; BTTs were thought to be used for hybrid seed production in barley; the American variety "Hembar" was the first commercial hybrid barley

Balanced Translocation: Occurs when two chromosomes break and exchange places leaving the same amount of genetic material. An individual with a balanced translocation will be unaffected, but children may be affected in a variety of ways.

Banding pattern: The linear pattern of deeply stained bands and weakly staining interbands that results from more or less defined local differences in the degree of DNA compactation along the chromosome

Banding: A special staining technique for chromosomes, which results in a longitudinal differentiation (Eg. Giemsa staining, which is a complex of stains specific for the phosphate groups of DNA)

Base analogues: A purine or pyrimidine base that differs slightly in structure from a normal base, but that because of its similarity to that base may act as a mutagen when incorporated into DNA (eg., uracil, 5-bromouracil, 5-fluorouracil, 5-methylcytosin, 5-bromocytosin, hypoxanthin)

Base collection: A collection of germplasm that is kept for long-term, secure conservation and is not to be used as a routine distribution source; seeds are usually stored at subzero temperatures and low moisture content

Base pair (bp): Two nitrogenous bases (adenine and thymine or guanine and cytosine) held together by weak bonds. Two strands of DNA are held together in the shape of a double helix by the bonds between base pairs.

Base population: The initial set of genotypes from which selections will be taken to establish a breeding population, for example, the wild forest; it is also sometimes referred to in the same meaning as recruitment population (in the first generation they are actually the same, and it can be referred to as one of the populations in a stratified tiered structure)

Base sequence analysis: A method, sometimes automated, for determining the base sequence based on some sets of postulates

Basic number: The haploid number of chromosomes in diploid ancestors of

Batch culture: A cell suspension grown in liquid medium of a set volume; inocula of successive subcultures are of similar size and cultures contain about the same cell mass at the end of each passage; cultures commonly exhibit five distinct phases per passage (a lag phase follows inoculation, then an exponential growth phase, a linear growth phase, a deceleration phase, and finally a stationary phase)

Binary vector system: A two-plasmid system in *Agrobacterium tumefaciens* for transferring into plant cells a segment of T-DNA that carries cloned genes; one plasmid contains the virulence gene (responsible for transfer of the T-DNA), and another plasmid contains the T-DNA borders, the selectable marker, and the DNA to be transferred

Biodiversity: The existence of a wide variety of species (species diversity), other taxa of plants or other organisms in a natural environment or habitat, or communities within a particular environment (ecological diversity), or of genetic variation within a species (genetic diversity

Biofortification: The process of breeding food crops that are rich in bioavailable micronutrients; these crops fortify themselves that is, they load high levels of minerals and vitamins in their seeds and roots, which are then harvested and eaten by animals or humans; through biofortification, breeders can provide farmers with crop varieties that naturally reduce anemia, cognitive impairment, and other nutritionally related health problems, potentially for use in third-world countries

Bioinformatics. Use of computer programs for searching and analysing electronic databases of DNA and protein sequences. This branch of science is an important component of molecular biology and genetics.

Biolistics. The process of introducing DNA into plant cells by shooting DNA-coated pellets into the cell using a compressed air delivery system know as a gene gun.

Biotechnology: A set of biological techniques developed through basic research and now applied to research and product development. In particular, the use by industry of recombinant DNA, cell fusion and new bioprocessing techniques.

Biparental inheritance: Plant zygotes that show traits indicating chloroplast chromosomes from

Biparental populations: The progeny derived after crossing two genotypes as male and female parents; such populations include F_2 genotypes generated from F_1 progeny, lines generated after doubling the haploids (DHs, obtained from F_1 plants through anther, egg cell or ovule culture or distant hybridisation), or recombinant inbred lines (RILs), which are derived by single seed descent for at least five or more generations by repeated selfing or sibling mating

Bisexual: Species comprises individuals of both sexes or a hermaphrotide organism in which an individual plant possesses both stamens and pistils in the flower

Bivalent formation: The association of two homologous chromosomes as a ring or rod configuration depending on chiasma formation

Bivalent: Two homologous chromosomes when they are paired during prophase-metaphase of the first meiotic division

Bivariate statistics: Statistics that describe the relationship between two variables

Blended variety: Multiline variety

Blunt ends: DNA fragments that are double-stranded paired over the whole length, usually produced by certain types of restriction enzymes

Blunt-end ligation: Ligation of DNA with blunt ends requires higher concentration of DNA ligase than sticky-end ligation; it is inhibited by ATP concentrations

Bollgard insect-protected cotton: A genetically improved cotton that offers protection against the cotton bollworm; the bollworm can cause significant damage to a cotton crop and require repeated applications of insecticides; Bollgard. cotton contains an insecticidal protein from a naturally occurring soil microorganism, *Bacillus thuringiensis,* that gives cotton protection from bollworms both parents are present and active

Bottleneck: A period when a population becomes reduced to only a few individuals; a form of genetic drift that occurs when a population is drastically reduced in size; some genes may be lost from the gene pool as a result of chance

Breakage-reunion hypothesis: The classical and generally accepted model of crossing-over by physical breakage and crossways reunion of broken chromatids during meiosis

Breakdown of resistance: The inability of a plant to maintain resistance when attacked by a pest biotype that has a gene for virulence at every locus corresponding to a gene for resistance in the host

Breeder's collection: Working collection

Breeder's rights: Varietal protection; the legal rights of a breeder, owner, or developer in controlling seed production and marketing of crop varieties

Breeder's seed: Seed or vegetative propagating material increased by the originating or sponsoring plant breeder or institution; it represents the true pedigree of the variety; it is used for the production of genetically pure, foundation, registered, and certified seeds

Breeding cycle: The shortest period between successive generations from germination of a seed to reproduction of the progeny (i.e. the seed-to-seed cycle)

Breeding line: A group of plants with similar traits that have been selected for their special combination of traits from hybrid or other populations; may be released or used for further breeding approaches

Breeding population: A group of individuals selected from a wild, experimental, or crossing population for use in a breeding program; usually phenotypically selected for desirable traits

Breeding system: The system by which a species reproduces; more specifically, the organisation of mating that determines the degree of similarity and/or difference between gametes effective in fertilisation

Breeding value: The value of an individual as defined by the mean value of its progeny, either on the basis of individual traits or a selection index

Breeding: The intentional development of new forms or varieties of plants or animals by crossing, hybridisation, and selection of offspring for desirable characteristics

Bridge cross: A method of bypassing an incompatibility barrier between two genotypes or species by using a third genotype or species, which is partly compatible with each of them, in an intermediate cross

Bridge-breakage-fusion-bridge cycle: A process that can arise from the formation of dicentric chromosomes; daughter cells are formed that differ in their content of genetic material due to duplications and/or deletions in the chromosomes

Broad-sense heritability: The ratio of total genetic variance to phenotypic variance; it is used to estimate the degree of genetic control of a trait in a population, and it is useful for predicting response to clonal selection

Bt crops: Crops that are genetically engineered to carry a gene from the soil bacterium Bacillus thuringiensis (Bt). The bacterium produces proteins that are toxic to some pests but non-toxic to humans and other mammals. Crops containing the Bt gene are able to produce this toxin, thereby providing protection for the plant. Bt corn and Bt cotton are examples of commercially available Bt crops.

Bud pollination: A procedure utilised in maintaining self-incompatible parent lines by self-pollination; hybrid seed production in plant species having a sporophytic self-incompatibility system is dependent upon the production of inbred lines homozygous for a self-incompatibility allele *S*

Bulk breeding: The growing of genetically diverse populations of self-pollinated crops in a bulk

Bulk generation: In the breeding process, one or more generations in which plants of a genetically diverse population of autogamous crops are grown in a bulk plot with or without mass selection

Bulk population selection: Selection procedure in self-pollinating crops; segregating populations are propagated as bulks until segregation has virtually ceased, at which time selection is initiated

Bulked segregant analysis (BSA)**:** A rapid mapping strategy suitable for monogenic qualitative traits; when DNA of a certain number of plants is bulked into one pool, all alleles must be present; two bulked pools of segregants, differing for one trait, will differ only at the locus harboring that trait

C banding: A cytological staining technique for chromosomes that labels regions around the centromere with Giemsa stain; usually a bandlike and darkly stained structure appears, which consists of heterochromatin; the technique is intensively used in chromosome identification and genome characterization, including structural changes and polymorphisms

C value: The DNA quantity per genome (i.e., per chromosome set); the content of diploids is referred to as the 2C; haploid cells contain the 1C amount of DNA

Callus culture: The *in vitro* culture of callus, often as first stage in the regeneration of whole plants in culture

Callus induction: Undifferentiated plant tissue is produced at wound edges; callus can also be induced and grown *in vitro* by varying the ratio of hormones (eg., auxin and cytokinin) in the growth medium

Callus: Undifferentiated plant cells that are multiplied under tissue culture.

Cancer genes: Mutant alleles of normal genes that lead to cancer

Candidate gene: A gene whose function suggests that it may be involved in the genetic variation observed for a particular trait (phenotype, disease, or condition); candidate genes can be divided into two categories: positional and functional

Candidate population: In forestry, trees, which are planted to serve as a base for forward selection or in some way selected from that for further studies; for example, selected phenotypic selections may serve as a candidate population, which is subject to further progeny testing before reselection to the breeding population; the genotypes taken into consideration for the breeding population

Carcinogen: Physical or chemical agent which induces cancer.

Carrier: (heterozygote) An individual who carries one copy of a recessive gene.

Cartagena Protocol on Biosafety: The first international treaty dealing with the movement of genetically modified organisms (GMOs) across country borders; the protocol was drawn up under the Convention of Biological Diversity and came into force in September 2003; more than 100 countries have ratified the agreement; although the biosafety protocol was pushed for by the South and drafted as a promise of legal protection against the introduction of GMOs, the weakness of its provisions means that the protocol and the national biosafety laws following its introduction are being steadily turned into tools to facilitate the introduction of GMOs

cDNA libraries: Libraries which store sequences copied into DNA from RNA transcripts; typically these sequences carry only the exon information for making proteins.

cDNA: A duplex DNA where one strand is identical in sequence (except for T in place of U) and one is complementary to a particular RNA.

Cell cycle: The sequence of events that occurs between the formation of a cell and its division into daughter cells; it is conventionally divided into G_0, G_1, (G standing for gap), S (synthesis phase during which the DNA is replicated), G_2, and M (mitosis)

Cell division: The reproduction of a cell by karyogenesis and cytogenesis

Cell envelope: The different surface components of the cell that are present outside the cytoplasmic membrane

Cell fusion: Fusion of two previously separate cells, occurs naturally in fertilization; it can be induced artificially by the use of fusogens such as polyethylene glycol; fusion may be restricted to cytoplasm, nuclei may fuse as well; a cell formed by the fusion of dissimilar cells is referred to as a heterokaryon

Cell generation time: The time span between consecutive divisions of a cell cycle

Cell hybridisation: The fusion of a somatic cell *in vitro* and formation of viable cell hybrids

Cell wall: The steadfast external coat that surrounds the cell

Cell: The basic structural and functional unit of a plant; it is a system surrounded by membranes and is compartmentalised into specific functional areas and/or organelles with special tasks

Center of diversity: A geographical location or local region where a particular taxon exhibits greater genetic diversity than it does anywhere else; N. I. VAVILOV developed this concept; he considered that the centers of diversity are also the centers of origin of a crop species

Center of origin: An area from which a given taxonomic group of plants has originated and spread and/or where wild-type species are found in greatest genetic variation; the theory was first published by V. I. VAVILOV in 1922

Centimorgan (cM): A unit of measure of recombination frequency. One cM is equal to 1% chance that a marker at one genetic locus will be separated from a marker at a second locus due to crossing-over in a single generation.

Central dogma: Refers to F. CRICK's seminal concept that in nature genetic information generally flows from DNA to RNA to protein

Centric constriction: The visible bight along a chromosome that bears the centromere

Centric fission: A chromosomal structural change that results in two acrocentric or telocentric chromosomes from one metacentric chromosome

Centric fusion: The whole-arm fusion of chromosomes by the joining together of two telocentric chromosomes to form one chromosome

Centric: Chromosomes having a centromere as opposed to acentric (having no centromere)

Centriole: In mitosis, this small spherical body forms the center of the astral rays

Centromere interference: An inhibitory influence by the centromere on crossing-over and the distribution of chiasmata in its vicinity

Centromere misdivision: A transverse instead of lengthwise division of the centromere resulting in telocentric chromosomes

Centromere orientation: The process of orientation of centromeres during prometaphase of mitosis and meiosis that contributes to a proper segregation of chromatids or chromosomes during anaphase

Centromere: The structure to which the two halves of a chromosome, the chromatids, are joined; the centromere is generally flanked by repetitive DNA sequences and it is late to replicate

Certified seed: Seed produced under an officially designated system of maintaining the genetic identity of, and provisions for, seed multiplication and distribution of crop varieties; is the progeny of breeder, select, foundation, or registered seed; it is grown in compliance with regulations determining standards of germination, freedom from diseases and weeds, and trueness to type

Chemical mutagen: A chemical capable of causing genetic mutation

Chemical-hybridizing agents (CHA): Compounds applied to plants prior to anthesis to selectively induce male sterility

Chiasma interference: The occurrence, less frequent or more frequent than expected by chance, of two or more crossing-over and chiasmata in a given segment of a chromosomal pairing configuration and/or chromosome

Chiasma terminalisation: The progressive shift of chiasmata along the arms of paired chromosomes from their points of origin toward terminal positions

Chiasma: A cross-shaped structure forming the points of contact between nonsister chromatids of homologous chromosomes, first seen in the diplotene stage of meiotic prophase I

Chimera: A tissue containing two or more genetically distinct cell types, or an individual composed of such tissue; chimeral plants may originate by grafting, spontaneous mutation, induced mutation, sorting-out from variegated seedlings, mixed callus cultures, or protoplast fusion;

Chi-squared test ($\chi 2$ test): A significance test used to statistically assess the goodness of fit of observed data to a prediction

Chromatid aberration: Chromosomal changes produced in one chromatid as a consequence of

Chromatid bridge: A bridgelike structure caused by a dicentric chromatid with the two centromeres passing to opposite poles during anaphase; the frequency of chromatid bridges in AII of meiosis is sometimes used as a measure for the level of cytological disturbances (E.g. in induced autopolyploids and allopolyploids)

Chromatid segregation: Segregation of two sister-chromatid segments of a chromosome

Chromatid: The two identical halves of a chromosome produced for cell division and meiosis.

Chromatin remodeling: Refers to a reshaping (at molecular scale) of chromatin, that alters specific genes so that DNA subsequently gets expressed; it can be caused by short interfering RNA (siRNA) or certain transcription activators

Chromatography: A technique used for separating and identifying the components from mixtures of molecules having similar chemical and physical properties; molecules are dissolved in an organic solvent miscible in water, and the solution is allowed to migrate through a stationary phase; since the molecules migrate at slightly different rates they are eventually separated

Chromocenter: A central aggregation of heterochromatic chromosomal elements of the cell nucleus; the euchromatic chromosome arms extend from the chromocenter

Chromosomal aberration: An abnormal chromosomal complement resulting from the loss, duplication, or rearrangement of genetic material

Chromosome banding: The experimental production of differentially stained regions because of the distribution of different chromatin constituents along a chromosome

Chromosome doubling: Induced or spontaneous doubling of chromosome sets leading to rediploidization or to polyploids

Chromosome elimination: The loss of chromosomes from nuclei during certain mitotic or meiotic stages; it is common in several artificial autopolyploids and allopolyploids

Chromosome engineering: Manipulation of whole chromosome sets, individual chromosomes, or even chromosome segments, by different means, for scientific analysis or improvement of performance of crop plants

Chromosome jumping: A technique that allows two segments of duplex DNA that are separated by thousands of base pairs (~200 kb) to be cloned together; after subcloning, each segment can be used as a probe to identify cloned DNA sequences that, at the chromosome level, are roughly 200 kb apart

Chromosome map: A map showing the location of genes on a chromosome, deduced from genetic recombination and cytological experiments

Chromosome number: The specific somatic chromosome number (2n) of a given species or a crop derivative of it

Chromosome painting: Fluorescent *in situ* hybridisation of a specifically labelled DNA probe or probes that hybridizes to the entire chromosome of the probe's origin; using different fluorescent dyes and probes, a pattern of multicolored chromosomes or chromosome segments appear

Chromosome pairing: The highly specific side-by-side association of homologous chromosomes during meiotic prophase

Chromosome reduplication: The synthesis of all compounds that result in an identical copy of the original chromosome

Chromosome segregation: The separation of the members of a pair of homologous chromosomes in a manner that only one member is present in any postmeiotic nucleus

Chromosome staining: The pretreatment and treatment of chromosomes with different dyes in order to make them more suitable for chromosome counting or specific analyses

Chromosome substitution: The replacement of one or more chromosomes by others from another source by spontaneous events or a crossing scheme

Chromosome theory of inheritance: States that the chromosomes, as the carriers of genetic information, represent the material basis of nuclear inheritance

Chromosome: The self-replicating genetic structure of cells, containing genes, which determines inheritance of traits. Chemically, each chromosome is composed of proteins and a long molecule of DNA.

Chromosome-walking technique: A procedure that is used for the determination of a gene on a particular DNA clone of a DNA library; the total DNA of a chromosome has to be available as a series of overlapping DNA fragments; such fragments are produced either by DNA shearing or by cleavage using restriction enzymes; the fragments are used for series of hybridizations; it starts with a cloned gene, which is already identified on the same chromosome; this known gene serves as a probe for detection of clones (fragments) that contain neighboring DNA sequences;

Cis configuration: Two sites on the same molecule of DNA.

Cis-acting locus: Locus that affects the activity only of DNA sequences on the same molecule of DNA; usually implies that the locus does not code for protein.

Cleaved amplified polymorphic sequences (CAPS): PCR-amplified DNA (STS, EST, or SCAR products) that is digested with restriction endonucleases to reveal polymorphisms in restriction sites

Cleistogamous: Designating a self-pollinated plant that produces inconspicuous flowers that never open (Eg. wheat or tomato)

Clone: A genetic replica of an organism created without sexual reproduction.

Cloning vector: DNA molecule originating from a virus, a plasmid, or the cell of a higher organism into which another DNA fragment of appropriate size can be integrated without loss of the vector's capacity for self-replication; vectors introduce foreign DNA into host cells, where it can be reproduced in large quantities. Examples are plasmids, cosmids, and yeast artificial chromosomes; vectors are often recombinant molecules containing DNA sequences from several sources.

Cloning: The process of asexually producing a group of cells (clones), all genetically identical, from a single ancestor. In recombinant DNA technology, the use of DNA manipulation procedures to produce multiple copies of a single gene or segment of DNA is referred to as cloning DNA.

Cluster analysis: A technique of statistical analysis in which similar variances are grouped or clustered; the results of statistical calculations are often shown as dendrograms; particularly, in cross breeding the cluster analysis is used in order to select most diverse parents for crossing

Coding sequence: The part of a gene that determines the sequence of amino acids of a protein, as opposed to noncoding sequences, such as promoter, operator, intron, or terminator regions

Coding strand: The strand of duplex DNA that is transcribed into a complementary Mrna molecule

Codominance: The expression of both alleles in the heterozygote with equal effect on the phenotype, as opposed to recessive and dominant

Codon bias: Although several codons code for a single amino acid, a plant may have a preferred codon for each amino acid; this is called codon bias

Codon: A triplet of nucleotides that represents an amino acid or a termination (STOP) signal.

Coefficient of correlation: A number that measures the linear dependence between two random variables; limiting values of −1 and +1 indicate perfect negative and perfect positive correlation, respectively; a correlation of zero suggests a complete lack of association between the two variables

Coefficient of variation: The standard variation expressed as percentage of the mean

Coenocytes: An organism or a portion thereof that is multinucleate; the nuclei are not each separate in one cell, such as in some protoplast or cell fusion products

Colchicine: A poisonous alkaloid drug ($C_{22}H_{25}NO_6$) that is obtained from meadow saffron *(Colchicum autumnale)*; it has a disruptive effect on microtubular activity; thus, it affects tissue metabolism generally, and mitosis and meiosis in particular; it is used for induction of polyploidy

Colchiploidy: Polyploidy induced by application of colchicines

Combination breeding: A breeding method that utilises the genetic diversity of individuals or varieties in order to create and to select new phenotypes on the basis of genetic recombination of useful characters of parental materialcombined in the hybrid

Combining ability: Capacity or ability of a genotype to transmit superior performance to its crosses

Comparative (gene) mapping: Localisation (mapping) of a common set of DNA probes onto linkage maps of different species; the results of these comparisons indicate substantial conservation of blocks of genes and even large segments of chromosomes between species; this approach shows synteny of markers among related species or genera; such maps have been established for cereals (wheat, maize, oats, rye, rice, sorghum, millet) and Solanaceae, such as potato, tomato, and paprika

Comparative genomic hybridisation (CGH): A molecular cytogenetic technique that allows detection of DNA sequence copy number changes throughout the genome in a single hybridisation

Comparative positional candidate gene: A gene that is likely to be located in the same region as a DNA marker that has been shown to be linked to a single-locus trait or to a quantitative trait locus (QTL), where the gene's likely location in the genome of the species in question is based on its known location in the map of another species (i.e. is based on the comparative map between the two species)

Compensating trisomic: An aneuploid individual with an extra chromosome in which a missing standard chromosome is compensated for by two novel interchange chromosomes; the two novel chromosomes carry two different arms of the missing chromosome

Complementarity: The chemical affinity between specific nitrogenous bases as a result of their hydrogen bonding properties. The property of two nucleic acid chains having base sequences such that an antiparallel duplex can form where the adenines and thymines (or uracils) are opposed to each other, and the guanines and cytosines are opposed to each other.

Complementary DNA (cDNA): DNA that is synthesised from a messenger RNA template; the single-stranded form is often used as a probe in physical mapping.

Complementary homopolymeric tailing: The process of adding complementary nucleotide extensions to different DNA molecules, for example, dG (deoxyguanosine) to the 3′-hydroxyl ends of one DNA molecule and dC (deoxycytidine) to the 3′-hydroxyl ends of another DNA molecule to facilitate, after mixing, the joining of the two DNA molecules by base pairing between the complementary extensions

Complementary sequences: Nucleic acid base sequences that can form a double-stranded structure by matching base pairs; the complementary sequence to G-T-A-C is C-A-T-G.

Complementation test: A genetic test to determine if two mutations with similar phenotypes are allelic (occur in the same gene) or are nonallelic (occur in separate genes); individuals homozygous for each mutation are crossed; if the mutations are allelic, the offspring will have the mutant phenotype; if they are nonallelic the offspring will be heterozygous for both alleles and will have the wildtype phenotype

Complementation: The production of a wild-type phenotype when two different mutations are combined in a diploid or a heterokaryon.

Complete diallel: A mating design and subsequent progeny test resulting from the crossing of a certain number of parents in all possible combinations including selfs and reciprocals; because of severe inbreeding depression in the selfs, these are often skipped, nevertheless the test is still called a full diallel

Complete penetrance: The situation in which a dominant gene always produces a phenotypic effect or a recessive gene in the homozygous state always produces a detectable effect

Completely randomized design: The structure of the experiment in a completely randomized design is assumed to be such that the treatments are allocated to the experimental units completely at random

Composite cross: A population derived from the hybridization of several parents, either by handpollination or by the use of male sterility

Composite transposon:: A transposable element formed when two identical or nearly identical transposons insert on either side of a nontransposable segment of DNA, such as the bacterial transposon Tn5

Composite variety: A plant population in which at least 70% of progeny result from cross of the parent lines

Composite: A plant of the immense family Compositae, regarded as comprising the most highly developed flowering plants *bot*; a mixture of genotypes from several sources, maintained by normal pollination

Composite-cross population: A population generated by hybridizing more than two varieties and/ or lines of normally self-fertilizing plants and propagatings successive generations of the segregating population in bulk in specific environments so that natural selection is the principal force acting to produce genetic change; artificial selection may also be imposed; the resulting population is expected to have a continuously changing genetic makeup

Compound chromosome: Formed by the union of two separate chromosomes

Compound cross: A combination of desirable genes from more than two inbred lines, breeding strains, or varieties

Conditional lethal genes: Genes that are expressed in response to a specific environmental stimulus such as a specific chemical; when expressed, they are lethal to the plant

Conditional mutation: A mutation that has the wild-type phenotype under certain environmental conditions (temperature, age, nutrition) and a mutant phenotype under other conditions

Conjugation: A process whereby organisms of identical species, but opposite mating types, pair and exchange genetic material (DNA); in molecular biology, natural process of DNA transfer between bacteria in which the DNA is never exposed

Consanguineous matings: Matings between two individuals who share a common ancestor in the preceding two or three generations

Consensus sequence: If a particular nucleotide sequence is always found with only minor variations, then the usual form of that sequence is called consensus sequence; the term is also used for genes that encode the same protein in different organisms

Conserved sequence: A base sequence in a DNA molecule (or an amino acid sequence in a protein) that has remained essentially unchanged throughout evolution.

Constitutive heterochromatin: The material basis of chromosomes or segments that exhibit heterochromatic properties under most conditions (e.g., centromeric or telomeric heterochromatin)

Constitutive mutation: Causes genes that usually are regulated to be expressed without regulation contain their own DNA which is separated from the cell's nuclear DNA; in transplastomic plants, the DNA in the chloroplasts has been genetically engineered; since only the nuclear DNA is inherited, the genetic modification in the chloroplasts will not be passed on to the next generation; therefore, transplastomic plants may be a solution for ensuring biocontainment

Contig map: A map depicting the relative order of a linked library of small overlapping clones representing a complete chromosomal segment.

Contig: A set of overlapping clones that provide a physical map of a portion of a chromosome

Contiguous map: The alignment of sequence data from large, adjacent regions of the genome to produce a continuous nucleotide sequence across a chromosomal region

Continuous culture: An *in vitro* suspension culture continuously supplied with nutrients by the inflow of fresh medium; the culture volume is normally constant

Conventional breeding: The selection process used to try and improve the characteristics of a plant; genetically superior parents are selected for breeding in a cyclic process until the desired trait is received; the goal of breeding is to produce genetically improved populations

Convergence breeding: A breeding method involving the reciprocal addition to each of two inbred lines of the dominant favorable genes lacking in one line and present in the other; backcrossing and selection are performed in parallel, each of the original lines serving as the recurrent parent in one series

Convergence-divergence selection: A breeding scheme in which selection of promising genotypes is made in a bulk population at different locations followed by massing of selection and allowing mating among them in a pollination field; the harvested bulk seeds constitute the basis for the next propagation cycle

Copy-choice hypothesis: The interpretation of intrachromosomal genetic recombination that is not regarded as a physical exchange of preformed genetic strands

Core collection: The basic sample of a germplasm collection; it is designed to represent the wide range of diversity in terms of morphology, geographic range, or genes

Corepressor: A metabolite that in conjugation with a repressor molecule binds to the operator gene present in an operon and prevents the synthesis of a repressible enzyme

Correlation: The degree to which statistical variables vary together; measured by the correlation coefficient, which has a value from zero (no correlation) to –1 or +1 (perfect negative or positive correlation)

Cosmid: Artificially constructed cloning vector containing the cos gene of phage lambda. Cosmids can be packaged in lambda phage particles for infection into E. coli; this permits cloning of larger DNA fragments (up to 45 kb) than can be introduced into bacterial hosts in plasmid vectors.

Cosuppression: Silencing of a gene by addition of transgenic DNA copies or infection by a virus; this term, which can refer to silencing at the post transcriptional (PTGS) or transcriptional

Covariance analysis: An analysis of the mean of the product of the deviation of two variates from their individual means; it measures the interrelationship between variables

CpG island: Repetitive CpG doublets creating a region of DNA greater than 200 bp in length with a G+C content of more than 0.5 and an observed/expected presence of CpG more than 0.6; usually associated with transcription-initiation regions of (housekeeping) genes transcribed at low rates that do not contain a TATA box; the CpG-rich stretch of 20–50 nucleotides occurs within the first 100–200 bases upstream of the start site region; a transacting transcription factor called SP1 recognizes the CpG islands

Criss-cross inheritance: The transmission of a gene from mother to son or father to daughter

Cross: The deliberate mating of selected parents based on particular genetic traits desired in the offspring.

Cross-fertilisation: The fusion of male and female gametes from different genotypes or individuals of the same species, as base of genetic recombination

Cross-hybridisation: In biotechnology, the hydrogen bonding of a single-stranded DNA sequence that is partially but not entirely complementary to a single-stranded substrate; often, this involves hybridising a DNA probe for a specific DNA sequence to the homologous sequences of different species

Crossing-over: The reciprocal exchange of material between homologous chromosomes during meiosis, which is responsible for genetic recombination. The process involves the natural breaking of chromosomes, the exchange of chromosome pieces, and the reuniting of DNA molecules

Cross-pollination: The transfer of pollen from the stamen of a flower to the stigma of a flower of a different genotype but usually of the same species, with subsequent growth of the pollen tube

Cross-protection: Plant protection conferred on a host by infection with one strain of, for example, a virus that prevents infection by a closely related strain

Cry1A protein: Derived from the bacterium *Bacillus thuringiensis* that is toxic to some insects when ingested; this bacterium occurs widely in nature and has been used for decades as an insecticide although it constitutes less than 2% of the overall insecticides used

Crystallography: The study of, for example, a protein structure by crystallising the protein and examining the crystals using X-rays; the diffraction angles of the X-rays are used to compute the relative positions of components of the protein, and thus its structure

Cybrid: The hybrid formed from the fusion of a cytoplast and a whole cell; the cytoplast may transmit cytoplasmic components independently of the cell genome

Cytogamy: The fusion or conjugation of cells

Cytogenetic breeding: Breeding techniques that involve the manipulation or alteration of genetic material in cells (E.g. exposure to radiation or application of chemicals such as colchicine)

Cytogenetic map: A map showing the locations of genes on a chromosome (i.e. the visual appearance of a chromosome when stained and examined under a microscope)

Cytogenetics: Scientific discipline that combines cytology with genetics; it includes (a) the organisation of chromosomes in the nucleus with the requirements of gene expression in different developmental and environmental contexts; (b) the behaviour of chromosomes with transmission of phenotypes to progeny; (c) changes in chromosomal structure and number with speciation; and (d) the evolution of the genome with the evolution of the species

Cytokinesis: During the division of a cell, the division of the constituents of the cytoplasm; it usually begins in early telophase with the formation of a cell plate, which is assembled within the

Cytoplasm: The part of a cell that is enclosed by the plasma membrane, but excluding the nucleus

Cytoplasmic inheritance: A non-Mendelian (extra-chromosomal) inheritance *via* genes in cytoplasmic organelles (mitochondria, plastids)

Cytoplasmic male sterility (CMS): Pollen abortion due to cytoplasmic factors, which are maternally transmitted, but which act only in the absence of pollen-restoring genes; this type of sterility can also be transmitted by grafting

Cytosine (C): A nitrogenous base, one member of the base pair G-C (guanine and cytosine).

Degenerate code: A genetic code in which some amino acids may be encoded by more than one codon each.

Degree of freedom (DF): The number of items of data that are free to vary independently; in a set of quantitative data, for a specified value of the mean, only

($n-1$) items are free to vary, since the value of the nth item is then determined by the values assumed by the others and by the mean

Deletion mapping: The use of overlapping deletions to localise the position of an unknown gene on a chromosome or linkage map

Deletion mutation: A mutation in which one or more bases are removed from the DNA sequence of a gene

Deletion: The loss of a chromosomal segment from a chromosome set; the size may vary from a single nucleotide to sections containing several genes

Denaturation: The separation of the two strands of a DNA double helix, or the severe disruption of the structure of any complex molecule without breaking the major bonds of its chains.

Density gradient centrifugation: The separation of macromolecules or subcellular particles by sedimentation through a gradient of increasing density under the influence of a centrifugal force; the density gradient may either be formed before the centrifugation run by mixing two solutions of different density (Eg. in sucrose density gradients), or it can be formed by the process of centrifugation itself (eg. in $CsCl_2$ and Cs_2SO_4 density gradients)

derived from a common ancestor

Diallel cross: The crossing in all possible combinations of a series of genotypes

Dichogamy: The condition in which male and female parts of a flower mature at different times

Differential selection: The difference between a selected plant, family, or clone, and the average of the population from which it is taken

Dihaploid: A haploid cell or individual containing two haploid chromosome sets - not to be confused with doubled-haploid

Dihybrid: A cross between individuals that differ with respect to two specified gene pairs

Dioecious: Possessing male and female flowers or other reproductive organs on separate, unisexual, individual plants (eg. in hemp, hops, asparagus, or spinach)

diploid number

Diploid parthenogenesis: A type of gametophyte apomixis by which a diploid embryo sac cell results in a diploid embryo; the diploid condition is the result of cytogenetic mechanisms occurring in the egg stage

Diploid: A cell with two chromosome sets or an individual with two chromosome sets in each cell; a diploid state is written as "2n" to distinguish it from the haploid state "n"

Diploidisation: In polyploids, a natural or induced mechanism in which the chromosomes pair completely or partially as bivalents, although polyploid sets of chromosomes are present; it may be caused by a structural differentiation of homologous chromosome sets or by genetic control; for example, in bread wheat three homoeologous genomes are available (AABBDD); they do not pair as hexavalents but exclusively as bivalents

Directional dominance: A type of dominance in which the majority of dominant alleles have positive effects in one direction

Directional mutation: A genetic change that favours a certain genotype or population

Directional selection: Selection resulting in a shift in the population mean in the direction desired by the breeder

Dirty crop approach: Partially resistant multilines where each component of the multiline carries a single gene for resistance but none of the resistances are effective against all known races of the pathogen (Marshall 1977); it has been advocated as a dynamic, natural biological system for the control of crowd diseases in crop plants; multilines could stabilise the race structure of the pathogen population

Disassortative mating: Occurs if the plants mating resemble each other less than plants belonging to pairs of random plants, with regard to some trait

Disease avoidance: Avoiding the disease by, for example, growing the crop sufficiently early so that the vulnerable part of the plant's growing cycle is over before the disease-causing organism arrives in the area; this stops the disease from starting; it is sometimes called "passive resistance"

Disease control: Several types of disease control can be classified: (1) disease resistance, (2) protection, (3) avoidance, (4) exclusion, (5) eradication, (6) therapy

Disease cycle: A cyclical sequence of host and parasite development and interaction that result in disease and in reproduction of the pathogen

Disease escape: For a variety of reasons, some individuals in a screening population may remain free of pests or disease; it also known as chance escape; this phenomenon can be very misleading because it is so easily confused with resistance

Disease resistance: The ability to resist disease or the agent of disease and to remain healthy

Disease triangle: The three conditions required for a disease to occur, namely: a susceptible host, a suitable pathogen, and an appropriate environment

Disease: A condition in which the use or structure of any part of a living organism is not normal; harmful deviation from normal functioning of physiological processes; six types of causal agents can be considered: (1) fungi, (2) bacteria, (3) viruses, (4) nematodes, (5) insects, and (6) plant parasites

Disruptive selection: A selection that changes the frequency of alleles in a divergent manner, leading to the fixation of alternative alleles in members of the population; the result after several generations of selection should be two divergent phenotypic extremes within the population

Distant hybridisation: The crossing and/or hybridization of members of different genera

Diversifying selection: Selection in which two or more genotypes show optimal adaptation under different environments

Diversity arrays technology (DArT) marker: A DArT marker is a segment of genomic DNA, the presence of which is polymorphic in a defined genomic representation; DArT markers are diallelic and behave in a dominant (present versus absent) or codominant (two doses versus one dose versus absent) manner; to identify the polymorphic markers, a complexity reduction method is applied on the metagenome, a pool of genomes representing the germplasm of interest; the genomic representation obtained from this pool is then cloned and individual inserts are arrayed on a microarray resulting in a "discovery array"; labelled genomic representations prepared from the individual genomes included in the pool are hybridised to the discovery array

Dizygotic: Two cells having been fertilised at the same time (often resulting in twins.)

DNA (deoxyribonucleic acid): The molecule that encodes genetic information. DNA is a double-stranded molecule held together by weak bonds between base pairs of nucleotides. the four nucleotides in DNA contain the bases stranded molecule held together by weak bonds between base pairs of nucleotides. The four nucleotides in DNA contain the bases: adenine (A), guanine (G), cytosine (C), and thymine (T). In nature, base pairs form only between A and T and between G and C; thus the base sequence of each single strand can be deduced from that of its partner.

DNA fingerprint(ing) (DFP**):** The unique pattern of DNA fragments identified by SOUTHERN hybridisation (using a probe that binds to a polymorphic region of DNA) or by polymerase chain reaction (PCR) using primers flanking the polymorphic region

DNA hybridisation: Base pairing of DNA from two different sources; in biotechnology, a technique for selectively binding specific segments of single-stranded (ss) DNA or RNA by base pairing to complementary sequences of ssDNA molecules that are trapped on a nitrocellulose membrane

DNA ligase (polynucleotide ligase)**:** An enzyme that creates a phosphodiester bond between the 5′-PO_4 end of one polynucleotide and the 3′-OH end of another, thereby producing a single, larger polynucleotide

DNA marker: A DNA sequence that exists in two or more forms that can be used to genotype individual organisms.

DNA micro arrays: A powerful, versatile and economical molecular technique for screening of genetic aberrations; high-density gene sequences are printed onto glass slides; fluorephorelabeled genomic or complimentary DNA (cDNA) is hybridized to slides with fixed signature patterns and resolved using computer driven fluorescent images

DNA polymerase: A group of enzymes mainly involved in copying a single-stranded DNA molecule to make its complementary strand

DNA replication: The process whereby a copy of a DNA molecule is made, and thus the genetic

DNA replication: The use of existing DNA as a template for the synthesis of new DNA strands. In humans and other eukaryotes, replication occurs in the cell nucleus.

DNA sequence: The relative order of base pairs, whether in a fragment of DNA, a gene, a chromosome, or an entire genome. See base sequence analysis.

DNA sequencing: The process of determining the order of DNA bases along a DNA strand.

DNA-amplified fingerprinting: A technology based on amplification of random genomic DNA sequences achieved by a single short (5–8 bases) oligonucleotide primer of arbitrary sequence; it produces a characteristic spectrum of short DNA pieces of varying complexity that are resolved on polyacrylamide gel (PAGE) following silver staining; it is used to detect genetic differences between genotypes as well as for detecting polymorphism even between organisms that are closely related, such as near isogenic lines

DNA-DNA hybridization: The annealing of two complementary DNA strands to produce hybrid

Domain of a protein: A discrete continuous part of the amino acid sequence that can be equated with a particular function.

Domain: A discrete portion of a protein with its own function. The combination of domains in a single protein determines its overall function.

Domestication: The process by which plants are genetically modified by selection over time by humans for traits that are more desirable or advantageous for humans

Dominance hypothesis: In hybrid breeding, dominant alleles of genes that should have stimulating effects on heterosis, while recessive ones should show inhibitory effects

Dominance variance: That portion of the genetic variance attributable to dominant gene effects

Dominance: The expression of a trait in the heterozygous condition.

Dominant gene: A gene whose phenotype is expressed when the allele is present as either one or two copies of the pair (Eg. AA or Aa).

Dominant-negative mutation: A (heterozygous) dominant mutation on one allele blocking the activity of wild-type protein still encoded by the normal allele causing a loss-of-function phenotype; the phenotype is indistinguishable from that of homozygous dominant mutation

Dosage compensation: A genetic process that compensates for genes that exist in two doses in the homozygous dominants, so that the heterozygotes produce the same amount of gene product as the homozygotes

Dosage effect: The influence upon a phenotype of the number of times a genetic element is present

Double cross: A cross between two F_1 hybrids; the method used for producing hybrid seed; four different lines (A, B, C, D) are used; A × B = AB hybrid and C × D = CD hybrid; the singlecross hybrids (AB and CD) are then crossed and the double-cross hybrid (ABCD) seed is used for the commercial crop

Double crossing-over: The situation in which two crossing-overs take place within a tetrad

Double fertilisation: The union of one sperm nucleus with the egg nucleus to form the diploid zygote and of the other sperm nucleus with the two polar nuclei to form a triploid endosperm nucleus

Double helix: The shape that two linear strands of DNA assume when bonded together.

Double monoisosomic: The presence of two isochromosomes, one for each arm; it can derive from a double monotelosomic individual

Double monotelosomic: The presence of two telocentrics, one for each arm

Double reduction: The genetic outcome of chromatid segregation, as opposed to chromosome segregation, whereby two sister-chromatid segments are included in the same meiotic product

Double-cross hybrids: Hybrids resulting from crossing two single cross hybrids

Doubled haploid: A diploid plant, which results from spontaneous or induced chromosome doubling of a haploid cell or plant, usually after anther or microspore culture by using different means

Down promoter mutations: Mutations that decrease the frequency of initiation of transcription; they lead to the production of less mRNA than is the case in the nonmutated state

Drought hardening: Adapting plants to survive periods of time with little or no water by stepwise reducing water supply or germinating and/or growing under insufficient moisture conditions

Drought resistance: the ability of a plant to withstand drought; this property can be very valuable in areas of uncertain rainfall, Eg. sorghum has greater drought resistance than maize, and is grown in many semi-arid areas for this reason

Duplication: Duplication of a sequence of DNA or section of chromosome.

Early generation test: Selection schemes in which poor recombinants are discarded already in F_2 and F_3 generations

Early selection: It is conducted on highly heterozygous individuals which become the parents of the next screening generation

Ear-to-row selection: A separate growing of progenies (i.e. the separate sowing of lines or families

EcoTILLING: A high throughput, low cost technique for rapid discovery of polymorphisms in natural populations of plants; it is similar to TILLING but it differs from TILLING in that natural polymorphisms are detected rather than polymorphisms induced through chemical mutagenesis; on the other hand, single nucleotide polymorphisms (SNPs), small insertions and deletions, and variations in microsatellite repeat number can be efficiently detected using the EcoTILLING technique

Ecovalence: This parameter is a quantitative measure for the evaluation of ecological adaptability

Electrophoresis: A method of separating large molecules (such as DNA fragments or proteins) from a mixture of similar molecules. An electric current is passed through a medium containing the mixture, and each kind of molecule travels through the medium at a different rate, depending on its electrical charge and size. Separation is based on these differences. Agarose and acrylamide gels are the media commonly used for electrophoresis of proteins and nucleic acids.

Emasculation: The removal of anthers from a flower before the pollen is shed. To produce F_1 hybrid seed in a species bearing monoecious flowers, emasculation is necessary to remove any possibility of self-pollination

Embryo culture: A method of inducing the artificial growth of embryos by excising the young embryos under septic conditions and placing them on suitable nutrient media; the method is often applied when postgamous incompatibility exists (Eg. in wide crosses)

Embryo rescue: When cross-pollination is made or occurs between genetically widely different plants, the resulting embryo may be aborted because of parental mutual incompatibility; such embryos may be excised during an early stage and grown on a congenial medium such as nutrient

Endomitosis: A doubling of the chromosomes within a nucleus that does not divide, thus producing a polyploid; the doubling may be repeated a number of times, which leads to endopolyploidy

Endonuclease: An enzyme that cleaves the phosphodiester bond within a nucleotide chain.

Endopolyploid: Diploid individuals whose cells contain 4C, 8C, 16C, 32C, etc., amounts of DNA in their nuclei

Endoreduplication: Chromosome reduplication in the interphase of the mitotic cell cycle

Endosperm: In grasses, the reserve food material in the caryopsis lying outside the embryo; it is a starchy tissue that is formed during grain development; it provides nourishment for the developing embryo and for the seedling after germination until it can establish itself

Enzyme: A protein catalyst which is essential to the correct functioning of biochemical reactions.

Enzyme-linked immunosorbent assay (ELISA): A technique using antibodies for detecting specific proteins. Used to test for the presence of a particular genetically engineered organism.

Epigenesis: The concept that an organism develops by the new appearance of structure and functions, as opposed to the hypothesis that an organism develops by the unfolding and growth

Epigenetic regulation: Regulation of gene activity mediated by a reversible change in DNA modification or chromatin structure

Epigenetic: Heritable variation caused by differences in the chemistry of either the DNA (methylation) or the proteins associated with the DNA (histone acetylation), rather than in the DNA sequence itself

Epigenetics: The study of the mechanisms by which genes bring about their phenotypic effects

Epimutation: A heritable change in gene activity that is not due to a change in base sequence; generally applied to an abnormal change in gene activity (Eg. in a tumor or a cell culture

Epistasis: The nonreciprocal interaction of non-allelic genes; the situation in which one gene makes the expression of another; more specifically, gene interaction in which one gene interferes with the phenotypic expression of another non-allelic gene so that the phenotype is determined effectively by the former; the latter is described as hypostatic

Equational division: The division of each chromosome during the metaphase of mitosis or meiosis into two equal longitudinal halves (i.e. sister chromatids), which are distributed to the two cell poles, and then incorporated into separate daughter nuclei

Error variance: Variance arising from unrecognised or uncontrolled factors in an experiment with which the variance of recognised factors is compared in tests of significance

EST: Expressed sequence tag. See sequence tagged site.

Estimation error: The amount by which an estimate differs from a true value; this error includes the error from all sources (Eg. sampling error and measurement error)

Eucentric: A chromosomal interchange by which the translocated segment does not change the relative position to the centromere

Euchromatin: Chromatin that shows the staining behaviour of the majority of the chromosome complement; it is uncoiled during interphase and condenses during mitosis, reaching a maximum density at metaphase

Euchromatization: The induced or spontaneous change of heterochromatin into euchromatin

Euchromosome: A chromosome showing the typical features of the standard complement of a given species

Eugenic: Favourable to the genetic quality of a population, as opposed to dysgenic

Euhaploid: A haploid genome showing no deviating number of chromosomes compared with the standard genome of the species

Eukaryon: The highly organised nucleus of a eukaryote

Eukaryote: An organism whose cells have a distinct nucleus

Euploid: Having the "correct" chromosome composition. Cells containing only complete sets of chromosomes.

***Ex situ* conservation:** A conservation method that entails the removal of seed, pollen, sperm, or individual organisms from their original habitat, keeping these resources of biodiversity alive outside of their natural environment; the storage of plant genetic resources as seeds in genebanks at subzero temperatures is the most widely applied method

***Ex situ*:** Out of place or not in the original environment (E.g. seeds stored in a genebank)

Exogenous DNA: DNA originating outside an organism.

Exon: Any segment of an interrupted gene that is represented in the mature RNA product. The protein-coding sequences of a gene.

Exonuclease An enzyme that cleaves nucleotides one at a time from an end of a polynucleotide chain.

Experimental design: The planning of a process of data collection; also used to refer to the information necessary to describe the interrelationships within a set of data; it involves considerations, such as number of cases, sampling methods, identification of variables and their scale-types, identification of repeated measures, and replications

Extrachromosomal inheritance: Inheritance that is not controlled by chromosomal determinants but by cytoplasmic components

Facilitated recurrent selection: A type of recurrent selection in which genetic male sterility is maintained in the population to maintain heterozygosity and genetic diversity and to permit the recombination and shifting of gene frequencies

Facultative heterochromatin: Heterochromatin that is present in only one of a pair of homologues or not permanently present

Familial trait: Any trait that is more common in relatives of an affected individual than in the general population; could be due to genetic and/or environmental causes.

Farmer selection (*syn* participatory plant breeding): An aspect of some plant breeding programs in which the farmers make the final selection of cultivar; each farmer is given a different group of new clones or pure lines of a crop, emerging from a breeding program

Farmer's privilege: A clause in the plant breeder's rights legislation of most countries that permits a farmer to use some of their own crop of a registered cultivar for seed on their own farm

Farmer's rights: The recognition of farmers (past, present, and future) as *in situ* agricultural innovators who collectively conserve and develop agricultural genetic resources around the world; as such, farmers are recognised as innovators entitled to intellectual integrity and to compensation whenever their innovations are commercialised

Fertilisation: The union of two gametes to produce a zygote that occurs during sexual reproduction

Field resistance: Synonymous with general resistance; it is under polygenic control (i.e. controlled by many genes with minor individual effects); in general, field resistance is longer lasting than race-specific resistance; field resistance slows down the rate at which disease increases in the field

Field trial: A test of a new technique or variety, including biotech-derived varieties, done outside the laboratory but with specific requirements on location, plot size, methodology, etc.

First division restitution (FDR): Results from an abnormal orientation of the spindles right before meiotic anaphase II; non-sister chromatids end up in the same nucleolus; it was found in several crop species, such as potato, rye, etc.

FISH (fluorescence *in situ* hybridisation): A physical mapping approach that uses fluorescein tags to detect hybridization of probes with metaphase chromosomes and with the less-condensed somatic interphase chromatin.

Fitness: The relative ability of a plant to survive and transmit its genes to the next generation

Flow cytometry: Analysis of biological material by detection of the light-absorbing or fluorescing properties of cells or subcellular fractions (i.e., chromosomes) passing in a narrow stream through a laser beam. An absorbance or fluorescence profile of the sample is produced. Automated sorting devices, used to fractionate samples, sort successive droplets of the analysed stream into different fractions depending on the fluorescence emitted by each droplet.

Flow karyotyping: Use of flow cytometry to analyse and/or separate chromosomes on the basis of their DNA content.

Footprinting: A method used to determine the length of nucleotide chains that are close to a protein (which bind to DNA); for example, certain types of drugs act by binding tightly to certain DNA molecules in specific locations

Foreign DNA: DNA that is not found in the normal genome concerned; usually it is directly or indirectly introduced into a recipient cell by several experimental means

Forward mutation: A mutation that alters (usually inactivates) a wild-type allele of a gene

Forward selection: Choosing good individuals out of a progeny test for possible use in seed orchards and/or subsequent generations of breeding

Foundation seed: Seed stocks increased from breeder seed, i.e. the progeny of breeder, select, or foundation seed; handled as to closely maintain the genetic identity and purity of the variety; it is a sort of certified seed, either directly or through registered seed

Founder effect: Genetic drift due to the founding of a population by a small number of individuals

Founder population: The first generation of a breeding population, e.g., in forest tree breeding often the initial plus trees; this is usually the starting point of calculations

Fragile Site: A site on a chromosome vulnerable to mutation to breakage.

Frame shift mutation: A mutation that is caused by a shift of the reading frame of the mRNA (usually by the insertion of a nucleotide) synthesized from the altered DNA template

Frequency distribution: A specification of the way in which the frequencies of members of a population are distributed according to the values they exhibit

Frequency-dependent selection: Selection where the fitness of a type varies with its frequency (i.e. whereby a genotype is at an advantage when rare and at a disadvantage when common) from application of pollen produced on the same plant

Functional genomics: The science of how the genes in organisms interact to express complex traits - that is, the field of research that aims to determine the function of newly discovered genes; attempts to convert the molecular information represented by DNA into an understanding of gene functions and effects; functional genomics also entails research on the protein function (proteomics) or, even more broadly, the whole metabolism (metabolics) of an organism

G banding: A special staining technique for chromosomes that results in a longitudinal differentiation by Giemsa stain, which is a complex of stains specific for the phosphate groups of DNA; the characteristic bands produced are called G bands; these bands are generally produced in AT-rich heterochromatic regions

Gamete: A specialised haploid cell whose nucleus and often cytoplasm fuses with that of another gamete in the process of fertilisation, thus forming a diploid zygote

Gametic (phase) disequilibrium: In relation to any two loci, the occurrence of gametes with a frequency greater than or less than the product of the frequency of the two relevant alleles

Gametic (phase) equilibrium: In relation to any two loci, the occurrence of gametes with a frequency equal to the product of the frequency of the two relevant alleles, Eg. loci *A* and *B* are in linkage equilibrium if the frequency of the gamete *A*n*B*n equals the product of the frequencies of alleles *A*n and *B*n

Gametic selection: The influences acting to cause differential reproductive success of one allele over another in a heterozygote

Gametocide: A chemical agent used to selectively kill either male or female gametes; it is used in hybrid seed production of autogamous crops (Eg. barley or wheat)

Gametoclonal variation: Variation among regenerants obtained from pollen and/or anther culture

Gametophytic self-incompatibility: Self-incompatibility is based on the genotypic and phenotypic relationship between the female and male reproductive system; alleles in cells of the pistil determine its receptivity to pollen; the phenotype of the pollen, expressed as its inability to effect fertilization, may be determined by its own alleles, referred to as gametophytic incompatibility

Gene (DNA) sequencing: Determining the exact sequence of nucleotide bases in a strand of DNA to better understand the behavior of a gene.

Gene amplification: The more or less specific production of multiple copies of a gene

Gene cloning: Insertion of a DNA fragment carrying a gene into a cloning vector; subsequent propagation of the recombinant DNA molecule in a host organism results in many identical copies of the gene (clones) in a form that is more easily accessible than the original chromosomal copy

Gene discovery: A process whereby the individual base nucleotides in an organism's DNA are identified in order to learn more about the genome of the organism, and to learn about or identify specific areas of interest on chromosomes. This process allows for the identification of specific genes that code for a particular trait or traits.

Gene dosage: The number of times a given gene is present in the nucleus of a cell and/or individual

Gene duplication: A process in evolution in which a gene is copied twice; the two copies lie side by side along the same chromosome

Gene expression profiling: The analysis of gene expression level of many genes at the same time; technologies that can be used to obtain this information include Northern blot, DNA chips, high-density array spotted on glass or membranes, and quantitative techniques

Gene expression: The phenotypic manifestation of a gene depending on the different levels of gene activation, or the process by which the information in a gene is used to produce a protein; in molecular genetics, the full use of the information in a gene *via* transcription and translation leading to production of a protein and hence the appearance of the phenotype determined by that gene; gene expression is assumed to be controlled at various points in the sequence leading to protein synthesis; this control is thought to be the major determinant of cellular differentiation

Gene families: Groups of closely related genes that make similar products.

Gene flow: The movement of genes from one individual or population to another genetically compatible individual or population.

Gene frequency: The number of loci at which a particular allele is found divided by the total number of loci at which it could occur for a given population, expressed as a proportion or percentage

Gene insertion: The addition of one or more copies of a normal gene into a defective chromosome

Gene interaction: Modification of gene action by a nonallelic gene or genes, generally the interaction between products of nonallelic genes

Gene library: In molecular genetics, a random collection of cloned DNA fragments in a number of vectors that ideally includes all genetic information of that species

Gene map: A graphic presentation of the linear arrangement of a chromosome or segment; it shows the relative distance between loci gained in linkage experiments

Gene mapping: Determining the relative physical locations of genes on a chromosome. Useful for plant and animal breeding.

Gene mutation: A heritable change of gene revealed by phenotypic modifications

Gene pool system: it consists of three informal categories in order to provide a genetic perspective and focus for cultivated plants

Gene pool: The reservoir of different genes of a certain plant species or lower and higher taxa available for crossing and selection; it may be differentiated between (1) primary gene pools (consists of those species that readily hybridize, produce viable hybrids, and have chromosomes that may freely recombine), (2) secondary gene pools (consists of those species with a certain degree of hybridization barrier due to ploidy differences, chromosome alterations, or incompatibility genes), and (3) tertiary gene pools (consists of distinct species or higher taxa with strong crossing barriers); in general, the total number of genes or the amount of genetic information that is possessed by all the reproductive members of a population of sexually reproducing organisms

Gene product: The biochemical material, either RNA or protein, resulting from expression of a gene. The amount of gene product is used to measure how active a gene is; abnormal amounts can be correlated with disease-causing alleles.

Gene silencing: The suppression of gene expression, for example, of the gene for polygalacturonase which causes fruit to ripen, or the gene for P34 protein in soybeans, *via* a variety of methods, for example, *via* RNA interference (RNAi), chemical genetics, effect of certain viruses, zinc finger proteins, or sense or antisense genes

Gene splicing: The enzymatic attachment (joining) of one gene or part of a gene to another; also removal of introns and splicing of exons during mRNA synthesis

Gene substitution: The replacement of one allele by another mutant allele in a population by natural or directed selection

Gene tagging: The labeling of a gene by a marker gene or specific DNA sequence closely linked

Gene therapy: Insertion of normal DNA directly into cells to correct a genetic defect.

Gene transfer: The physical transfer of a gene by crossing, chromosomal manipulation, and molecular means; in biotechnology, different methods are described, such as (1) microinjection, (2) insertion *via* microprojectiles (particle gun, particle bombardment) using silicon fibers as carriers of the DNA, (3) direct transfer, (4) electroporation, (5) liposome fusion, or (6) a vector-mediated transfer

Gene translocation: The transfer or movement of a gene or gene fragment from one chromosomal location to another; often it alters or abolishes expression

Gene: The fundamental physical and functional unit of heredity, which carries information from one generation to the next; a segment of DNA, composed of a transcribed region and a regulatory sequence that makes transcription possible.

Genebank: An establishment in which both somatic and hereditary genetic material are conserved (seeds, pollen, whole plants, extracted DNA); it stores, in a viable form, material from plants that are in danger of extinction in the wild and cultivars that are not currently in popular use; the stored genetic material can be called up when required; the normal method of storage is to reduce the water content of seed material to around 4% and keep it at 0°C or less (–20°C); all stored stocks are periodically checked by germination tests

Gene-for-gene theory: In certain plant-pathogen interactions, a gene for resistance in the host corresponds to and is directed against a gene for virulence in the pathogen; in flowering plants, four families of disease resistance genes confer gene-for-gene resistance to a wide array of pathogens by recognising the products of the corresponding pathogen avirulence genes

General combining ability: The average performance of a strain or genotype in series of hybrid combinations with other genotypes is termed as General Combining Ability

General resistance: Resistance against all biotypes of a pathogen; nonspecific host-plant resistance general, a nontypical form or function

Generalised lattice design: Each replicate is subdivided into incomplete blocks typically of 5 or 6 plots; sometimes referred to as "alpha design" (Patterson and Williams , 1976)

Genetic (linkage) map: The linear arrangement of gene loci on a chromosome, deduced from genetic recombination experiments; a genetic map unit is defined as the distance between gene pairs for which one product of meiosis out of a hundred is recombinant (i.e. it equals a recombination frequency of 1%)

Genetic code: The set of correspondences between base triplets in DNA and amino acids in protein; these base triplets carry the genetic information for protein synthesis

Genetic correlation: The correlation between the genotypic values of two characters with respect to the genetic character

Genetic counseling: Information and support provided by a specialist doctor, usually a geneticist, to parents who have known conditions in their families or who are concerned about the future possibility of genetically transmitted conditions.

Genetic drift: The random fluctuations of gene frequencies in a population such that the genes amongst offspring are not a perfectly representative sampling of the parental genes

Genetic engineering: Manipulation of an organism's genes by introducing, eliminating or rearranging specific genes using the methods of modern molecular biology, particularly those techniques referred to as recombinant DNA techniques.

Genetic equilibrium: An equilibrium in which the frequencies of two alleles at a given locus are maintained at the same values generation after generation; a tendency for the population to equilibrate its genetic composition and resist sudden change is called genetic homoeostasis

Genetic erosion: The loss of genetic information that occurs when highly adaptable cultivars are developed and threaten the survival of their more locally adapted ancestors, which form the genetic base of the crop

Genetic heterogeneity: A similar phenotype being caused by different mutations. Most commonly used for a similar phenotype being caused by mutations in different genes. Allelic heterogeneity refers to different mutations in the same gene.

Genetic load: The average number of lethal mutations per individual in a population

Genetic mapping: The process of determination of a genetic map; in former time, genetic mapping was usually on a gene by gene basis; until molecular markers, there were no specific projects to map entire genomes; therefore, genetics will never be the same because of greater precision (<0.5 cM), fewer crosses, use of F_2 data, fewer breeding generations, more reliance on maps, mapping kits, etc.

Genetic marker: A gene or DNA sequence at a known location on a chromosome that can be used to identify individuals or species. A genetic marker may be a short or long DNA sequence. Some common types of genetic markers are RFLPs (Restriction Fragment Length Polymorphism), SSLPs (Simple Sequence Length Polymorphism), SNPs (Single Nucleotide Polymorphism), and SSRs (Simple Sequence Repeat DNA marker). Genetic markers play a role in genetic engineering because they allow breeders to locate and utilize genes of interest for genetic gains.

Genetic modification: The production of heritable improvements in plants or animals for specific uses, via either genetic engineering or other more traditional methods. Some countries other than the United States use this term to refer specifically to genetic engineering.

Genetic polymorphism: Differences between DNA sequences.

Genetic variance: A portion of phenotypic variance that results from the varying genotypes of the individuals in a population; together with the environmental variance, it adds up to the total phenotypic variance observed amongst individuals in a population; it is divided into additive and dominance variance

Genetic vulnerability: The susceptibility of genetically uniform crops to damage or destruction caused by outbreaks of a disease or pest or unusually poor weather conditions or climatic change

Genetically modified organism (GMO): A plant or animal whose genetic material (DNA) has been altered through genetic engineering/biotechnology techniques (insertion/deletion of genes) to produce a genotype that possesses a modified trait that is not found in naturally occurring plants of that species. When genes are inserted, they usually come from a different species. The principle of producing a GMO is to add new genetic material into an organism's genome.

Genetics. The study or science of genes and heredity.

Genome mapping: Determining a set of landmarks or genetic markers–genes or short DNA sequences–in the genome that will enable researchers to find new genes. A genome map is one-dimensional–a straight line with landmarks that stand for genes or DNA sequences. They guide a researcher toward a gene that is suspected to be involved in some process that is of interest.

Genome mutation: Spontaneous or induced changes in the number of complete chromosomes that result either in polyploids or aneuploids

Genome projects: Research and technology development efforts aimed at mapping and sequencing some or all of the genome of human beings and other organisms.

Genome sequencing: Determining the order of every DNA base in the genome. More detailed than a genome map and usually done on short segments of the genome. Done to establish the distinctive order of genes of an organism.

Genome: The total genetic material of an organism, i.e. an organism's complete set of DNA sequences.

Genomic imprinting: The phenomenon whereby genes function differently depending on whether they are inherited from the maternal or paternal parent; this is thought to be caused by information superimposed on DNA sequences, which is different in male and female gametes; such information is transmitted, or inherited, in somatic cells but usually erased and reset in the germ line; it is due to methylation of one of the alleles depending on its origin

Genomic *in situ* hybridisation (GISH): An *in situ* hybridisation technique that uses total genomic DNA of a given species as a probe and total genomic DNA of another species as a blocking DNA; it is based on fluorescence *in situ* hybridisation; it is a useful method to detect interspecific or intergeneric genome differentiation, chromosome rearrangements (translocations), and substitutions or additions

Genomic library: A collection of clones made from a set of randomly generated overlapping DNA fragments representing the entire genome of an organism.

Genomic selection (GS): A method in which the number of polymorphic bands resembling the

Genomics: The discipline of genetics that encompasses gene mapping, gene sequencing, and determining gene function.

Genotype x Environment (GxE): The interaction of a plant's genotype with the environment in which it is grown that contributes to its performance

Genotype: The genetic identity of an individual. Genotype often is evident by outward characteristics, but may also be reflected in more subtle biochemical ways not visually evident.

Genotyping: The process of identifying the genetic make-up of an organism, by using molecular markers, DNA sequencing, etc.

Germ cells: Specialised cells that form the reproductive organ where they ultimately undergo meiosis, thereby producing gametes that contain half the number of chromosomes as there body cells. Germ cells are responsible for transmitting genes to the next generation of an organism.

Germplasm enhancement: Any activity that includes gene transfer *via* sexual and asexual means from germplasm accessions, or increasing the frequencies of desirable genes in crop gene

Germplasm: The genetic material that carries the inheritance traits of an organism. Specifically, a collection of diverse genetic resources that is available to be used in the development of improved breeding lines and varieties.

Golden rice: A biotechnology-derived rice (*Oryza sativa*) created in the 1990s by I. Potr ykus and P. Beyer , which contains large amounts of beta-carotene, i. e. precursor of vitamin A, in its seeds; the researchers utilised *Agrobacterium tumefaciens* bacteria to genetically engineer the rice plant, i.e. by inserting the following genes from daffodil and from the bacterium *Erwinia uredovora*: (a) phytoene synthase - from daffodil (narcissus), which converts geranylgeranyl-diphosphate into phytoene; (b) CRTL gene - from *Erwinia uredovora*, which codes for phytoene desaturase, which causes the rice plant to convert phytoene (a "light harvesting" carotenoid involved in photosynthesis) into lycopene (a carotenoid which is then utilised by the rice plant in the production of beta-carotene); and (c) lycopene beta-cyclase - from daffodil, which converts lycopene into beta-carotene

Goodness of fit: Methods to test the conformity of an observed empirical distribution function of data with a posited theoretical distribution function (Eg. Chi-square test) by comparing observed

Green fluorescent protein (GFP): Protein from the jellyfish, *Aequorea victoria* (Scyphozoa); the gene is used as a reporter gene; it fluoresces in UV light; several variants have been developed, which each exhibit characteristic spectra; significant advantages are that the protein

Grid design: For the grid design, plants or variants are divided into blocks and the best ones chosen from each

Guanine (G): A nitrogenous base, one member of the base pair G-C (guanine and cytosine).

GUS gene: a gene that codes for production of beta-glucuronidase (GUS protein) in *Escherichia*

Gynandromorph: An individual exhibiting both male and female sexual differentiation

Gynodioecy: Plant species or population in which female plants as well as hermaphroditic plants

Gynoecium: The collective term for the female reproductive organs of a flower, comprising one or more carpels

Half sibs: Progeny with only one common parent

Haploid parthenogenesis: The situation in which the unfertilised egg hatches and develops normally to produce a viable male adult whose cells contain only the haploid number of chromosomes

Haploid: A single set of chromosomes (half the full set of genetic material), present in the egg and sperm cells of animals and in the egg and pollen cells of plants.

Haploidization: The process whereby diploid somatic cells become haploid during a parasexual cycle or by experimental means, for example, in barley by crossing with wild (bulb) barley (*Hordeum bulbosum*), or in wheat by crossing with pearl millet (*Pennisetum glaucum*)

Haplo-insufficiency: Situation where one normal copy of a gene alone is not sufficient to maintain normal function; it is observed as a dominant mutation on one allele (or deletion of it) resulting in total loss-of-function in a diploid cell because of the insufficient amount of the wild-type protein encoded by the normal allele on the other haplotype

Haplontic: Organisms in which meiosis occurs in the zygote resulting in four haploid cells

Haplotype: A set of closely linked alleles (genes or DNA polymorphisms) inherited as a unit. A contraction of the phrase "haploid genotype."

Hardy -Weinberg law (equilibrium): The stable frequency distribution of genotypes, *AA*, *Aa*, and aa, in the proportions p^2, 2pq, and q^2, respectively (where p and q are the frequencies of the alleles, *A* and *a*); that is, a consequence of random mating in the absence of mutation, migration, natural selection, or random drift; it states that in an infinitely large, interbreeding population in which mating is random and in which there is no selection, migration, or mutation, gene and genotype frequencies will remain constant from generation to generation

Heat shock protein: When certain plants are exposed to high temperature, heat shock proteins are synthesized; they provide thermal protection to subsequent heat stress

Herbicide-tolerant crops: Crops that have been developed to survive application(s) of particular herbicides by the incorporation of certain gene(s) either through genetic engineering or traditional breeding methods. The genes allow the herbicides to be applied to the crop to provide effective weed control without damaging the crop itself.

Hereditary determinant: Any genetically acting unit of an organism that is replicated and conserved, transferred from generation to generation

Hereditary: Transmissible from parent to offspring or progeny

Heritability: A measure of the degree to which a phenotype is genetically influenced and can be modified by selection; it is represented by the symbol h^2;

this equals Vgen/Vphe where Vgen is the variance due to genes with additive effects and Vphe is the phenotypic variance;

Herkogamy: Pollination by the neighbour individual, population, or species

Hermaphrodite: A plant having both female and male reproductive organs in the same flower, eg. cotton, sugarbeet, alfalfa, rapeseed, rye, or sunflower

Heterochromatin: The darkly stained regions in chromosomes

Heteroduplex analysis: A method of detecting gene mutation by mixing PCR-amplified mutant and wild-type DNA followed by denaturation and reannealing; the resultant products are resolved by gel electrophoresis, with single base substitutions detectable under optimal electrophoretic conditions and gel formulations; large base pair mismatches may also be analysed by using electron microscopy to visualise heteroduplex regions

Heteroduplex DNA: A double-stranded DNA molecule formed by the annealing of strands from two different sources, as opposed to homoduplex, which has homologous strands; as a result, there are regions noncomplementary and showing abnormalities in the form of extra loops

Heterosis: The increased vigor of growth, survival, and fertility of hybrids, as composed with the two homozygotes; it usually results from crosses between two genetically different, highly inbred lines

Heterostyly: A polymorphism among flowers that ensures cross-fertilisation through pollination by visiting insects; flowers have anthers and styles of different length

Heterozygosity: The presence of different alleles at one or more loci on homologous chromosomes.

Heterozygous: Having two diffeent alleles (eg. Aa) governing a single phenotypic trait. A hybrid corn plant, for example, is heterozygous at every locus (eg. Aa Bb Cc Dd...).

High-throughput DNA sequencing: The industrialised process of determining the exact order of the four letter code (bases, GAT or C) that composes all genetic material; because a typical higher organism contains > 20,000 genes, each with several thousand bases of DNA, data storage and analysis is one of the most important parts of this process

Histone: One of a group of globular, simple proteins that have a high content of the amino acids arginine and lysine; it forms part of the chromosomal material of eukaryotic cells and appears to play an important role in gene regulation; highly conserved basic proteins that are involved in the packing of DNA; histone proteins and the nucleosomes form, with DNA, the fundamental building blocks of eukaryotic chromatin; they bind to the phosphate groups of DNA by their amino termini; there are five major types of histone proteins; two copies of H_2A, H_2B, H_3, and H_4 bind to about 200 base pairs of DNA to form the repeating structure of chromatin (nucleosome), with H1 binding to the linker sequence

Holocentric: Applied to chromosomes with diffuse centromeres such that the properties of the centromere are distributed over the entire chromosome

Homeotic mutation: A mutation that causes one body structure to be replaced by a different body structure during development

Homoeologous: Partially homologous; chromosomes or genomes that are believed to have originated from ancestral homologous chromosomes

Homogamy: The preference of individuals to mate with others of a similar genotype or phenotype; in botany, the condition in which male and female parts of the flower mature simultaneously

Homologies: Similarities in DNA or protein sequences between individuals of the same species or among different species.

Homologous chromosomes: Chromosomes that pair with each other at meiosis.

Homopolymer-tailing: Attachment of identical nucleotides to the 3′ end of a DNA molecule that can be achieved with terminal deoxynucleotide transferase

Homozygosity: The presence of identical alleles at one or more loci in homologous chromosomal segments

Homozygote: An individual possessing a pair of identical alleles at a given locus on a pair of homologous chromosomes.

Honeycomb design: In a honeycomb design, the plant at the center of the hexagon is compared with every other plant within the hexagon; a plant is chosen only if it is superior to every other plant in the hexagon; it was developed for selecting individual plants in a population; seeds or plants are usually spaced equidistantly from one another in a hexagon pattern; plants are spaced far enough apart that they cannot compete with adjacent individuals; homogeneous checks can be included; the size of the hexagon determines the selection intensity; it is used to minimise adverse effects of interplant competition

Hordeum bulbosum **procedure:** A method for producing zygotic haploids in barley by crossing *Hordeum bulbosum* with *Hordeum vulgare* genotypes; after formation of zygotes, the wild *H. bulbosum* chromosomes are subsequently eliminated during embryogenesis, which results in haploid *H. vulgare* plants

Horizontal resistance: Resistance conditioned by polygenes or quantitative genes; it is race nonspecific in nature and does not reveal a gene-for-gene hypothesis; this type of resistance is difficult to identify

Host plant resistance: A method of pest control in which resistant, tolerant, or unattractive host organisms are used; the inherited qualities of resistance influence the extent of pest damage

Housekeeping gene: Gene that is expressed in virtually all cells since it is fundamental to the any cell's functions.

Human Genome Initiative: Collective name for several projects begun in 1986 by DOE to (1) create an ordered set of DNA segments from known chromosomal locations, (2) develop new computational methods for analysing genetic map and DNA sequence data, and (3) develop new techniques and instruments for detecting and analysing DNA. This DOE initiative is now known as the Human Genome Program. The national effort, led by DOE and NIH, is known as the Human Genome Project.

Hybrid breeding (*syn* heterosis breeding): The discovery of heterosis has been recognised as one of the major landmarks of plant breeding; in comparison to inbred lines and homozygous material, the phenotypic superiority of heterozygotes is the basis of hybrid breeding; it is exploited for production of hybrid, synthetic, and composite varieties; hybrid breeding can be performed using traditional breeding techniques, or the process can be hastened using gene marker technology to rapidly identify parents with desired genes for certain attributes; numerous commercial crops are hybrids with increasing tendency; seeds from a hybrid variety, if planted, will not deliver the same benefits as the original seeds and after several offspring will have lost the desired qualities from the original hybridisation

Hybrid inviability: Reduced vigour of hybrid plants compared to their crossing parents

Hybrid seed production: The production of hybrid seeds by combination of more or less defined parental forms; usually those hybrid seeds are more productive or more suitable than pure lines; they are used for subsequent growing and commercial production of a crop

Hybrid selection: The process of choosing plants possessing desired traits among a hybrid population

Hybrid variety: A variety produced from the cross-fertilisation of inbred lines with favourable combining ability; the progeny is homogeneous and highly heterozygous; it can be produced by

Hybrid: The offspring of any cross between two organisms of different genotypes.

Hybridisation: A method of breeding new varieties that applies crossing to obtain genetic recombination; in genetics, the fusion of unlike genetic material, such as sexual organs or DNA; in molecular biology, pairing of complementary DNA and/or RNA

Hydrogen bond: A weak bond involving the sharing of an electron with a hydrogen atom; hydrogen bonds are important in the specificity of base pairing in nucleic acids and in the determination of protein shape.

Hypersensitivity: The response to attack by a pathogen of certain host plants in which the invaded cells die promptly and prevent further spread of infection

Hypoploidy: Having missing chromosome complements compared to the standard chromosome set

Identical by descent: Two genes that are identical in nucleotide sequence because they are both

Identity preservation: The segregation of one crop type from another at every stage from production and processing to distribution. This process is usually performed through audits and site visits and provides independent third-party verification of the segregation.

Ideotype breeding: A method of breeding to enhance genetic yield potential based on modifying individual traits where the breeding goal for each trait is specified

Ideotype: Crop plant with model characteristics known to influence photosynthesis, growth, and grain production; "ideotype" was coined by Donald (1968); it means literally a form denoting an idea, and in the broadest sense it is a biological model which is expected to perform or behave in a predictable manner within a defined environment; more specifically a crop ideotype is a plant model which is expected to yield a greater quantity of grain, oil or other useful product when developed as a cultivar

Idiogram: A diagrammatic representation of the karyotype of a plant

Idiotype: The sum of the hereditary determinants of a cell or plant consisting of the genotype and plasmotype; idiotype is an established term, first used by Siemens (1921); "idio" in idiotype is derived from the Greek word "*idios*", meaning one's own, personal, separate, distinct

Imprinting: Patterns of inheritance affected by whether the inheritance was from the mother or father.

***In silico*:** Modern term used to characterise biological experiments carried out entirely in a computer

***In situ* conservation:** Refers to the maintenance of plant populations in the habitats where they naturally occur and have evolved; in crop species, conservation can be achieved in home gardens or on-farm, in wild species through the designation of protected areas, such as natural parks and nature reserves

***In situ* hybridisation:** Use of a DNA or RNA probe to detect the presence of the complementary DNA sequence in cloned bacterial or cultured eukaryotic cells.

***In situ*:** in place; where naturally occurring

***In vitro* culture:** The cell, organ, or tissue culture performed under artificial conditions in tubes, glasses, dishes, etc.

***In vitro* fertilisation:** Pollination performed aseptically *in vitro* by direct application of the pollen to the ovule; it is used to overcome prezygotic incompatibility

***In vitro* mutagenesis:** Methods for altering DNA outside the host cells; mutagenesis can be random or specific for the site and base change depending on the technique used

***In vitro* selection:** Used to screen large numbers of plants or cells for a certain characteristic before growing them in the field or in glasshouses, eg. salt tolerance

***In vitro*:** Literally, "in glass;" but applied more generally to studies and propagation of living plant material that are performed under artificial conditions in tubes, glasses, dishes, etc.

***In vivo*:** Literally, "in life"; applied to studies and propagation of whole, living organisms, on intact organ systems therein, or on populations of microorganisms

Inbred line: A line produced by continued inbreeding; usually a nearly homozygous line originating by continued self-fertilisation, accompanied by selection

Inbred pure lines: Involves inbreeding of annual seed-propagated material; homogeneous and homozygous lines are isolated by selection of desired recombinants or segregates in F_2 to F_7 generations of crosses between parental pure lines (generally monogenotypic lines can be blended to form multilines, eg. in tomato, lettuce, soybean, pea, cowpea, snapbean, field bean, Arabian coffee, *Capsicum* pepper, eggplant, okra, lentil, and papaya)

Inbred: A plant resulting from successive self-fertilisation of parents throughout several generations

Inbred-variety cross: The F_1 cross of an inbred line with a variety

Inbreeding coefficient: The probability that the two genes at any locus in a diploid individual are identical by descent (i.e. they originated from the replication of one gene in a previous generation)

Inbreeding depression: The loss of vigour when some crops have reduced heterozygosity due to enforced self-pollination

Inbreeding load: The extent to which a population is impaired by inbreeding

Inbreeding: The crossing of closely related plants; one important purpose of induced inbreeding is the development of genotypes that can be maintained through multiple generations of seed production; self-pollinated cultivars are reproduced for many generations by inbreeding; inbreeding is also used to reduce the frequency of deleterious recessive alleles in genotypes that serve as parents of a synthetic or a vegetatively propagated cultivar; inbreeding increases the genetic and phenotypic variability among individuals in a population; four mating systems are used to increase the homozygosity in a breeding population

Incompatibility (homomorphic or heteromorphic): A genetically determined inability to obtain fertilisation and seed formation after self-pollination or cross-pollination; there are several types of progamous or postgamous incompatibility; in contrast to heteromorphic incompatibility (eg. heterostyly in *Primula* sp.), homomorphic incompatibility is not associated with morphological differences

Incompatibility group: Plasmids that are incompatible with each other belong to the same incompatibility group

Incomplete block design: A block design in which not every treatment occurs in a block; it is preferable where large numbers of cultivars are compared in a single yield trial; the entries in each replication are subdivided into smaller blocks, in a manner designed to reduce the error caused by soil variation; usually, it refers to a lattice design, considering the restriction that (1) the number of testers must be harvested, (2) inferior strains cannot be discarded prior to harvest to reduce harvest expenses, and (3) researchers still must analyze the experiment as a lattice design

Independent assortment: The random distribution in the gametes of separate genes; if an individual has one pair of alleles *A* and *a*, and another pair *B* and *b* then it should produce equal numbers of four types of gametes: *AB*, *Ab*, *aB*, and *ab;* it is asserted in Mendel 's second law—the law of independent assortment

Indirect embryogenesis: Embryoid formation on callus tissues derived from zygotic or somatic embryos, seedling plants, or other tissues in culture

Induced mutation: A change in a gene caused by a treatment

Induced resistance: A form of resistance activated in plants by an external stimulus - for example, elicitor treatment or preinoculation with a microorganism; it can be a local or systemic acquired resistance

Informatics: The study of the application of computer and statistical techniques to the management of information. In genome projects, informatics includes the development of methods to search databases quickly, to analyse DNA sequence information, and to predict protein sequence and structure from DNA sequence data.

Inheritance: The passing of familial elements from one generation to the next.

Insecticide resistance: The development or selection of heritable traits (genes) in an insect population that allow individuals expressing the trait to survive in the presence of levels of an insecticide (biological or chemical control agent) that would otherwise debilitate or kill this species of insect. The presence of such resistant insects makes the insecticide less useful for managing pest populations.

Insect-resistance management: A strategy for delaying the development of pesticide resistance by maintaining a portion of the pest population in a refuge that is free from contact with the insecticide. For Bt crops this allows the insects feeding on the Bt toxin to mate with insects not exposed to the toxin produced in the plants.

Insect-resistant crops: Plants with the ability to withstand deter or repel insects and thereby prevent them from feeding on the plant. The traits (genes) determining resistance may be selected by plant breeders through cross-pollination with other varieties of this crop or through the introduction of novel genes such as Bt genes through genetic engineering.

Intellectual property rights: The legal protection for inventions, including new technologies or new organisms (such as new plant varieties). The owner of these rights can control their use and earn the rewards for their use. This encourages further innovation and creativity for the benefit of us all. Intellectual property rights protection includes various types of patents, trademarks, and copyrights.

Intellectual property: All patent applications, patents, plant variety protection (PVP) certificates, or trade secrets that make up proprietary information.

Intercalating agent: A chemical that can insert itself between the stacked bases at the center of the DNA double helix, possibly causing a frameshift mutation

Interchange trisomic: An additional chromosome to the diploid set, which is composed of two different chromosomes *via* translocation

Interference: The effect of recombination in one interval on the probability of recombination in an adjacent interval *gene*; in patenting biological material, when two or more patent applications

Intergeneric cross: Spontaneous or experimental crosses of individuals of different genera, for example, wheat *(Triticum aestivum)* and rye *(Secale cereale)*; this cross even resulted in a human-made new crop plant "triticale"

Interphase: The period in the cell cycle when DNA is replicated in the nucleus; followed by mitosis.

Interspecific cross: A crossing between two species

Intervarietal: Effects and phenomenons between varieties (cultivars)

Introgression library: Increasing the genetic diversity of elite breeding materials with exotic germplasm requires techniques that minimize negative side effects attributable to genetic interactions between recipient and donor; this seems achievable by an introgression library approach involving the systematic transfer of a limited and/or restricted number of short donor chromosome segments from an agricultural unadapted source (donor) into an elite line (recipient or recurrent parent); established introgression libraries represent a dynamic resource that can substantially foster breeding programs and provide an opportunity to proceed towards functional genomics

Introgression: The incorporation of genes of one species into the gene pool of another; if the ranges of two species overlap and fertile hybrids are produced, the hybrids tend to backcross with the more abundant species; it results in a population in which most individuals resemble the

Introgressive hybridisation: Crossbreeding of plants from different species that results in introgression

Intron: A segment of DNA of unknown function within a gene; it may be transcribed in precursor RNA, but cannot be found in functional mRNA

Inversion: Occurs where a chromosome breaks in two and becomes reattached in reverse orientation. This may or may not affect gene function.

Isochromosome: A chromosome with two identical arms; it usually derives from telocentric chromosomes

Isodicentric chromosome: A structurally abnormal chromosome containing a duplication of part of the chromosome including the centromere; the resulting structure contains two centromeres and a point of symmetry that depends on the position of the breakpoint

Isogenic lines: Two or more lines differing from each other genetically at one locus only

Karyotype: The entire chromosomal complement of an individual cell or individual, which may be observed during mitotic metaphase

Kilobase (kb): Unit of length for DNA fragments equal to 1000 nucleotides.

Kinetochore: A dense, plaquelike area of the centromere region of a chromatid, to which the microtubules of the spindle attached during cell division

***Lac* operon:** A cluster of structural genes specifying the enzymes acetylase, permease, and betagalactosidase

Lagging strand: DNA strand growing in the 3′ to 5′ direction, synthesized discontinuously

Lampbrush chromosome: A particular type of chromosome shape, usually found in the diplotene stage of animals including flies; it is a type of puffed chromosome; the loops of DNA strands form a lampbrush-like shape; puffed chromosome regions may also occur in plants (eg. in *Phaseolus* beans)

Landraces: A local variety that has been developed by adaptation to its local environment; usually heterogeneous varieties

Lattice design: An incomplete block design in which the number of treatments must be an exact square; there are several types of lattice designs, for example, alpha lattice, partially balanced square lattice; the lattice design is "simple" if $r = 2$; square lattice designs are also called "nets"; they are partially balanced with respect to an association scheme of Latin square type

Leader sequence: The sequence at the 5′ end of an mRNA that is not translated into protein.

Lethal doses: the concentration of a poison that kills a certain amount of cells, individuals, etc.; for example, LD50 = 50% of the cells or individuals are killed

Lethal gene: A gene whose expression results in the premature death of the organism carrying it; dominant alleles kill heterozygotes, whereas recessive alleles kill homozygotes only

Lethal mutation: A gene mutation whose expression results in the premature death of the organism carrying it

Library: A set of cloned fragments together representing the entire genome, created then placed into storage.

Ligase: DNA ligase; an enzyme that can rejoin a broken phosphodiester bond in a nucleic acid; requires a 5′ phosphate and a 3′ OH.

Line breeding: A system of breeding in which a number of genotypes, which have been progeny tested in respect to some characteristics, are composited to form a variety; examples of line varieties are from normally self-fertilised crops, for example, "Gaines" wheat,

Linkage (of genes): the association of genes that results from their being on the same chromosome; linkage is detected by the greater association in inheritance of two or more non-allelic genes than would be expected from independent assortment; the nearer such genes are to

Linkage disequilibrium: The non-random association of alleles at two or more loci, i.e. they occur together more often than would be expected by chance

Linkage drag: The inheritance of undesirable genes along with a beneficial gene due to their close linkage

Linkage group: All genes in one chromosome form one linkage group.

Linkage map: A map of the relative positions of genetic loci on a chromosome, determined on the basis of how often the loci are inherited together. Distance is measured in centimorgans (cM).

Linkage, genetic linkage: When two chromosomal regions are located physically near each other such that there is a high likelihood they will be inherited together

Locus: A specific location on a chromosome.

Lod score: The logarithm of the ratio of the odds that two loci are linked with a recombination fraction equal to or greater than 0 and less than 0.5 to the likelihood of independent assortment. Also called "Z."

Luxuriance: Hybrids that are larger, faster growing, or otherwise exceed the parental forms in some traits; usually occurs due to complementary gene action present in the parents and

Macro restriction map: Map depicting the order of and distance between sites at which restriction enzymes cleave chromosomes.

Macromutation: a mutation that results in a profound change in an organism, such as a change in a regulatory gene that controls the expression of many structural genes, as opposed to micromutation

Maintainer: Used for maintaining and multiplication of a cytoplasmic male sterile line; usually genotypes containing the normal cytoplasm and recessive at the restorer locus

Male gametocide: Any substance that kills the male reproductive cells of a plant (pollen or pollen mother cells), rendering it male-sterile

Male sterility: Producing no functional pollen

Map distance: The distance between any two markers on a genetic map, based on the percentage of crossing-over; the minimum distance between linked genes is 1% and maximum 50%

Map-based cloning: The isolation of important genes by cloning the gene in question on the basis of molecular maps, where the biochemical function is unknown; it involves the identification of a mutant phenotype for the trait of interest (obtained by mutagenesis or from natural variation) and genetic fine mapping using many progeny plants; this map is then used for chromosome walking or landing, with the help of large-insert DNA libraries or physical maps to isolate the gene

Mapping: The process of identifying the location of a gene or DNA segment along a chromosome. In *genetic* mapping, this is done by analysing patterns of inheritance in segregating populations (measured in recombinational units, commonly centimorgans). In*physical* mapping, this describes the actual location of a sequence in a particular genomic region (measured in bp)

Marker-assisted introgression: The use of DNA markers to increase the speed and efficiency of introgression of a new gene or genes into a population; the markers will be closely linked to the genes in question

Marker-assisted selection (MAS): Use of genetic/DNA markers to guide in choosing specific plants or lines with targeted traits for new variety development.

Mass pedigree selection: A system of breeding in which a population is propagated in bulk until conditions favorable for selection occur; usually, after mass pedigree selection, pedigree selection follows

Mass selection: A form of breeding in which individual plants are selected on their individual advantages and the next generation propagated from the aggregate of their seeds; the easiest method is to select and multiply together those individuals from a mixture of phenotypes, which correspond to the breeding aim (positive mass selection); it is still applied in cross-pollinating of vegetable species, such as carrots, radishes, or beetroots, in order to improve the uniformity; when all undesired off-types are rouged in grown crop population, and the remaining individuals are propagated further, the method is termed negative mass selection; negative mass selection is no longer an adequate breeding method for highly advanced varieties; it is usually applied in multiplication of established varieties (i.e., for seed production in order to remove diseased plants, casual hybrids, or other defects)

Maternal effect: Any nonlasting environmental effect or influence of the maternal genotype or phenotype on the immediate offspring

Maternal inheritance: Phenotypic differences found between individuals of identical genotype due to an effect of maternal inheritance

Megabase (Mb): Unit of length for DNA fragments equal to 1 million nucleotides and roughly equal to 1 cM.

Megagametogenesis: The development of the female gametophyte from a functional megaspore

Megasporogenesis: The development of the megaspore from the archesporial cell

Meiocyte: The sporocyte giving rise to the embryo sac and to pollen grains

Meiosis: A type of nuclear division that occurs at some stage in the life cycle of sexually reproducing organisms; by a specific mechanism, the number of chromosomes is halved to prevent doubling in each generation

Melting: Denaturation of DNA.

Mendel 's laws of inheritance: The inheritance of chromosomal genes on the basis of the chromosome theory of heredity; three laws are considered: (1) law of dominance or of uniformity of hybrids, (2) law of segregation, (3) law of independent assortment

Mendelian population: A group of (potentially) interbreeding plants (cross-fertilizing crops), which may occur in a certain area or in an experimental design

Mendelian ratio: The segregation ratios according to Mendel 's laws of inheritance

Messenger DNA (mDNA): A single-stranded DNA that acts as a messenger of the protein biosynthesis

Messenger RNA (mRNA): A single-stranded RNA molecule responsible for the transmission to the ribosomes of the genetic information contained in the nuclear DNA; it is synthesized during transcription and its base sequencing exactly matches that of one of the strands of the double-stranded DNA molecule

Metabolomics: An extended discipline of biochemistry that involves the analysis (usually high throughput or broad scale) of small-molecule metabolites and polymers, such as starch; it involves descriptions of biological pathways and current metabolomic databases, such as the Kyoto Encyclopedia of Genes and Genomes (KEGG)

Metacentric: Applied to a chromosome that has its centromere in the middle

Metaphase: A stage in mitosis or meiosis during which the chromosomes are aligned along the equatorial plane of the cell.

Microarrays: A tool to examine the expression levels and intertwined interactions among genes and their products; complementary DNA of interest is affixed to a glass slide in an ordered array; the expression level is determined by the binding to cDNA; a small glass, or filter, square may contain probes for thousands of gene products; in statistics, a microarray is a block design with blocks of size 2; the concept arose in DNA testing, where each array is probed with two DNA samples

Micropropagation: *In vitro* clonal propagation; the multiplication of genetically identical copies of a cultivar by asexual means is called clonal propagation

Microsatellite marker: Microsatellites or simple sequence repeats are a type of molecular marker; microsatellites consist of tandem repeats of 1–6 nucleotide motifs; the repeats usually are in units of ten or more, although repeats as small as six units have been found

Microsatellite-primed PCR (MP-PCR): A technique resulting in RAPD-like patterns after agarose gel electrophoresis and ethidium bromide staining; the MP-PCR technique is more reproducible than RAPD analysis because of higher stringency

Minichromosome: A very small chromosome, usually as a result of chromosome aberrations; engineered minichromosomes offer an opportunity to improve crop performance; unlike conventional gene transformation technologies, minichromosomes can be used simultaneously to transfer and to stably express (multiple) sets of genes; because they segregate independently of host chromosomes, they provide a platform for accelerating plant breeding; strategies for producing artificial chromosomes may consider engineering of endogenous chromosomes or *de novo* assembly from chromosomal constituents

Minisatellite: Highly polymorphic DNA markers comprised of a variable number of tandem repeats that tend to cluster near the telomeric ends of chromosomes; the repeats often contain a repeat of 10 nucleotides; they are useful tools for genetic mapping

Missense mutation: A single DNA base change which leads to a codon specifying a different amino acid.

Mitotic recombination: The recombination of genetic material during mitosis and the process of asexual reproduction; the mechanism for the production of variation in heterokaryons

Model organisms: Creatures used in genomic analysis because they have many genetic mechanisms in common with each other and with humans. These organisms lend themselves well to classical breéding experiments and direct manipulation of the genome.

Molecular biology: The study of the structure and function of proteins and nucleic acids in biological systems.

Molecular breeding: Plant breeding assisted by using DNA markers or protein markers.

Molecular farming: The application of biotechnology for the selected production of pharmaceutical compounds or other health or industrial compounds within a living organism (eg. microbe or agricultural crop)

Molecular genetics: The study of genes, the base components of heredity. Genes are made up of DNA, whose bases are placed in a unique order that determines the hereditary traits in living organisms.

Molecular marker: A gene or DNA sequence that identifies a particular locus on the chromosome (whether the actual location is known or not) and whose inheritance can be followed

Monocentric chromosome: A chromosome with only one centromere

Monohaploid: A haploid cell or individual possessing only one chromosome set in the nucleus

Monosomic: A genome that is basically diploid but that has only one copy of one particular chromosome type, so that its chromosome number is $2n-1$; monosomic series were developed

Monosomy: The total loss of one of a pair of chromosomes. This occurs, for example, in Turner Syndrome where one X chromosome is lost leaving a total of 45 chromosomes.

Monotelocentric: A cell or individual lacking one chromosome pair but showing one telocentric chromosome for one arm of the two missing homologues

Monotelodisomic: A cell or individual lacking one chromosome pair but showing two homologous telocentric chromosomes for one arm of the two missing homologues

Monotelomonoisosomic: A cell or individual lacking one chromosome pair but showing a telocentric chromosome for one arm of the missing homologous pair and an isochromosome for

Monotelotrisomic: A cell or individual showing an additional telocentric chromosome to a certain pair of chromosomes more abundant parents but also possess some of the characters of the other parent species

Mosaicism: Where a genetic or chromosomal abnormality does not occur in all body cells. Often related to X chromosome inactivation.

Multiplexing: A sequencing approach that uses several pooled samples simultaneously, greatly increasing sequencing speed.

Mutagen: Any agent that is capable of increasing the mutation rate.

Mutant allele: An allele differing from the allele found in the standard, or wild type.

Mutant: A gene having undergone genetic change or mutation.

Mutation: Any heritable change in DNA sequence. Compare polymorphism.

Natural selection: A complex process in which the total environment determines which members of a species survive to reproduce and so pass on their genes to the next generation

Near-isogenic lines (NIL)**:** Not fully isogenic; for example, in maize, two distinct composites of F_3 lines from a single cross, one consisting of lines homozygous recessive and the other consisting of lines homozygous dominant for a certain gene (i.e. there is same genetic background), however, differing only in being homozygous dominant versus recessive for the genes; in wheat, near-isogenic lines have been produced for different *Rht* (reduced height) genes causing different straw length

Negative assortative mating: A type of nonrandom mating in which individuals of unlike phenotype mate more often than predicted under random mating conditions

Nested association mapping (NAM)**:** An integrated mapping strategy that allows genome-wide high-resolution mapping in a cost-effective way; it greatly facilitates complex trait dissection in many species; it is used as a large-scale maize mapping resource and as a genomics and bioinformatics tool; it requires large amounts of SNPs and/or individuals, and it is sensitive to genetic heterogeneity

Nondisjunction: The failure of separation of paired chromosomes at metaphase, resulting in one daughter cell receiving both and the other daughter cell none of the chromosomes in question; it can occur both in meiosis and mitosis

Nonhomologous chromosomes: The different chromosomes of a haploid chromosome set, which usually cannot pair with one another

Non-Mendelian inheritance: An unusual ratio of progeny phenotypes that does not reflect the simple operation of Mendel 's law, for example, mutant:wild-type ratios of 3:5, 5:3, 6:2, or 2:6, indicate that gene conversion has occurred; in general, it refers to extrachromosomal and/or non-chromosomal inheritance

Nonpreference: A term used to describe a resistance mechanism where parasites prefer to be on some host genotypes more than others; the less preferred genotypes are resistant

Nonrandom mating: A mating system in which the frequencies of the various kinds of matings with respect to some trait or traits are different from those expected according to chance

Nonrecurrent apomixis: Refers to occasional apomixis, usually caused by haploid parthenogenesis

Nonsense codon: (also called STOP codon) Any one of three triplets (UAG, UAA, UGA) that cause termination of protein sysnthesis.

Non-sense mutation: A mutation that alters a gene so that a non-sense codon is inserted; such a codon is one for which no normal tRNA molecule exists, therefore it does not code for an amino acid; usually non-sense codons cause the termination of translation; several non-sense codons are recognised (eg. amber, ochre, opal)

Normalising selection: The removal of genes and/or alleles that produce deviations from the normal phenotype of a population

Northern blotting: A procedure analogous to Southern blotting, but transferring RNA, instead of DNA, from a gel to a carrier (like nitrocellulose) and then hybridizing to a radioactively labelled DNA probe complementary to the desired sequence, and visualised by autoradiography; the technique can therefore be used to locate and identify an RNA fragment containing a specific sequence gene

Nucleic acid: A large molecule composed of nucleotide subunits.

Nucleotide: A subunit of DNA or RNA consisting of a nitrogenous base (adenine, guanine, thymine, or cytosine in DNA; adenine, guanine, uracil, or cytosine in RNA), a phosphate molecule, and a sugar molecule (deoxyribose in DNA and ribose in RNA). Thousands of nucleotides are linked to form a DNA or RNA molecule.

Nucleus: The cellular organelle in eukaryotes that contains the genetic material.

Null allele: An allele that makes no gene product or whose product has no activity of any kind; a deletion of a gene is necessarily a null allele.

Null hypothesis: A hypothesis that there is no discrepancy between observation and expectation

Null mutation: A mutation, which eliminates all enzymatic activity; usually deletion mutations

Nullihaploid: A cell or individual that possesses a haploid chromosome set plus a missing single

Nullisome: A plant lacking both members of one specific pair of chromosomes

Nullisomic analysis: In nullisomic analysis, observations are made for phenotypic or other differences between the nullisomic for each chromosome and the disomic condition within the same variety; the method is applied for localisation of genes within a given genome; the method can only be used in polyploids, while diploids commonly do not tolerate the loss of both homologous chromosomes

Nullitetrasomic: A cell or individual, usually an allopolyploid, that lacks one pair of chromosomes, which is partially compensated for by a tetrasomic (four-fold dosage) of another, usually homoeologous, chromosome; a whole series of nulli-tetrasomics has been produced in hexaploid wheat, and successfully used in numerous genetic and molecular studies

Numeric aberration: The variation of the number of genomes or chromosomes, for example, ploidy variation, aneuploids (nullisomics, monosomics, trisomics, tetrasomics, etc.), substitutions, or additions

Off-type: An individual differing from the population norm in morphological or other traits; the term also includes escapes and contaminants (eg. seeds that do not conform to the characteristics of a variety, uncontrolled self-pollination during production of hybrid seed, segregates from plants, etc.)

Oligonucleotide ligation assay (OLA): A PCR-based method for SNP typing; it is a ligase mediated gene detection system which uses exact 3′ matching of a primer to one of the SNP allele; if this happens, the other labeled oligonucleotide that binds to the nucleotide immediately next to the SNP on the other side would be joined to the primer by ligase; the resulting sample can then be tested for the presence of the label

Oligonucleotides: Small single-stranded segments of DNA typically 20-30 nucleotide bases in size which are synthesized *in vitro*.

Oncogene: A gene, one or more forms of which is associated with cancer. Many oncogenes are involved, directly or indirectly, in controlling the rate of cell growth.

Open-reading frame (ORF): The mRNA region between the start and stop codon

Operator: A region of DNA at one end of an operon that acts as the binding site for a specific repressor protein and so controls the functioning of adjacent cistrons

Operon: A set of adjacent structural genes whose mRNA is synthesized in one piece, together with the adjacent regulatory genes that affect the transcription of the structural genes; it is under the control of an operator gene, lying at one end of it

Optical mapping: An enabling technology for whole genome analysis, which involves the capture of individual DNA molecules, obtained directly from genomic DNA, followed by digestion *in situ* by selected restriction endonucleases; the resulting fragments are then visualised directly to produce detailed optical restriction maps; this methodology allows patterns of sequence variation to be detected across entire genomes, without the need for DNA amplification, and, unlike other genomewide scanning methods, provides detailed haplotype information by analysing individual DNA molecules or issued patents claim the same invention

Organ culture: The growth in aseptic culture of plant organs, such as roots or shoots, beginning with organ primordia or segments and maintaining the characteristics of the organ

Organic agriculture: A concept and practice of agricultural production that focuses on production without the use of synthetic inputs and does not allow the use of transgenic organisms. USDA's National Organic Program has established a set of national standards for certified organic production which are available online.

Organic conservation breeding: Varieties that are of conventional origin are conserved and propagated by cultivation under organic standards; relevant to older varieties which should be conserved as a genepool for future breeding efforts and to varieties of high importance

Orthologous genes: Homologous genes that have become differentiated in different species derived from a common ancestral species; as opposed to paralogous genes

Outbreeding: The crossing of plants that are not closely related genetically, in contrast to inbreeding, in which the individuals are closely related

Outcrossing: Cross-pollination between plants of different

Overdominance: The phenomenon in which the character of the heterozygotes is expressed more markedly in the phenotype than in that of either homozygote; usually the heterozygote is fitter than the two homozygotes; this can give rise to monohybrid heterosis when the hybrid vigor is obtained by crossing parents differing in a single specified pair of allelic genes

Overlapping DNA: A special type of gene organization; one DNA sequence may code for different proteins; it is performed by two open reading frames, which subsequently act

Palindrome: Adjacent inverted DNA repeats; the identical base sequences are on the opposite strands; long (>130 bp) uninterrupted palindromes occur in eukaryotic DNA but are lethal in bacteria

Panmixis: Random mating; a mode of sexual reproduction in which male and female gametes encounter each other incidentally

P**arallel mutation:** A mutation that causes similar phenotypes but in different species or genotypes

Paralogous genes: Homologous genes that have arisen through gene duplication and that have evolved in parallel with the same organism; as opposed to orthologous genes

Paramutable allele: An unstable allele where the phenotypic consequences are enhanced by the

Paramutagenic allele: An allele possessing the ability to cause paramutation

Paramutation: A mutation in which one allele in the heterozygous condition permanently changes the partner allele

Parent-offspring analysis: The linear regression of the performance of offspring on that of the parents was proposed as a method of estimating heritability

Parthenocarpy: Production of fruit without seeds, as in bananas and some grapes

Parthenogenesis: The development of an individual from an egg without fertilisation; it occurs in some plants

Partial correlation: Displays the numerical relationship between a single Xi and Y in multiple correlation; in contrast to the simple correlation, the partial correlation accounts for interaction between the Xi variables and illustrates the true relationship between the Xi and Y

Partial dominance: Incomplete dominance; the production of an intermediate phenotype in individuals heterozygous for the gene concerned; it is generally considered to be a type of incomplete dominance, with the heterozygote resembling one homozygote more than the other

Partial self-incompatibility: Self-incompatibility in some species results in a lower percentage of seed set than occurs with cross-pollination

Partially balanced design: A design is partially balanced with respect to an association scheme if the number of blocks containing two points depends only on which class of the partition contains the given pair of points

Partially balanced incomplete block design (PBIB)**:** Refers to an experimental design where the residual variance of the difference between the candidates may adopt one out of only two different values

Participatory plant breeding: Approaches that involve close collaboration between researchers and farmers, and potentially other stakeholders, to bring about plant genetic improvements within a species; participatory plant breeding covers the whole research and development cycle of activities associated with plant genetic improvement, such as identifying breeding objectives, generating genetic variability or diversity, selecting within variable populations to develop experimental materials, evaluating these materials, release of materials, diffusion, seed production, and distribution; it also may include assessing existing policy or legislative measures, or both, and designing new ones where needed

Participatory variety selection: Selection of fixed lines (including landraces) by farmers in their target environments using their own selection criteria; it consists of four methodological steps: (a) situation analysis and identification of farmer's varietal needs, (b) search for suitable genetic materials, (c) farmer's experimentation with new crop varieties in their own fields and with their own crop-management practices, and (d) wider dissemination of farmer-preferred crop varieties

Particulate inheritance: The model proposing that genetic information is transmitted from one generation to the next in discrete units (particles) so that the character of the offspring is not a smooth blend of factors from the parents

Path-coefficient analysis: A method for analysing regular and irregular breeding systems; the path coefficient is a measure of the importance of a given path of influence from cause to effect; it is the ratio of the standard deviation of the effect to the total standard deviation

Pathogen: A simple organism which damages a crop plant. The most important crop pathogens are fungi, bacteria and viruses. Larger pathogens, such as nematodes, insects, rats, birds etc. are usually referred to as *pests* rather than as *pathogens*

Pedigree breeding: A system of breeding in which individual plants are selected in the segregating generations from a cross on the basis of their desirability judged individually and on the basis of a pedigree record; the advantages are (1) if selection is effective, inferior genotypes may be discarded before lines are further evaluated; (2) selection in generation involves a different environment, which provides good genetic variability; and (3) the genetic relationship of lines is estimated and can be used to maximise genetic variability

Pedigree: The record of the ancestry of an individual, genetic line or variety

Penetrance: The proportion of individuals with a specific genotype who manifest that genotype at the phenotype level.

Pericentric inversion: The inversion of a chromosome segment containing a block of genes that involves the centromere

Periclinal chimera: A plant made up of two genetically different tissues, one surrounding the other

Pest-resistant crops: Plants with the ability to withstand, deter or repel pests and thereby prevent them from damaging the plants. Plant pests may include insects, nematodes, fungi, viruses, bacteria, weeds, and other.

Phage: Short for bacteriophage; a virus for which the natural host is a bacterial cell. Literally "bacteria eaters."

Phasmid: A cloning vector that has the capabilities to replicate as a plasmid or as a phage; the two modes of replication are usually functional in different bacterial species

Phenotype: The visible appearance of an organism. The phenotype reflects the combined action of the genotype and the environment where the individual exists

Phenotypic segregation: The phenotypic differentiation patterns of cells or individuals in segregating populations; as opposed to genetic segregation

Phenotypic selection: Development of a variety based on its physical appearance without regard to its genetic constitution

Phenotypic variance: The total variance observed in a character; it includes experimental error, genotype × environment interaction, and the genotypic variance

Physical map: Map of the locations of identifiable landmarks on DNA (eg. restriction enzyme cutting sites, genes) regardless of inheritance; the distance is measured in base pairs

Phytosanitary certificate: A certificate from a recognised plant quarantine service that indicates that a sample is substantially free from diseases or pests

Plant breeder's rights: The intellectual property rights that are legally accorded to plant breeders by various laws or international treaties

Plant breeding: The application of genetic principles and practices to development of individuals or cultivars more suited to the needs of humans; it uses knowledge from agronomy, botany, genetics, cytogenetics, molecular genetics, physiology, pathology, entomology, biochemistry, or statistics

Plant introduction (PI): Germplasm brought to the United States from other parts of the world to provide new genes for potential improvement of crop productivity.

Plant quarantine: The isolation of newly imported plants to ensure that they are not introducing any foreign parasites; the term quarantine may refer to the quarantine station itself, or to the process of testing and purifying the plant material

Plant Variety Protection Act: Enacted in 1970 in the United States to provide patentlike protection for seed plant species; prior to 1970, breeders in the industry worked primarily with maize and sorghum, with some efforts directed to alfalfa, cotton, sugarbeet, and certain vegetables; the purpose of the PVPA was to stimulate private plant breeding research and to provide better seed cultivars to farmers and gardeners

Plant variety rights: The legal rights of a plant breeder (not necessarily as a person) who has developed a new variety

Plant-incorporated protectants (PIPs): Pesticidal substances introduced into plants by genetic engineering that are produced and used by the plant to protect it from pests. The protein toxins of Bt are often used as PIPs in the formation of Bt crops.

Plaque: In bacterial virus analysis, a clear area of a petri dish, devoid of bacterial cells, indicating the presence of viral particles.

Plasmid: Autonomously replicating, extra chromosomal circular DNA molecules, distinct from the normal bacterial genome and nonessential for cell survival under nonselective conditions. Some plasmids are capable of integrating into the host genome.

Plastid DNA: Organelle DNA that is present in a plastid

Plastid inheritance: Non-Mendelian inheritance that is caused by hereditary factors present in the plastids plot with or without mass selection, generally followed by a single-plant selection; it is a procedure for inbreeding a segregating population until the desired level of homozygosity is achieved

Point mutation: A change in a single base pair.

Polarity: An overall direction.

Polyacrylamide gel electrophoresis (PAGE): A method for separation of proteins and amino acids on the basis of their molecule size; the molecules move through the gel under the influence of an electric field

Polycistronic RNA: mRNA that codes for more than one polypeptide

Polycross: Open pollination of a group of genotypes (generally selected) in isolation from other compatible genotypes in such a way as to promote random mating *inter se*; it is a widely used procedure for intercrossing parents by natural hybridization

Polygenes: One of a group of genes that together controls a quantitative character; individually each gene has little effect on the resulting phenotype, which instead requires the interaction of

Polygenic resistance: Resistance to parasites based on many genes, i.e. a series of genes controlling a quantitative character

Polyhaploid: Haploid plant derived from a polyploid individual

Polyhybrid: Individuals that are heterozygous with respect to the alleles of many gene loci or of crosses involving parents that differ with respect to the alleles of more loci

Polymerase chain reaction (PCR): A method for amplifying a DNA base sequence using a heat-stable polymerase and two 20-base primers, one complementary to the (+)-strand at one end of the sequence to be amplified and the other complementary to the (-)-strand at the other end. Because the newly synthesized DNA strands can subsequently serve as additional templates for the same primer sequences, successive rounds of primer annealing, strand elongation, and dissociation produce rapid and highly specific amplification of the desired sequence. PCR also can be used to detect the existence of the defined sequence in a DNA sample.

Polymerase, DNA or RNA: Enzymes that catalyse the synthesis of nucleic acids on preexisting nucleic acid templates, assembling RNA from ribonucleotides or DNA from deoxyribonucleotides.

Polymorphic site: A chromosome site with two or more identifiable alleleic DNA sequences. Also called a polymorphic locus.

Polymorphism: A Difference in DNA sequence among individuals. Genetic variations occurring in more than 1% of a population would be considered useful polymorphisms for genetic linkage analysis. Compare mutation.

Polyploidization: The spontaneous or induced multiplication of a haploid or diploid genome of a cell or individual polyploids, represented by "x"

Polyploidy: A state in which multiple copies of a complete genome are present. Polyploidy is rare in animals, but common in plants. In animals (and also plants) some tissues within a diploid organism can be polyploid. The polyploid series is haploid (1 copy), diploid (2 copies), triploid (3 copies), tetraploid (4 copies), pentaploid (5 copies), hexaploid (6 copies) etc.

Polysomy: The reduplication of some but not all of the chromosomes of a set beyond the normal

Polytene chromosome: A chromosome that is formed by repeated reduplication of single chromatids; sections may appear to puff or swell due to differential gene activation; it is visible through the light microscope

Population genetics: The study of inherited variation in populations and its modulation in time and space; it relates the heritable changes in populations to the underlying individual processes of inheritance and development

Population variance: A measure of dispersion defined as the average of the squared deviations between the observed values of the elements of a population and the corresponding mean of those values

Positional candidate gene: A gene known to be located in the same region as a DNA marker that has been shown to be linked to a single-locus trait or to a quantitative trait locus and whose function suggests that it could be the source of genetic variation in the trait in question

Positional cloning: The process where researchers obtain the clone of a gene without prior knowledge of its protein product or function; it uses large-scale physical and formal genetic linkage maps to find specific genes.

Positive assortative mating: A type of nonrandom mating in which individuals of similar phenotype mate more often than predicted under random mating conditions

Post-transcriptional gene silencing (PTGS): Silencing of an endogenous gene caused by the introduction of a homologous dsRNA, transgene, or virus; in PTGS, the transcript of the silenced gene is synthesized but does not accumulate because it is rapidly degraded; this is a more general term than RNAi

Prebreeding: All research and screening activities before a plant material enters the directed breeding process (eg. the development of germplasm to a state where it is viable for breeder's use); primarily, it involves the evaluation of traits from exotic material and their introduction into more cultivated backgrounds

Primary gene pool: Includes the cultivated species of a crop and related species from which useful genes can be most readily obtained for breeding; in general, it is the total sum of all the genetic variation in the breeding population of a species and closely related species that commonly interbreed with, or can be routinely crossed with, the species

Primer: Short, pre-existing oligonucleotide or polynucleotide cahin to which new DNA can be added by DNA polymerase.

Primer-dimers: Occur when single stranded primer oligonucleotides bind to each other rather than the DNA template

Probe: Single-stranded DNA or RNA molecules of specific base sequence, labeled either radioactively or immunologically, that are used to detect the complementary base sequence by hybridisation.

Processed pseudogene: A copy of a functional gene which has no promoter, no introns, and which, consequently, is not itself transcribed; pseudogenes are thought to originate from the integration into the genome of cDNA copies synthesized from mRNA molecules by reverse transcriptase; pseudogenes therefore have a poly (dA) sequence at their 5′ ends; because they are not subject to any evolutionary pressure to maintain their coding potential, pseudogenes accumulate mutations and often have stop codons in all three reading frames

Progeny selection: Selection based on progeny performance

Progeny test: A test of the value of a genotype based on the performance of its offspring produced in some definite system of mating

Prokaryote: An organism without a nucleus; eubacteria, archaebacteria, and blue-green algae.

Promoter: A region of DNA involved in binding of RNA polymerase to initiate transcription.

Protandry: The maturation of anthers before carpels (eg. in sugarbeet, sunflower, or carrot)

Protein: A large molecule composed of one or more chains of amino acids in a specific order; the order is determined by the base sequence of nucleotides in the gene coding for the protein. Proteins are required for the structure, function, and regulation of the bodys cells, tissues, and organs, and each protein has unique functions. Examples are hormones, enzymes, and antibodies.

Proteome: The complete set of proteins detectable in a tissue; it is the functional representation of the genome that includes the types, functions, and interactions of proteins that are present in a cell; it is not a fixed characteristic of a cell, but variable depending on developmental stage, hormonal status, etc.

Proteomics: A system for identifying proteins within a cell or tissue sample, i.e. systematic characterization of proteins, which are found in a tissue or in a specific physiological condition; the proteins can be identified using mass spectroscopy

Protoclonal variation: Variability of somatic cells derived from protoplast culture

Protogyny: A condition in which the female parts develop first (eg. in bajra)

Proto-oncogene: A gene that can mutate to an allele, an oncogene, that causes a cell to become cancerous.

Protoplast culture: The isolation and culture of plant protoplasts by mechanical means or by enzymatic digestion of plant tissue, organs, or cultures derived from these; protoplasts are utilised for selection or hybridisation at the cellular level and for a variety of other purposes

Protoplast fusion: A technique used in somatic hybridisation experiments; it is used for overcoming crossing barriers; protoplasts are placed together and induced to fuse, applying fusogenic agents, such as polyethylene glycol or physical means; subsequent regeneration of the cell wall allows the propagation and regeneration of a somatic hybrid plant

Protoplast: Cell without a cell wall; protoplasts are produced by enzymes, which digest the wall; they are used for production of hybrid cells by protoplast fusion or for injection of foreign DNA

Pseudoallele: Genes that behave as alleles in the allelism test but that can be separated by crossingover

Pseudodominance: The apparent dominance of a recessive gene (allele), owing to a deletion of the corresponding gene in the homologous chromosome

Pseudogamy: A type of apomixis in which the diploid egg cell develops into the embryo without fertilisation of the egg cell, although only after fertilization of the polar nuclei with one of the sperm cells from the male gamete to form a normal triploid endosperm

pUC18 vector: A plasmid cloning vector; size: 2.7 kb; ~100 copies per chromosome; it shows ampicillin resistance for selection and alpha complementing fragment of beta-galactosidase

Pulsed-field gel electrophoresis: An electrophoretic technique in which the gel is subjected to electrical fields alternating between different angles, allowing very large DNA fragments to move through the gel, and hence permitting efficient separation of mixtures of such large fragments

Pure line: A number of individuals of a successive, self-pollinated crop, which derives from a single plant; a strain homozygous at all loci

Purine: A nitrogen-containing, single-ring, basic compound that occurs in nucleic acids. The purines in DNA and RNA are adenine and guanine.

Pyramiding of genes: The process of bringing together several disease resistance or agronomically important genes from different sources into one genetic background (genotype)

Pyrimidine: A nitrogen-containing, double-ring, basic compound that occurs in nucleic acids. The pyrimidines in DNA are cytosine and thymine; in RNA, cytosine and uracil.

Q banding: A chromosomal staining technique using the fluorescence dye quinacrin mustard; under UV light a characteristic light and dark banding is induced

Qualitative character: A character in which variation is discontinuous

Qualitative inheritance: An inheritance of a character that differs markedly in its expression amongst individuals of a species; variation is discontinuous; such characters are usually under the control of major genes

Qualitative trait: Traits that are easily classified into distinct phenotypic categories. These phenotypes are under the genetic control of only one or a few genes with little or no environmental modification to obscure the genetic effects.

Quantitative character: A character in which variation is continuous so that classification into discrete categories is arbitrary

Quantitative genetics: A branch of genetics that deals with the inheritance of quantitative traits; sometimes it is also called biometrical or statistical genetics

Quantitative inheritance: An inheritance of a character that depends upon the cumulative action of many genes, each of which produces only a small effect; the character shows continuous variation (i.e. a graduation from one extreme to the other)

Quantitative trait loci (QTL): Regions of the chromosome associated with the inheritance of polygenic traits

Quarantine: The official confinement of plants subject to phytosanitary regulations for observation and research or for further inspection and/or testing; more general, a legal ban on the export or import of certain noxious weeds or insects that may be attached to the plants

Random amplified polymorphic DNA (RAPD) technique: A comparative study (among individuals, populations, or species) of the DNA fragment length produced in controlled DNA synthesis reactions started with short sequences of DNA (primers); as a genetic mapping methodology, it utilises as its basis the fact that specific DNA sequences (polymorphic DNA) are repeated (i.e., appear in sequence) with a gene of interest; thus, the polymorphic DNA sequences are linked to that specific gene; their linked presence serves to facilitate genetic mapping within a genome

Random genetic drift: Variation in gene frequency from one generation to another due to chance fluctuations

Random mating: For a given population, where an individual of one sex has an equal probability

Randomisation: The process of making assignments at random; in field trials, randomization of entries is required to obtain a valid estimate of experimental error; each entry must have an equal chance of being assigned to any plot in a replication and an independent randomization is required for each replication

Randomised block design: A randomised block analysis of variance design (eg. one-way blocked ANOVA) is created by first grouping the experimental subjects into blocks; the subjects in each block are as similar as possible (eg. littermates); there are as many subjects in each block as there are levels of the factor of interest; randomly assigning a different level of the factor to each member of the block, such that each level occurs once and only once per block; the blocks are assumed not to interact with the factor

Reading frame: A sequence of sense codons such that each suceeding triplet generates the correct order of amino acids resulting in a polypeptide chain.

Realised genetic gain: Refers to the observed difference between the mean phenotypic value of the offspring of the selected parents and the phenotypic value of the parental generation before selection

Realised heritability: Refers to heritability measured by a response to selection; it is the ratio of the single-generation progress of selection to the selection differential of the parents

Recalcitrant seed: Seed that does not survive drying and freezing; in particular, seed that cannot withstand either drying or temperatures of less than +10°C and, therefore, cannot be stored for long periods, as compared to orthodox seeds

Recessive allele: An allele whose phenotypic effect is not expressed in a heterozygote.

Recessive gene: A gene whose phenotype is expressed only when it is present as two copies of both alleles (eg. aa).

Reciprocal cross: One of a pair of crosses in which the two opposite mating types are each coupled with each of two different genotypes and mated with the reciprocal combination

Reciprocal full-sib selection: A method of interpopulation improvement for species in which the commercial product is hybrid seed; a cycle of selection is completed in the fewest number of seasons by use of plants from which both selfed and hybrid seed can be obtained

Reciprocal recurrent selection: A breeding method used to achieve an accumulation of genes that are valuable for specific traits but also for combining ability; in practice, two populations form the basis of selection; reciprocally, one population serves as a tester for the investigation of the selections deriving from the other population; from the first population, usually a greater number of plants is selfed and at the same time crossed with several plants of the second population; the same procedure is realised with the second population; during the second year, the progenies of the test-crosses are subjected to performance trials; the progenies of the cross of one female parent, derived from population 1, are combined with several male parents from population 2; based on the performance testing, progenies from selfed seed, produced by the best plants, are grown during the third year; in the same year, the best progenies of selfings are crossed in many combinations; seeds obtained from those crosses form the improved populations for growing during the fourth year; within such populations, individual plants may be selfed again and crossed for utilisation in a new cycle

Reciprocal translocation: A translocation that involves an exchange of chromosomes segments between two non-homologous chromosomes

Recombinant DNA molecules: A combination of DNA molecules of different origin that are joined using recombinant DNA technologies.

Recombinant DNA technology: Procedures used to join together DNA segments in a cell-free system (eg. in a test tube outside living cells or organisms). Under appropriate conditions, a recombinant DNA molecule can be introduced into a cell and copy itself (replicate), either as an independent entity (autonomously) or as an integral part of a cellular chromosome.

Recombinant inbred lines (RILs): A population of fully homozygous individuals that is obtained through the repeated selfing of an F_1 hybrid, and that comprises 50% of each parental genome in different combinations; they derive from multiple inbred strains (either by selfing or by full-sib mating) and can serve as a powerful resource for the genetic dissection of complex traits; however, the use of such multiple-strain requires a detailed knowledge of the haplotype structure in such lines

Recombination frequency: The number of recombinants divided by the total number of progeny, expressed as a percentage or fraction; such frequencies indicate relative distances between loci on a genetic map

Recombination: The formation of a new combination of alleles through independent assortment or crossing-over.

Recurrent full-sib selection: A method of intra population improvement that involves the testing of paired-plant crosses; it is the only method of recurrent selection in which the seeds from two individuals, rather than one, are used for testing and to form the new population

Recurrent half-sib selection: A method of intra population improvement that includes the evaluation of individuals through the use of their half-sib progeny; the general procedure for a cycle of selection is (1) to cross the plants being evaluated to a common tester, (2) evaluate the half-sib progeny from each plant, and (3) intercross the elected individuals to form a new population

Recurrent mass selection: A breeding method designed to increase the levels of desirable qualities, which are quantitative variables, by changing the frequency of polygenes; in each screening generation, the best individuals are selected, and they become the parents of the next screening generation; the process is repeated for as many generations as necessary, but the rate of progress declines dramatically after a few generations recurrent parent is used as the selection criterion

Recurrent parent: The parent to which a hybrid is crossed in a backcross; it replaces the dragged alleles step-by-step with the alleles of the original variety

Recurrent reciprocal selection: A recurrent selection breeding system in which genetically different groups are maintained and, in each selection cycle, individuals are mated from the different groups to test for combining ability

Recurrent selection: A method designed to concentrate favourable genes scattered among a number of individuals; it is performed by repeated selection in each generation among the progeny produced by matings *inter se* of the selected individuals of the previous generation

Reduction division: The two nuclear divisions in meiosis that produce daughter nuclei, each of which has half as many chromosomes as the parental nucleus

Redundant DNA: DNA that does not appear to be genetically active and hence is not translated or transcribed; it often consists of repeated sequences

Redundant gene: A gene that is present in many functional copies, so that one copy can complement the loss of another copy

Reduplication: Doubling of the genetic matter of a haploid chromosome set

Regeneration: The embryonic development of shoots and roots from callus tissue, producing a plantlet.

Regression analysis: The process of fitting a regression equation to a set of data by using the method of least squares; it includes the various statistical tests and estimates associated with the use of the fitted equation

Regression coefficient: The rate change of the dependent variable with respect to the independent variable

Regulator gene: In the operon theory of gene regulation, a gene that is involved in switching on or off the transcription of structural genes; when transcribed, the regulator gene produces a repressor protein, which switches off an operator gene and hence the operon that this controls; the regulator gene is not part of the operon and may even be on a different chromosome

Regulatory regions or sequences: A DNA base sequence that controls gene expression.

Renaturation: The reassociation of denatured complementary single strands of a DNA double helix.

Resolution: Degree of molecular detail on a physical map of DNA, ranging from low to high.

Restorer line (R line): A pollen parent line; it contains the restorer gene or genes, which restore cytoplasmic male sterile plants to pollen fertility; it is crossed with an A line in the production of hybrid seeds

Restriction enzyme cutting site: A specific nucleotide sequence of DNA at which a particular restriction enzyme cuts the DNA. Some sites occur frequently in DNA (E.g. every several hundred base pairs), others much less frequently (rare-cutter; e.g., every 10,000 base pairs).

Restriction enzyme, endonuclease: A protein that recognises specific, short nucleotide sequences and cuts DNA at those sites. Bacteria contain over 400 such enzymes that recognise and cut over 100 different DNA sequences. See restriction enzyme cutting site.

Restriction enzyme: An enzyme that functions in a bacterial modification-restriction system and recognises specific nucleotide sequences and breaks the DNA chain at these sites; there is a great number of them, each with different recognition and/or cutting sites; they are intensively used as a tool in molecular genetics and also in producing chromosomal banding patterns in cytogenetics

Restriction fragment length polymorphism (RFLP): Variation between individuals in DNA fragment sizes cut by specific restriction enzymes; polymorphic sequences that result in RFLPs are used as markers on both physical maps and genetic linkage maps. RFLPs are usually caused by mutation at a cutting site. See marker.

Restriction map: Representation of DNA with the position of restriction sites indicated

Restriction site: A certain nucleotide sequence within the double-stranded DNA; it is recognised by a restriction endonuclease; the enzyme cuts the double strand within the recognition sequence; the restriction sites are usually composed of four to six base pairs and are bilaterally symmetric; both strands are cut either on exactly opposite positions (blunt ends) or alternated ones (sticky ends); the type of cutting depends on the enzyme used

Retrotransposon: Retrotransposons are a ubiquitous and major component of plant genomes; those with long terminal DNA repeats (LTRs, Ty1-copia-like family) are widely distributed over the chromosomes of many plant species

Reverse genetics: Using linkage analysis and polymorphic markers to isolate a disease gene in the absence of a known metabolic defect, then using the DNA sequence of the cloned gene to predict the amino acid sequence of its encoded protein; in general, a technology aiming at isolating mutants of a given sequence; it is also applied for identification of gene function

Reverse mutation: The production, by further mutation, of a premutation gene from a mutant gene; it restores the ability of the gene to produce a functional protein; strictly, reversion is the correction of a mutation (i.e. it occurs at the same site)

Reverse transcriptase: An enzyme from retroviruses for the synthesis of a DNA complementary to a RNA molecule (i.e., cDNA); it is used for (1) filling-in reactions, (2) for DNA sequencing, and (3) for cDNA synthesis

Ribonucleic Acid (RNA): A chemical substance made up of nucleotides compound of sugars, phosphates, and derivatives of the four bases adenine (A), guanine (G), cytosine (C), and uracil (U). RNAs function in cells as messengers of information from DNA that are translated into protein or as molecules that have certain structural or catalytic functions in the synthesis of proteins. RNA is also the carrier of genetic information for certain viruses. RNAs may be single or double stranded.

Ribosomal RNA (rRNA): A class of RNA found in the ribosomes of cells.

Ribosomes Small cellular components composed of specialised ribosomal RNA and protein; site of protein synthesis.

Ring Formation: Occurs where the ends of a chromosome bend over and fuse together with a loss of genetic material.

RNA interference (RNAi): The use of double stranded RNA to interfere with gene expression; RNAi is usually mediated by approximately 21-nt small interfering RNAs

RNA polymerase: An enzyme that transcribes an RNA molecule from the template strand of a DNA molecule; it adds to the 3′ end of the growing RNA molecule one nucleotide at a time using ribonucleotide triphosphates (rNTPs) as substrates; RNA polymerase I is dedicated to the synthesis of only one type of RNA molecule (pre-rRNA); RNA polymerase II is required for general transcription reactions; RNA polymerase III produces small RNAs such as tRNAs and 5S rRNA

RNA transcriptase: The enzyme responsible for transcribing the information encoded in DNA into RNA; it is also called transcriptase or RNA polymerase

Robertsonian translocation: A chromosomal mutation due to centric fusion or centric fission (i.e., a reciprocal translocation with breakpoints within the centromeric regions); in wheat, Robertsonian translocations arise from centric misdivision of univalents at anaphase/ telophase I, followed by segregation of the derived telocentric chromosomes to the same nucleus, and fusion of the broken ends during the ensuing interkinesis root disease in certain plants

Satellite chromosome: Each chromosome with a secondary constriction; this constriction divides a satellite part from the rest of the chromosome (arm)

Satellite DNA: A highly repetitive DNA, composed of repeated hepta- to deca-nucleotide sequences; DNA of different buoyant density; a minor DNA fraction that has sufficiently different base composition from the bulk of the DNA in order to separate distinctly during cesium chloride density gradient centrifugation; it can derive from nucleus, plastid, or mitochondrial

Saturation mutagenesis: Induction and recovery of large numbers of mutations in one area of a genome, or in one biological function, in order to identify all the genes in that area, or affecting that function

Second division restitution (SDR): Due to premature cytokinesis before the second meiotic division takes place; sister chromatids end up in the same nucleus

Secondary gene pool: Species in the secondary gene pool include those from which genes can be transferred to the cultivated species, however, with more difficulties as compared to species

Seed certification: A procedure developed as a means of assuring that seeds have a high standard of purity and quality

Segregating population: A population, generally the progeny of a cross, in which genetic differences are detectable, thus permitting identification of individuals having a desired trait and

Segregation: The process whereby alleles are separated from one another as a result of meiosis. Segregation can only occur with respect to genes which are in the heterozygous state in the parental plant

Selectable marker: A gene, often encoding resistance to an antibiotic or an herbicide, introduced into a group of cells to allow identification of those cells that contain the gene of interest from the cells that do not. Selectable markers are used in genetic engineering to facilitate identification of cells that have incorporated another desirable trait that is not easy to identify in individual cells.

Selection differential: In artificial selection, the difference in mean phenotypic value between the individuals selected as parents of the following generation and the whole population

Selection intensity: The ratio of the number of genotypes selected divided by the number of genotypes tested

Selection response: The difference between the mean of the individuals selected to be parents and the mean of their offspring; it is expressed by the formula $R = h^2 \times S$ (h = heritability, S = selection coefficient, R = phenotypic difference between the mean of all selected fractions and the mean of total population)

Selection: The process determining the relative share allotted individuals of different genotypes in the propagation of a population; natural selection occurs if zygotic genotypes differ with regard to fitness

Selective advantage: An advantage for survival of a genotype in a population, and for production of viable progeny as compared to other genotypes, which may show a selective disadvantage with respect to fitness and viability

Selective breeding: Making deliberate crosses or matings of organisms so the offspring will have particular desired characteristics derived from one or both of the parents.

Selective disadvantage: Inferior fitness of one genotype compared to others in the population

Self-incompatibility: Controlled physiological hindrance to self-fertilization; inability to set seed

Selfish DNA: A segment of the genome with no apparent function other than to ensure its own replication

Self-pollination: The transfer of pollen from anther to stigma of the same plant

Self-sterility: The inability of some hermaphrodites to form viable offspring by self-fertilisation

Semiconservative replication: Replication of DNA in which the molecule divides longitudinally, each half being conserved and acting as a template for the formation of a new strand

Sense mutation: A mutation that changes a termination (stop) codon into one that codes for an

Sense strand DNA: The DNA strand with the same sequence as Mrna

sequence characterised amplified region (SCAR): A molecular marker; the DNA amplified as a PCR derived marker (eg. RAPD) is sequenced; the sequence is used to develop a new PCR marker comprising a single DNA species, unique to the target organism or genotype

Sequence tagged site (STS): Short (200 to 500 base pairs) DNA sequence that has a single occurrence in the human genome and whose location and base sequence are known. Detectable by polymerase chain reaction, STSs are useful for localising and orienting the mapping and sequence data reported from many different laboratories and serve as landmarks on the developing physical map of the human genome. Expressed sequence tags (ESTs) are STSs derived from cDNAs.

Sequence-tagged-microsatellites (STMs) marker: Primers constructed from the flanking regions of microsatellite DNA which can be used in PCR reactions to amplify the repeat region

Sequencing: Determination of the order of nucleotides (base sequences) in a DNA or RNA molecule or the order of amino acids in a protein.

Sex chromatin: A condensed mass of chromatin representing an inactivated X chromosome

Sex chromosomes: The X and Y chromosomes in human beings that determine the sex of an individual. Females have two X chromosomes in diploid cells; males have an X and a Y chromosome. The sex chromosomes comprise the 23rd chromosome pair in a karyotype. Compare autosome.

Sex linkage: Genes located and inherited on a sex chromosome

Shotgun method: A way of preparing DNA for genetic studies; random-cut fragments of DNA are cloned into a vector; the fragments resemble a clone library; using this collection of clones, several molecular techniques are applicable

Shuttle breeding: A program that shuttles seed between two (or more) locations to be grown at each; this method was introduced by N. Borlaug and is a proven contributor to the success of CIMMYT (Mexico); shuttle breeding allows two breeding cycles per year instead of one - for example, a winter cycle in the northern desert of Sonora, and a summer crop in the central Mexican highlands; this not only fast-forwards selection, but also exposes test varieties to radically different day lengths, temperatures, altitudes, and diseases; resulting plants are broadly adapted; they grow well in numerous environments

Shuttle vector: A vector constructed in such a way that it can replicate in at least two different host species recombined into such a vector can be tested or manipulated in several cell types

Sib mating: A form of inbreeding in which progeny of the same parents (siblings) are crossed

Silent mutation: Mutation in which the function of the protein product of the gene is unaltered.

Single sequence repeat (SSR) DNA marker technique: A genetic mapping technique that utilises the fact that microsatellite sequences repeat (appear repeatedly in sequence within the DNA molecule) in a manner enabling them to be used as markers

Single-gene disorder: Hereditary disorder caused by a mutant allele of a single gene (e.g., Duchenne muscular dystrophy, retinoblastoma, sickle cell disease). Compare polygenic disorders.

Single-nucleotide polymorphisms (SNPs): Variations (in individual nucleotides) that occur within DNA at the rate of approximately one in every 1,300 bp in most organisms; SNPs usually occur in the same genomic location in different individuals

Single-seed descent (SSD): Derivation of plants by a selection procedure in which F_2 plants and their progeny are advanced by single seeds until genetic purity is achieved; single-seed descent methods (single-seed, single-hill, multiple-seed) are easy ways to maintain populations during inbreeding; natural selection cannot influence the population, unless genotypes differ in their ability to produce viable seeds; artificial selection is based on the phenotype of individual plants, not on the progeny performance

Single-strand conformational polymorphism (SSCP): This feature relies on secondary and tertiary structural differences between denatured and rapidly cooled amplified DNA fragments that differ slightly in their DNA sequences; different SSCP alleles are resolved on non-denaturing acrylamide gels, usually at low temperatures; the ability to resolve alleles depends on the conditions of electrophoresis and this requires DNA sequence data

Site-specific mutagenesis: The use of recombinant DNA technology to create specific deletions, insertions, or substitutions *in vitro* in a particular gene; the technique allows the production of proteins having any desired amino acid at any position

Site-specific recombination: A crossover event that requires homology of only a very short region and uses an enzyme specific for that recombination; recombination occurring between two specific sequences that need not be homologous, mediated by a specific recombination system

Small interfering RNAs (siRNAs): Models of PTGS indicate that 21–23 nucleotide dsRNAs mediate PTGS; introduction of siRNAs can induce PTGS, e.g. in mammalian cells; siRNAs are apparently produced *in vivo* by cleavage of dsRNA introduced directly or *via* a transgene or virus; amplification by an RNA-dependent RNA polymerase (RdRP) may occur in some organisms; siRNAs are incorporated into the RNA-induced silencing complex (RISC), guiding the complex to the homologous endogenous mRNA where the complex cleaves the transcript

SNP: An abbreviation for "single nucleotide polymorphism", pronounced "snip". A SNP which distinguishes two sequences can be used as a genetic marker

Somaclonal variation: Variation among the tissues or plant derived from the *in vitro* somatic cell culture i.e., callus or suspension cultures.

Somatic cells: All the cells of an organism except those of the germ line.

Somatic crossing-over: Crossing-over during mitosis of somatic cells such that parent cells heterozygous for a given allele, instead of giving rise to two identical heterozygous daughter cells, give rise to non-identical daughter cells, one of which is homozygous for one of these alleles, the other being homozygous for the other allele

Somatic embryo: An organised embryonic structure morphologically similar to a zygotic embryo but initiated from somatic cells; somatic embryos develop into plantlets *in vitro* through developmental processes that are similar to those of zygotic embryos

Somatic hybridisation: The fusion of genetically different somatic cells by different means, usually to overcome natural crossing (incompatibility) barriers

Somatic mutation: A mutation occurring in a somatic cell; if the mutated cell continues to divide, the individual will develop a patch of tissue with a genotype different from the cells of the rest of the body

Southern blotting: Transfer by absorption of DNA fragments separated in electrophoretic gels to membrane filters for detection of specific base sequences by radiolabeled complementary probes. SPERM (abbreviation of Spermatozoon) The male reproductive cell carrying 23 chromosomes.

Splicing: The removal of the intron (noncoding sequence) from an mRNA gene sequence and the fusion of the exons (coding sequence) during the mRNA processing

Split plot design: Allows the testing of two factors in combination; one factor (the main effect) serves as a replication for the second factor (the split effect); there are many split plot design options, but the basic principle involves assigning one set of treatments to the main plots that are arranged in randomised complete blocks; the second set of treatments is assigned to subplots within each main plot; the statistical analysis is similar to that used with the randomized complete block design, i.e., an ANOVA set up for a split plot is followed by a means comparison test such as the Dunn's test or Duncan's

Spontaneous mutation: A naturally occurring mutation, as opposed to one artificially induced by chemicals or irradiation; usually such mutations are due to errors in the normal functioning of cellular enzymes

Sporophytic self-incompatibility: Self-incompatibility is based on the genotypic and phenotypic relationship between the female and male reproductive system; alleles in cells of the pistil determine its receptivity to pollen; the phenotype of the pollen, expressed as its inability to effect fertilization, may be determined by the maternal plant, referred to as sporophytic incompatibility

Stabilizing selection: A type of selection that removes individuals from a population (i.e., from both ends of a phenotypic distribution divided by deviation of a sample of means)

Stacked traits: Term used to describe seed products or GMO's with more than one biotechnology trait. These GMO's have had more than one gene from another organism or organisms transferred into their genome, thus creating a GMO with stacked "biotech" traits.

Standard deviation (SD): A measure of the variability in a population of items, i.e., a set of *n* measurements *x1, x2, x3, ..., xn* is equal to the positive square root of the variance of the measurements

Standard error (SE): A measure of variation of a population of means

Statistic: A quantity that is calculated from a sample of data in order to give information about unknown values in a corresponding population

Statistical analysis: Analysing collected data for the purposes of summarising information to make it more usable and/or making generalizations about a population based on a sample drawn from that population

Statistical significance: The degree to which a value is greater or smaller than would be expected by chance; typically, a relationship is considered statistically significant when the probability of obtaining that result by chance is less than 5% if there were, in fact, no relationship in the population

Stratified mass selection: Mass selection in which the population is split into subpopulations that are grown under different environmental conditions (i.e., in different fields or in different parts of a field); plants for next generation seed are selected from the different subpopulations

Subcloning: Transplantation of a piece of DNA from one vector to another

Subculture: The aseptic transfer of a part of a stock culture to a fresh medium

Substitution line: A line in which one or more chromosomes are replaced by one or more chromosomes of a donor variety or species; it is used for genetic analysis and gene transfer

Supernumerary chromosomes: Chromosomes present, often in varying numbers, in addition to the characteristic invariable complement of chromosomes

Suppression: Changes that eliminate the effects of a mutation without reversing the original change in DNA.

Suppressor mutation: A mutation that counteracts the effects of another mutation. A suppressor maps at a different site than the mutation it counteracts, either within the same gene or at a more distant locus. Different suppressors act in different ways.

Synaptonemal complex (SC): Ribbonlike structures observed in electron micrographs of nuclei in the synaptic stages of meiosis; the ribbon represents a system that promotes synapsis of homologous chromosomes; the SC seems to be the essential prerequisite for homologous chromosome pairing and crossing-over

Synthetic variety: A variety produced by crossing *inter se* a number of genotypes selected for good combining ability in all possible hybrid combinations, with subsequent maintenance of the variety by open pollination; usually the first generation of a synthetic variety is obtained by a polycross involving a certain number of components with a good general combining ability; the components are maintained by identical reproduction, either by vegetative propagation (clones) or by continued sib mating (inbred populations)

Tandem repeat sequences: Multiple copies of the same base sequence on a chromosome; used as a marker in physical mapping.

Tandem repeat: A chromosomal mutation in which two identical chromosome segments lie adjacent to each other, with the same gene order; the DNA, which codes for the rRNA, contains many tandem repeats

Tandem selection: In the case of successive multiple selection, the selection concerns other traits in the first few generations than in later generations

Taq polymerase: A DNA-dependent RNA polymerase from phage T7, which recognises a very specific promoter sequence; it is used in many expression vectors

Targeting-induced local lesions in genomes (TILLING): A technique to screen a population that has either been deliberately mutagenized or possesses natural biodiversity (EcoTILLING) to identify those plants with a change in a specific gene of interest; the technique uses high-throughput methods useful to plant breeders, including a reverse molecular method that combines random chemical mutagenesis with PCR-based screening of gene regions of interest; it provides a range of allele types, including mis-sense and knockout mutations; by comparing the phenotypes of isogenic genotypes differing in single sequence motifs, TILLING provides direct proof of function of both induced and natural polymorphisms without the use of transgenic modifications

TATA box: A canonic DNA sequence; part of a plant promoter; promotes the transcription of DNA

Technology transfer: The process of converting scientific findings from research laboratories into useful products by the commercial sector.

Telocentric (chromosome): Refers to the chromosomal centromere which lies at the end

Telomerase: A reverse transcriptase (h*TERT*) containing an RNA molecule (h*TR*) that functions as the template for the tandem repeat at the telomere; it synthesizes the telomere to maintain its length after each cell division; it is active in young cells and gametes, inactive in differentiated somatic cells, and reactivated in malignant cells; telomerase can add one base at a time to the telomeric end of a chromosome; this maintenance work is required for cells to escape from replicative senescence

Telomere: The ends of chromosomes. These specialized structures are involved in the replication and stability of linear DNA molecules. See DNA replication.

Telotrisomic: In allopolyploids, such as hexaploid wheat, a cell or individual with one missing chromosome but having a telocentric and an isochromosome for the same arm of the missing one

Temperature-sensitive mutation: A class of conditional mutations; the mutant phenotype is observed in one temperature range and the wild-type phenotype is observed in another temperature range.

Template strand: The strand of the DNA double helix that is copied by base pair complementarity to make an RNA. The other, non-template strand of the DNA duplex has a sequence that is identical to the synthesized RNA (except in RNA, U replaces T).

Template: The DNA single strand, complementary to a nascent RNA or DNA strand, which serves to specify the nucleotide sequence of the nascent strand

Terminalisation: A progressive shift of chiasmata from their sites of origin to more terminal positions

Termination codon: Stop codon

Terminator genes: A set of transgenes designed to arrest the final stages of seed development in a crop plant; the resulting trait is the production of seeds that will not germinate; it is also called the technology protection system

Terminator technology: Plants are genetically engineered so when the crops are harvested, all new seeds produced from these crops are sterile

Tertiary gene pool: Gene transfer from a species of the tertiary gene pool to the cultivated species, since the primary gene pool usually requires special crossing and embryo rescue techniques in order to get viable hybrids

Test cross: A cross between a heterozygote of unknown genotype and an individual homozygous for the recessive genes

Tetrad analysis: The use of tetrads to study the behavior of chromosomes and genes in crossingover during meiosis; particularly used in studies of fungi

Tetraploid: Having four sets of chromosomes in the nucleus

Tetraploidisation: The mitotic or meiotic procedure which produces tetraploids

Tetrazolium test: A quick test to determine seed viability; tetrazolium is a class of chemicals that have the ability to accept hydrogen atoms from dehydrogenase enzymes during the respiration process in viable seeds; this ability is the basis of the test during which the tetrazolium chemical undergoes a color change, usually from colorless to red (formazan)

Three-point test cross: Cross involving one parent with three heterozygous gene pairs and another (tester) with three homozygous recessive gene pairs

Three-way cross: A first generation hybrid between a single-cross and an inbred line or pure line

Thymine (T): A nitrogenous base, one member of the base pair A-T (adenine-thymine).

Ti plasmid: Established tumors in plants contain only a part of the total plasmid

Topcross hybrid: The first generation of a cross between an inbred line and an open pollinated variety

Topcross: A cross between a selection, line, clone, etc., and a common pollen parent, which may be a variety, inbred line, single cross, etc.; the common pollen parent is called the "topcross" or "tester" parent; in maize, a topcross is commonly called an "inbred-variety cross"; usually, it is used in order to test the "general combining ability"

Topoisomerase: Enzyme that catalyses the reduction of supercoiling of DNA; it is important during DNA replication to relieve the negative supercoiling that occurs when the two strands of the DNA double helix are separated

Totipotency: The potential ability of a cell to express all its genetic information under appropriate conditions and to proceed through all the stages of development to produce a fully differentiated adult; A. Hansen first postulated in 1879 that plants are totipotent, referring to the power of regeneration of *Begonia* that can develop new plantules of leaf cuttings, a feature still used by gardeners today; Hansen 's statement was generalised by G. Haberland

Traditional breeding: Modification of plants and animals through selective breeding. Practices used in traditional plant breeding may include aspects of biotechnology such as tissue culture and mutational breeding.

Trait: A recognisable, measurable character in a plant, which is under some genetic control

Trans configuration: The configuration of two sites refers to their presence on two different molecules of DNA (chromosomes).

Transcription factors: Proteins that are directly involved in regulation of transcription initiation by binding to the control elements and allowing RNA polymerase to act; there are ubiquitous transcription factors as well as cell- and tissue-specific ones; several families have been identified including helix-loop-helix proteins, helix-turn-helix proteins, leucine zipper proteins and zinc finger proteins

Transcription unit: The distance between sites of initiation and termination by RNA polymerase; may include more than one gene.

Transcription: The polymerisation of ribonucleotides into a strand of RNA in a sequence complementary to that of a single strand of DNA; by this means, the genetic information contained in the latter is faithfully matched in the former; the process is mediated by a DNA-dependent RNA polymerase

Transcriptome: DNA sequences of the expressed genome

Transcriptomics: The study of global gene expression and changes caused by genetics, environmental effects, drugs or chemicals, or tissue location; it is the application of micro- or macroarrays and sequence-based methods to conduct expression profiling to determine gene expression at a global (genome wide) level

Transfection of eukaryotic cells: The acquisition of new genetic markers by incorporation of added DNA.

Transfer RNA (tRNA): A class of RNA having structures with triplet nucleotide sequences that are complementary to the triplet nucleotide coding sequences of mRNA. The role of tRNAs in protein synthesis is to bond with amino acids and transfer them to the ribosomes, where proteins are assembled according to the genetic code carried by mRNA.

Transformation A process by which the genetic material carried by an individual cell is altered by incorporation of exogenous DNA into its genome.

Transgene: A modified native gene or a gene from another species that is inserted into plants to bring about a desired change or introduce a trait not native to the natural genome of that species.

Transgenic organism: An organism resulting from the insertion of genetic material from another organism using recombinant DNA techniques.

Transgressive segregation: The segregation of individuals in the F_2 or a later generation of a cross that shows a more extreme development of a character than either parent

Translation The process in which the genetic code carried by mRNA directs the synthesis of proteins from amino acids.

Translocation: A change in the arrangement of genetic material, altering the location of a chromosome segment; the most common forms of translocation are reciprocal, involving the exchange of chromosome segments between two non-homologous chromosomes

Transplastomic plants: Plants with genetic modifications in the DNA of their chloroplasts; plant cells contain chloroplasts, cell organelles responsible for photosynthesis; these chloroplasts

Transposable element: A chromosomal locus that may be transposed from one spot to another within and among the chromosomes of the genome; it happens through breakage on either side of these loci and their subsequent insertion into a new position either on the same or a different chromosome

Transposition: In molecular biology, the process of moving a transposon or other inserts from one position to another within a genome

Transposon tagging: The blocking activity of functional genes by insertion of foreign DNA

Transposon: Chromosomal loci capable of being transposed from one spot to another within and among the chromosomes of a complement

Transverison: A mutation in which a purine is replaced with a pyrimidine or vice versa

Trihybrid: Progeny resulting from a cross of parents differing in three genes

Triisosomic: In allopolyploids, such as hexaploid wheat, when a cell or individual lacks one chromosome pair while three homologous isosomes for the same arm are present

Triploid fusion nucleus: The result of the fusion of one male gamete (pollen) with two haploid female nuclei (polar cells) in the second polar event; this nucleus divides to form the cells of the endosperm

Triploidy: A state in which three chromosome sets are present

Trisomic analysis: A method for mapping gene loci on individual chromosomes by comparing disomic and trisomic segregation patterns of a series of individuals

Trisomic: A genome that is diploid but that contains an extra chromosome, homologous with one of the existing pairs, so that one kind of chromosome is present in triplicate

True breeding: A situation in which a group of identical individuals always produce offspring of the same phenotype when intercrossed; a true-breeding plant is that when self-fertilized, only produces offspring with the same traits; the alleles for this type of plants are homozygous

Twins: May be genetically identical (monozygous) when they arise from a single fertilised egg or non-identical (dizygous) when they arise from two separate eggs.

Ultracentrifugation: Centrifugation carried out at high rotor speeds (<100,000 rpm) and therefore under high centrifugal forces (<750,000 g)

Unequal crossing-over: A crossing-over after improper pairing between chromosome homologues that are not perfectly aligned; the result is, for example, one crossing-over chromatid with one copy of the segment and another with three copies

Union for Protection of New Varieties of Plants (UPOV): An intergovernmental organisation with headquarters in Geneva, Switzerland; it is based on the International Convention for the Protection of New Varieties of Plants, as revised since its signature in Paris on December 2, 1961; on April 16, 1993, the union consisted of 23 member States; the objective of the convention is the protection of new varieties of plants by an intellectual property right

Uniparental inheritance: A pattern of inheritance in which only one parent provides genes to the progeny; a phenomenon, usually exhibited by extranuclear genes, in which all progeny have the phenotype of only one parent

Unique DNA sequence: A DNA sequence that is present only once per genome, with no repetitive nucleotide sequence

Univalent shift: A spontaneous change in monosomy from one chromosome to another; it is caused by partial asynapsis or desynapsis during meiosis

Univalent: A single chromosome observed during meiosis when bivalents are also present; it has no pairing mate

Unreduced gametes: Gametes not resulting from common meiosis, and so showing the number of chromosomes per cell that is characteristic of a sporophyte; they spontaneously arise as consequence of irregular division in anaphase I of meiosis; they may contribute to spontaneous (meiotic) polyploidisation; in rye, they were used for production of tetraploids *via* valence crosses

Variability: The sum of different genetic or phenotypic characters within different taxa

Uracil: A nitrogenous base normally found in RNA but not DNA; uracil is capable of forming a base pair with adenine.

Variable expression: A variation in phenotype between affected members of the same family (i.e.individuals carrying identical mutations); it occurs in many dominant conditions and may be associated with reduced penetrance

Variable number tandem repeat (VNTR): Genetic markers that consist of DNA segments that are duplicated end to end; the number of copies present at a locus can vary, giving rise to a large number of alleles; these genetic markers are commonly used for DNA fingerprinting

Variance: When all values in a population are expressed as plus and minus deviations from the population mean, the variance is the mean of the squared deviations; it is a measure of variation of a population; it can be divided into phenotypic variance, genotypic variance, and environmental variance

Variety: A plant differing from other members of the species to which it belongs by the possession of some hereditary traits; breeding varieties can be classified according to the manner of propagation, such as clone varieties (maintained by vegetative propagation), line varieties (maintained by self-fertilisation), panmictic varieties (propagated by cross-fertilisation), or hybrid varieties (produced by directed crosses)

Vector In cloning, the plasmid, phage, or yeast chromosomal sequences used to propagate a cloned DNA segment.

Vertical resistance: The existence of differential levels of resistance to different races of a given pathogen conditioned by one or a few qualitative genes

Virus: A noncellular biological entity that can reproduce only within a host cell. Viruses consist of nucleic acid covered by protein; some animal viruses are also surrounded by membrane. Inside the infected cell, the virus uses the synthetic capability of the host to produce progeny virus.

Western blotting: A technique similar to Southern blotting but for the analysis of proteins instead of DNA; a technique in which proteins are separated by gel electrophoresis and transferred by capillary action to a nylon membrane or nitrocellulose sheet; a specific protein can be identified through hybridization to a labelled antibody

Wide hybridisation: Cross combinations between taxonomically remote species or genera

Wild type: The genotype or phenotype that is found most commonly in nature or in the standard laboratory stock for a given organism.

Working collection: A collection of germplasm kept under short-term storage conditions, commonly used by breeders or researchers

X chromosome: A sex chromosome found in a double dose in the homogametic sex and in a single dose in the heterogametic sex

Xenia: A situation in which the genotype of the pollen influences the developing embryo of the maternal tissue (endosperm) of the fruit to produce an observable effect on the seed

X-linked-recessive: Abnormal gene carried on the X chromosome.

X-ray crystallography: A technique for deducing molecular structure by aiming a beam of X rays at a crystal of the test compound and measuring the scatter of rays.

Y chromosome: Plays a role in sex determination

Yeast artificial chromosome (YAC): A vector used to clone DNA fragments (up to 400 kb); it is constructed from the telomeric, centromeric, and replication origin sequences needed for replication in yeast cells.

Z chromosome: A sex chromosome that is limited to the male sex

zDNA: A left-handed, alternate form of DNA in which the backbone phosphates zigzag

Zygote: The product of the fusion between one of the pollen sperm nuclei and the egg cell of the female gamete at fertilisation. Following a number of mitotic divisions, the zygote differentiates into the embryo.